METHODS IN CELL BIOLOGY

VOLUME XIV

Contributors to This Volume

STUART A. AARONSON

DENNIS ALEXANDER

ALISON M. BADGER

WILLIAM I. BENSINGER

GEORGE BOLCSFOLDI

ANDRÉ BOUÉ

MICHAEL BRENNER

WILLIAM A. CASSEL

ROLAND A. COOK

P. H. COOPER

SIDNEY R. COOPERBAND

FREDERICK T. COUNTER

SUSAN CURE

T. M. DEXTER

LOIS H. DICKERMAN

RANDALL L. DIMOND

EVA ELIASSON

ROSANN A. FARBER

P. J. FISCHINGER

MITSURU FURUSAWA

ROBERT D. GOLDMAN

DAVID J. GRDINA

JOEL S. GREENBERGER

J. STEPHEN HASKILL

J. FRANK HENDERSON

KIYOSHI HIGUCHI

ROGER B. HOWARD

TOBI L. JONES

FREDERICK H. KASTEN

HAJIM KATSUTA

DOUGLAS R. LaBRECQUE

CHRISTOPHER A. LOMAX

WILLIAM F. LOOMIS

JOYCE K. McCOLLEY

JOHN J. MONAHAN

ALFRED I. NEUGUT

TOSHIKAZU NISHIMURA

S. NOMURA

YOSHIO OKADA

LLOYD C. OLSON

TADAMASA OOKA

ROBERT B. OWENS

DEMETRIOS PAPAHADJOPOULOS

GEORGE POSTE

C. R. W. RAYNER

PETER REEVE

PAUL SCHRATTER

D. R. STANWORTH

TOSHIKO TAKAOKA

N. G. TESTA

WILLIAM G. THILLY

TSUYOSHI UCHIDA

PAUL UNRAU

WILLIAM J. VAIL

I. BERNARD WEINSTEIN

D. N. WHEATLEY

MICHAEL H. WIGLER

GARY M. WILLIAMS

JOHN A. WYKE

MASARU YAMAIZUMI

DOMINIC K. YIP

Methods in Cell Biology

Edited by

DAVID M. PRESCOTT

DEPARTMENT OF MOLECULAR, CELLULAR AND
DEVELOPMENTAL BIOLOGY
UNIVERSITY OF COLORADO
BOULDER, COLORADO

VOLUME XIV

1976

ACADEMIC PRESS • New York San Francisco London
A Subsidiary of Harcourt Brace Jovanovich, Publishers

ACADEMIC PRESS, INC.
111 Fifth Avenue, New York, New York 10003

United Kingdom Edition published by
ACADEMIC PRESS, INC. (LONDON) LTD.
24/28 Oval Road, London NW1

LIBRARY OF CONGRESS CATALOG CARD NUMBER: 64-14220

ISBN 0-12-564114-1

PRINTED IN THE UNITED STATES OF AMERICA

CONTENTS

7. Enucleation of Mammalian Cells in Suspension
Michael H. Wigler, Alfred I. Neugut, and I. Bernard Weinstein

8. Cell Culture with Synthetic Capillaries
Paul Schratter

9. Magnesium Acetate Treatment of Glass Roller Bottles to Facilitate Cell Attachment
John J. Monahan

10. Use of Plastics for Suspension Culture Vessels
Roland A. Cook, Frederick T. Counter, and Joyce K. McColley

27. *Hypertonicity and the Synchronization of Mammalian Cells in Mitosis*
D. N. Wheatley

28. *Selection of Synchronized Populations of HeLa Cells*
Alison M. Badger and Sidney R. Cooperband

29. *The Preparation and Characterization of Intact Isolated Parenchymal Cells from Rat Liver*
Douglas R. LaBrecque and Roger B. Howard

LIST OF CONTRIBUTORS

Numbers in parentheses indicate the pages on which the authors' contributions begin.

STUART A. AARONSON, Laboratory of RNA Tumor Viruses, National Cancer Institute, Bethesda, Maryland (237)

DENNIS ALEXANDER, Central Veterinary Laboratory, Ministry of Agriculture, New Haw, Weybridge, England (1)

ALISON M. BADGER, Departments of Microbiology, Surgery, and Medicine, and the Cancer Research Center, Boston University School of Medicine, Boston, Massachusetts (319)

WILLIAM I. BENSINGER, Laboratory of RNA Tumor Viruses, National Cancer Institute, Bethesda, Maryland (237)

GEORGE BOLCSFOLDI, The Wenner-Gren Institute, University of Stockholm, Stockholm, Sweden (159)

ANDRÉ BOUÉ, Centres d'Etudes de Biologie Prenatale, Paris, France (125)

MICHAEL BRENNER, Biological Laboratories, Harvard University, Cambridge, Massachusetts (187)

WILLIAM A. CASSEL, Department of Microbiology, Division of Basic Health Sciences, Emory University, Atlanta, Georgia (181)

ROLAND A. COOK, Eli Lilly and Company, Indianapolis, Indiana (113)

P. H. COOPER, Department of Experimental Pathology, Medical School, University of Birmingham, Birmingham, England (365)

SIDNEY R. COOPERBAND, Departments of Microbiology, Surgery, and Medicine, and the Cancer Research Center, Boston University School of Medicine, Boston, Massachusetts (319)

FREDERICK T. COUNTER, Eli Lilly and Company, Indianapolis, Indiana (113)

SUSAN CURÉ, Centres d'Etudes de Biologie Prenatale, Paris, France (125)

T. M. DEXTER, Paterson Laboratories, Christie Hospital, and Holt Radium Institute, Manchester, England (387)

LOIS H. DICKERMAN, Department of Biology, Case Western Reserve University, Cleveland, Ohio (81)

RANDALL L. DIMOND, Department of Biology, Massachusetts Institute of Technology, Cambridge, Massachusetts (187)

EVA ELIASSON, The Wenner-Gren Institute, University of Stockholm, Stockholm, Sweden (159)

ROSANN A. FARBER, Division of Human Genetics, Children's Hospital Medical Center, Boston, Massachusetts (265)

P. J. FISCHINGER, Laboratory of Viral Carcinogenesis, National Cancer Institute, Bethesda, Maryland (229)

MITSURU FURUSAWA, Laboratory of Embryology, Department of Biology, Faculty of Science, Osaka City University, Osaka, Japan (73)

ROBERT D. GOLDMAN, Department of Biological Sciences, Carnegie-Mellon University, Pittsburgh, Pennsylvania (81)

DAVID J. GRDINA, Section of Experimental Radiotherapy, The University of Texas System Cancer Center, M.D. Anderson Hospital and Tumor Institute, Houston, Texas (213)

JOEL S. GREENBERGER,[1] Laboratory of RNA Tumor Viruses, National Cancer Institute, Bethesda, Maryland (237)

J. STEPHEN HASKILL, Department of Basic and Clinical Immunology and Microbiology, Medical University of South Carolina, Charleston, South Carolina (195)

J. FRANK HENDERSON, Cancer Research Unit (McEachern Laboratory) and Department of Biochemistry, University of Alberta, Edmonton, Alberta, Canada (205)

KIYOSHI HIGUCHI,[2] Microbiological Associates, Walkersville, Maryland (131)

ROGER B. HOWARD, Michael Reese Hospital and Medical Center, Chicago, Illinois (327)

TOBI L. JONES, Department of Surgery, Michigan State University, East Lansing, Michigan (195)

FREDERICK H. KASTEN, Department of Anatomy, Louisiana State University Medical Center, New Orleans, Louisiana (165)

HAJIM KATSUTA, Department of Cancer Cell Research, Institute of Medical Science, University of Tokyo, Shirokanedai, Minato-ku, Tokyo, Japan (145)

DOUGLAS R. LaBRECQUE, Department of Internal Medicine, Liver Study Unit, Yale University, New Haven, Connecticut (327)

CHRISTOPHER A. LOMAX, Départment de Microbiologie, Centre Hospitalier Universitaire, Université de Sherbrooke, Sherbrooke, P. Q., Canada (205)

WILLIAM F. LOOMIS, Department of Biology, University of California, San Diego, La Jolla, California (187)

JOYCE K. McCOLLEY, Eli Lilly and Company, Indianapolis, Indiana (113)

JOHN J. MONAHAN, Department of Cell Biology, Baylor College of Medicine, Houston, Texas, (105)

ALFRED I. NEUGUT, College of Physicians and Surgeons, Department of Pathology, and Institute of Cancer Research, Columbia University, New York, New York (87)

TOSHIKAZU NISHIMURA,[3] Laboratory of Embryology, Department of Biology, Faculty of Science, Osaka City University, Osaka, Japan (73)

S. NOMURA, Laboratory of Viral Carcinogenesis, National Cancer Institute, Bethesda, Maryland (229)

YOSHIO OKADA, Department of Animal Virology, Research Institute for Microbial Disease, Osaka University, Osaka, Japan (73)

LLOYD C. OLSON, Department of Microbiology, Indiana University School of Medicine, Indianapolis, Indiana (11)

TADAMASA OOKA, Départment de Biologie Générale et Appliquée, Laboratoire Associé au C.N.R.S., Université Claude Bernard (Lyon-I), Villeurbanne, France (287)

ROBERT B. OWENS, Cell Culture Laboratory, School of Public Health, University of California, Berkeley, California (341)

DEMETRIOS PAPAHADJOPOULOS, Department of Experimental Pathology, Roswell Park Memorial Institute, Buffalo, New York (23, 33)

GEORGE POSTE, Department of Experimental Pathology, Roswell Park Memorial Institute, Buffalo, New York (1, 23, 33)

[1] *Present address:* Joint Center for Radiation Therapy, Department of Radiation Therapy, Harvard Medical School, Boston, Massachusetts.
[2] *Present address:* 199 Meadowdale Lane, Frederick, Maryland.
[3] *Present address:* Department of Anatomy, Aichi Medical University, Nagakute-cho, Aichi-ken, Japan.

C. R. W. RAYNER,[4] St. George's Hospital and Westminster Hospital, London, England (379)

PETER REEVE,[5] Department of Bacteriology, University College Hospital Medical School, London, England (1)

PAUL SCHRATTER, Amicon Corporation, Lexington, Massachusetts (95)

D. R. STANWORTH, Department of Experimental Pathology, Medical School, University of Birmingham, Birmingham, England (365)

TOSHIKO TAKAOKA, Department of Cancer Cell Research, Institute of Medical Science, University of Tokyo, Shirokanedai, Minato-ku, Tokyo, Japan (145)

N. G. TESTA, Paterson Laboratories, Christie Hospital, and Holt Radium Institute, Manchester, England (387)

WILLIAM G. THILLY, Toxicology Group, Department of Nutrition and Food Science, Massachusetts Institute of Technology, Cambridge, Massachusetts (273)

TSUYOSHI UCHIDA, Department of Animal Virology, Research Institute for Microbial Disease, Osaka University, Osaka, Japan (73)

PAUL UNRAU, Biology Branch, Chalk River Nuclear Laboratories, Chalk River, Ontario, Canada (265)

WILLIAM J. VAIL, Department of Microbiology, University of Guelph, Guelph, Ontario, Canada (33)

I. BERNARD WEINSTEIN, College of Physicians and Surgeons, Department of Medicine, and Institute of Cancer Research, Columbia University, New York, New York (87)

D. N. WHEATLEY, Department of Pathology, University of Aberdeen, Foresterhill, Aberdeen, Scotland (297)

MICHAEL H. WIGLER, College of Physicians and Surgeons, Departments of Microbiology, Pathology, and Medicine, and Institute of Cancer Research, Columbia University, New York, New York (87)

GARY M. WILLIAMS, Naylor Dana Institute for Disease Prevention, American Health Foundation, Valhalla, New York (357)

JOHN A. WYKE, Department of Tumour Virology, Imperial Cancer Research Fund, London, England (251)

MASARU YAMAIZUMI, Department of Animal Virology, Research Institute for Microbial Disease, Osaka University, Osaka, Japan (73)

DOMINIC K. YIP, Department of Anatomy, Louisiana State University Medical Center, New Orleans, Louisiana (165)

[4] *Present address:* c/o Plastic Surgery Unit, Withington Hospital, West Didsbury, Manchester, England.

[5] *Present address:* Sandoz Forshungsintitut Gmbh, Vienna, Austria.

PREFACE

Volume XIV of this series continues to present techniques and methods in cell research that have not been published or have been published in sources that are not readily available. Much of the information on experimental techniques in modern cell biology is scattered in a fragmentary fashion throughout the research literature. In addition, the general practice of condensing to the most abbreviated form materials and methods sections of journal articles has led to descriptions that are frequently inadequate guides to techniques. The aim of this volume is to bring together into one compilation complete and detailed treatment of a number of widely useful techniques which have not been published in full detail elsewhere in the literature.

In the absence of firsthand personal instruction, researchers are often reluctant to adopt new techniques. This hesitancy probably stems chiefly from the fact that descriptions in the literature do not contain sufficient detail concerning methodology; in addition, the information given may not be sufficient to estimate the difficulties or practicality of the technique or to judge whether the method can actually provide a suitable solution to the problem under consideration. The presentations in this volume are designed to overcome these drawbacks. They are comprehensive to the extent that they may serve not only as a practical introduction to experimental procedures but also to provide, to some extent, an evaluation of the limitations, potentialities, and current applications of the methods. Only those theoretical considerations needed for proper use of the method are included.

Finally, special emphasis has been placed on inclusion of much reference material in order to guide readers to early and current pertinent literature.

DAVID M. PRESCOTT

METHODS IN CELL BIOLOGY

VOLUME XIV

Chapter 1

Enhancement of Virus-Induced Cell Fusion by Phytohemagglutinin

GEORGE POSTE,[1] DENNIS ALEXANDER,[2] AND
PETER REEVE[3]

I. Introduction

Studies in several laboratories have shown that the binding of plant lectins to the cell surface can significantly alter the cellular response to virus infection (Poste, 1975). A striking example of the ability of lectins to alter virus–host cell interactions concerns their effect on virus-induced cell fusion. Treatment of cultured mammalian cells with concanavalin A (Con A), wheat germ agglutinin, and soybean agglutinin has been shown to inhibit their subsequent fusion by both RNA- (Poste et al., 1974) and DNA–containing viruses (Rott et al., 1975). In contrast, treatment of cells with phytohemagglutinin (PHA), a lectin obtained from the red kidney

[1] Department of Experimental Pathology, Roswell Park Memorial Institute, Buffalo, New York.
[2] Central Veterinary Laboratory, Ministry of Agriculture, New Haw, Weybridge, England.
[3] Department of Bacteriology, University College Hospital Medical School, London, England. *Present address:* Sandoz Forshungsintitut Gmbh, Vienna, Austria.

bean (*Phaseolus vulgaris*), results in significant enhancement of virus-induced cell fusion (Poste *et al.*, 1974; Reeve *et al.*, 1974; Sullivan *et al.*, 1975; Yoshida and Ikeuchi, 1975). In this chapter we describe how this property of PHA can be exploited to enhance the efficiency of cell fusion by inactivated viruses to increase the yield of hybrid cells produced by fusion of different cell types.

II.   Phytohemagglutinin (PHA)

Commercially available preparations of PHA from Difco Laboratories Inc. (Detroit, Michigan) and the Wellcome Co. (Triangle Park, North Carolina and Beckenham, England) are able to induce significant enhancement of virus-induced cell fusion (Poste *et al.*, 1974; Reeve *et al.*, 1974; Yoshida and Ikeuchi, 1975). Preparations from both manufacturers are supplied as a lyophilized powder, which can be reconstituted and diluted in phosphate-buffered saline (PBS) or culture medium as outlined in Section IV.

Large amounts of PHA with cell fusion-enhancing properties can also be isolated directly from red kidney beans by affinity chromatography on a thyroglobulin–Sepharose column (Matsumoto and Osawa, 1972; Felsted *et al.*, 1975), and we now routinely employ a modification of this procedure to generate PHA for cell fusion studies. Red kidney beans (100 gm) are homogenized in an ice-bath to produce a fine powder, which is then added to 500 ml of PBS; the mixture is incubated overnight at 4° C with continuous stirring. The solution is then filtered through cheesecloth to remove large particulate matter, and the filtrate is centrifuged at 45,000 g for 30 minutes at 4°C. The supernatant, which is viscous, slightly turbid, and off-brown in color, provides the starting extract for use with the affinity column.

Affinity columns are prepared by mixing porcine thyroglobulin (Sigma, type III) with cyanogen bromide-activated Sepharose 4B (Pharmacia) (200 mg of thyroglobulin per gram of activated Sepharose) at room temperature for 20 hours in 0.1 M NaHCO$_3$ buffer (pH 8.0) containing 0.5 M NaCl. The resin is then filtered and washed several times with buffer. Protein determinations are made on the combined filtrate and the washes to determine the amount of protein coupling, which is usually between 30 and 60%. The washed resin is then incubated with 1 M ethanol (pH 7.6) for 2 hours at room temperature and then sequentially washed with 0.1 M acetate buffer (pH 4.0) containing 1 M NaCl and 0.1 M borate buffer (pH 8.0) containing 1 M NaCl. The washing procedure should be done at least three times. The resin is then finally washed with 6.5 mM potassium phosphate (pH 7.2) containing 0.15 M NaCl, after which the resin is ready for use.

For affinity chromatography the saline extract described above is diluted with an equal volume of PBS to reduce its viscosity, and the diluted extract is added to the thyroglobulin–Sepharose resin (typically, 100–120 ml of resin per 100 gm of bean extract); the suspension is mixed at room temperature for 2 hours, after which the mixture is centrifuged at 8000 g for 10 minutes. The resin is then resuspended in a limited volume of PBS and packed into a column (30 ml of resin for a 2.5-cm × 6-cm column). Fractions can be collected from the column during settling of the resin. The column is first washed with PBS to elute early fractions containing UV (280 nm)-absorbing material, then fractions are eluted with 1 mM potassium phosphate (pH 7.2) containing 1 M NaCl until elution of UV-absorbing

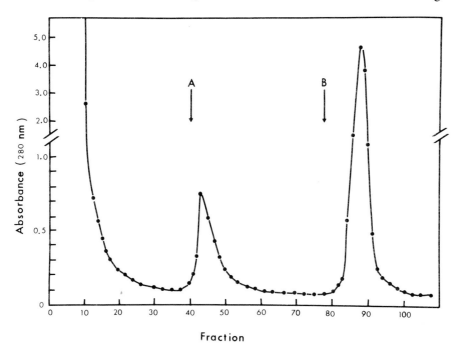

FIG. 1. Elution profile for saline extract of red kidney beans (*Phaseolus vulgaris*) on a porcine thyroglobulin-Sepharose affinity column. The saline extract from 25 gm of beans was mixed with 30 ml of thyroglobulin–Sepharose, packed into a 2.4-cm × 6-cm column, and 10-ml fractions were collected during settling of the resin. Initial washes until elution of UV (280 nm)–absorbing material had stabilized were made with phosphate-buffered saline; the wash was then changed (arrow A) to 1 mM potassium phosphate (pH 7.2) in 1 M NaCl, and elution of UV-absorbing material was again allowed to stabilized; after this (arrow B) the remaining fractions were all eluted with 0.05 M glycine HCl (pH 3.0) containing 0.5 M NaCl.

material again stabilizes; after this all subsequent fractions are eluted with 0.05 M glycine HCl (pH 3.0) containing 0.5 M NaCl. A typical elution profile is shown in Fig. 1. As described later in Section V, the high-salt elute from the column (fractions 40 to 50 in Fig. 1) are significantly more active on a per unit weight basis in enhancing virus-induced cell fusion than are the later glycine-eluted fractions, even though these contain virtually all the erythroagglutinin activity. Both the high-salt and the glycine eluates are concentrated before being used to treat cells by ultrafiltration through a PM-10 membrane (Amicon Corp., Lexington, Massachusetts). Two absorptions of a saline extract from 100 gm of beans by the above method yields between 500 and 600 mg of purified PHA proteins. The thyroglobulin–Sepharose resin can be regenerated and used a number of times without apparent loss of activity.

III. Virus-Induced Cell Fusion and Quantitation of Hybrid Cell Formation

In this chapter we will be concerned only with the effect of PHA on cell fusion induced by inactivated viruses (so-called fusion from without), which results from the direct interaction of virus particles with the cell surface and does not require entry of virus into cells. The use of inactivated Sendai virus to induce cell fusion is the best-known example of this phenomenon, but a number of other paramyxoviruses are equally effective in producing this type of cell fusion (see Poste, 1972; Poste and Waterson, 1975). In the present study, Sendai virus, and an avian paramyxovirus, Newcastle disease virus (NDV), will be used for inducing cell fusion from without. A second type of virus-induced cell fusion, so-called fusion from within, in which cell fusion occurs as a consequence of the intracellular replication of the virus, will not be considered here, though PHA also enhances this type of virus-induced cell fusion (Poste *et al.*, 1974; Reeve *et al.*, 1974; Sullivan *et al.*, 1975).

Interspecific mammalian somatic cell hybrids were produced by virus-induced fusion of thymidine kinase-deficient CI-ID mouse cells (Dubbs and Kit, 1964) and hypoxanthine guanine phosphoribosyltransferase-deficient WI-18Va2 human cells (Weiss *et al.*, 1968) with subsequent selection of hybrid cells in HAT medium (Littlefield, 1966). Egg-grown stocks of Sendai virus and NDV (strain Texas) were inactivated with β-propiolactone as described previously (Poste, 1973). Equal numbers (5 $\times$ 10^6) of each cell type were suspended in 1.0 ml of serum-free Eagle's basal medium with or without PHA, and the medium was adjusted to the required pH (see Section IV). Except where stated otherwise cells were incubated with the indicated concentrations of PHA for 90 minutes at 37°C before final addition of virus [2000 HAU of inactivated Sendai virus or 2000 egg infectious dose $(EID)_{50}$ per cell of NDV]. The cell suspension was incubated with virus for 1 hour at 37°C.

To determine the extent of virus-induced cell fusion, the mixed cell population was inoculated onto coverslips and incubated for a further 16 hours at 37°C; the coverslips were then fixed and stained with May–Grünwald–Giemsa. The extent of cell fusion, expressed as the percentage of polykaryocytosis, was measured by counting the number of nuclei present in multinucleate cells (polykaryocytes) and expressing this as a percentage of the total number of nuclei present in the same microscope field. At least 20 microscope fields were counted from each coverslip, and four replicate coverslips were examined for each sample.

To determine the yield of hybrid cells produced by fusion of mouse and human cells, 1×10^6 cells from the virus-treated mixed cell population were inoculated into 35-mm plastic petri dishes in 2.0 ml of HAT medium buffered to the required pH. The medium was changed and replaced with fresh HAT medium at the same pH after 1, 3, 6, 9, and 12 days. The culture dishes were finally fixed and stained with May–Grünwald–Giemsa 14 days after fusion, and the number of colonies of surviving hybrid cells were counted. At least 10 dishes were used for each determination. The hybrid nature of cells surviving in HAT was confirmed by their expression of both mouse-specific and human-specific enzymes (demonstrated by starch gel electrophoresis).

IV. Enhancement of Virus-Induced Cell Fusion and Hybridization by PHA

The enhancing effect of different preparations of PHA on virus-induced fusion of a mixed population of CI–ID and WI–18Va2 cells is shown in Fig. 2.

Cell fusion induced by inactivated viruses has a well defined pH optimum of 7.6 to 8.0 (Bratt and Gallaher, 1969; Croce et al., 1972). This effect is shown in Table I, where the extent of cell fusion induced by both Sendai virus and NDV in the absence of PHA is significantly greater at pH 7.6 to 8.0. Although a similar pH optimum is also found for fusion of cells with

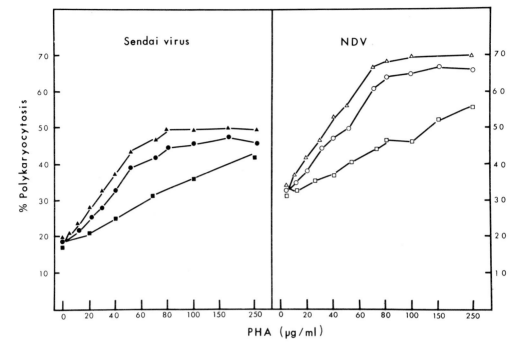

FIG. 2.   Effect of different phytohemagglutinin (PHA) preparations on fusion of mixed populations of human (WI-18Va2) and mouse (Cl-ID) cells at pH 8.0 by inactivated Sendai virus (▲, ●, ■) and inactivated Newcastle disease virus (△, ○, □). Cells were preincubated for 90 minutes at 37°C with the indicated concentrations of affinity-purified PHA (high salt eluate-fractions 40–50 in Fig. 1) (▲—▲ and △—△); Wellcome PHA (●—● and ○—○); or Difco phytohemagglutinin-P (■—■ and □—□) before treatment with virus. The extent of cell fusion, expressed as percentage of polykaryocytosis, was measured as described in text.

PHA (Table I), PHA treatment significantly enhances the extent of cell fusion at all pH values (Table I).

PHA not only enhances the extent of virus-induced cell fusion (Table I), but also significantly increases the yield of hybrid cells from fused mixed-cell populations (Table II). Similar enhancement of virus-induced hybrid cell formation by PHA has been described by Yoshida and Ikeuchi (1975) for fusion between human and mouse cells and between mouse and rat cells.

These results indicate that pretreatment of cells with PHA before exposure to cell-fusing viruses can be profitably employed to enhance cell fusion and to increase the efficiency of present methods used in the production of mammalian somatic cell hybrids.

TABLE I

EFFECT OF AFFINITY-PURIFIED PHYTOHEMAGGLUTININ (PHA) (100 μG/ML) ON FUSION
OF MOUSE CI-ID AND HUMAN WI-18Va2 CELLS BY INACTIVATED VIRUSES
AT DIFFERENT pH VALUES

	Percent polykaryocytosis[b]					
	Untreated control		Sendai virus		NDV (strain Texas)	
pH[a]	No PHA	PHA[c]	No PHA	PHA[c]	No PHA	PHA[c]
6.4	2.8	3.5	8.3	22.6	16.8	39.3
7.2	3.1	3.5	10.4	30.7	21.3	40.6
7.6	3.6	3.9	20.6	42.5	30.7	63.5
8.0	4.3	4.5	27.2	45.8	39.4	65.9
8.4	3.8	3.5	22.3	43.6	N.D.	N.D.

[a] Culture medium buffered by organic buffers at the following concentrations: pH 6.4 PIPES buffer (10 mM); pH 7.2 BES buffer (15 mM); pH 7.6 and 8.0 HEPES buffer (15 mM); and pH 8.4 Tricine buffer (15 mM).

[b] Measured 16 hours after treatment with virus as described in Section III.

[c] Cells were incubated with PHA for 90 minutes at 37°C before treatment with virus. PHA preparation was obtained as the high-salt eluate from porcine thyroglobulin–Sepharose column (fractions 40–50 in Fig. 1).

TABLE II

EFFECT OF AFFINITY-PURIFIED PHYTOHEMOGGLUTININ (PHA) (100 μG/ML) ON FORMATION
OF HYBRID CELL COLONIES PRODUCED BY VIRUS-INDUCED FUSION OF
HUMAN WI-18Va2 CELLS AND MOUSE CI-ID CELLS

	Average number of hybrid colonies per 35-mm dish 14 days after fusion					
	Cocultivation		Sendai virus		NDV (strain Texas)	
pH[a]	No PHA	PHA[b]	No PHA	PHA[b]	No PHA	PHA[b]
7.2	0	2.7	2.9	10.5	12.1	30.7
7.6	3.8	6.9	15.7	50.8	37.2	98.3
8.0	8.5	11.4	41.4	95.3	62.5	154.5

[a] Culture medium buffered by organic buffers as outlined in footnote a to Table I.

[b] Cells were incubated with PHA for 90 minutes at 37°C before treatment with virus (or plating in case of cocultivated cells). The PHA preparation was identical to that described in footnote c in Table I.

V.   Mechanism of PHA-Induced Enhancement of Cell Fusion

The mechanism by which PHA enhances virus-induced cell fusion remains to be elucidated. Since PHA-conjugated to Sepharose is equally as effective as the free lectin in enhancing cell fusion (Reeve et al., 1974), it is clear that the effect of PHA on the fusion process is mediated entirely at the cell surface and does not require entry of lectin into the cells.

The most straightforward explanation for the ability of PHA to enhance virus-induced cell fusion is that agglutination of cells by PHA creates conditions of close cell-to-cell contact that facilitate subsequent fusion. Despite the obvious simplicity and attractiveness of this interpretation, the role of agglutination in promoting cell fusion is confused by the observation that while PHA enhances fusion, other lectins such as Con A, wheat germ agglutinin, and soybean agglutinin inhibit virus-induced cell fusion even though they are equally as effective as PHA in agglutinating cells (Poste *et al.*, 1974). A possible explanation for the conflicting effects of different lectins on virus-induced cell fusion concerns the question of whether or not a particular lectin is able to bind to virus-receptor sites on the cell surface. Binding of lectin to virus receptors on the cell surface would prevent subsequent attachment of virus and consequently also prevent virus-induced cell fusion. Recent work in this laboratory has shown that lectins, such as Con A and wheat germ agglutinin, that inhibit paramyxovirus-induced cell fusion do in fact prevent binding of virus to the cell surface whereas PHA does not alter either the rate or the extent of virus particle binding (G. Poste, unpublished observations). These results suggest that the enhancement of virus-induced cell fusion by PHA might indeed result merely from these ability to agglutinate cells without affecting the subsequent binding of virus to its receptors on the cell surface. If this interpretation is correct, then other agents that promote cell agglutination without impairing virus binding might also be effective in enhancing virus-induced cell fusion. It is therefore of interest to note that De Boer and Loyter (1971) found that agglutination of erythrocytes by pretreatment with polylysine enhanced their subsequent fusion by inactivated Sendai virus and phospholipase C.

PHA is a mixture of five isolectins, which possess varying degrees of erythroagglutinating, leukoagglutinating, and mitogenic activities (see Weber *et al.*, 1972; Yachnin and Svenson, 1972; Miller *et al.*, 1973; Felsted *et al.*, 1975). Ion-exchange chromatography and polyacrylamide gel electrophoresis have been used to isolate two major biological fractions, designated L-PHAP and H-PHAP (Yachnin and Svenson, 1972; Miller *et al.*, 1973). The former is a homogeneous single protein with potent leukoagglutinin activity but low erythroagglutinin activity. The H-PHAP fraction is a complex protein mixture with high erythroagglutinin activity but little or no leukoagglutinin activity. Both L-PHAP and H-PHAP possess mitogenic activity. Characterization of the cell fusion-enhancing activity of these various PHA fractions isolated from affinity chromatography-purified PHA is currently in progress in this laboratory. Preliminary results indicate that cell fusion-enhancing activity resides largely in the L-PHAP fraction, which contains a singleprotein corresponding to the most anodal protein of the five PHA proteins resolved by gel electrophoresis. This finding also provides a possible explanation for the greater fusion-enhancing activity of commercial PHA from Wellcome compared with Difco phytohemagglutinin-P (Fig. 1). Characterization of these preparations by polyacrylamide gel electrophoresis reveals that Wellcome PHA contains only one major protein band, which has a mobility similar to the most anodal protein component of the five PHA proteins resolved by gel electrophoresis of complete PHA. In contrast, Difco phytohemagglutinin-P is a much more heterogeneous preparation containing five distinct proteins.

The above results suggest that the cell fusion-enhancing activity of PHA is predominantly a property of a single protein, which corresponds to the so-called L-PHAP of Yachnin and Svenson (1972) and Miller *et al.* (1973). In addition, assay of PHA isolated on thyroglobulin–Sepharose has shown that more than 80% of the cell fusion-enhancing activity found in PHA is associated with material in the high-salt eluate (Fig. 1). This material possesses only very low erythroagglutinating activity and by gel electrophoresis is found to be largely devoid of cathodal-migrating proteins. These observations lend further support to the conclusion that cell fusion-enhancing activity of PHA is a property of the most anodal of the PHA isolectins, which, in addition to its effect on cell fusion, is also a potent leukoagglutinin and mitogen.

REFERENCES

Bratt, M. A., and Gallaher, W. R. (1969). *Proc. Natl. Acad. Sci. U.S.A.* **64**, 536–543.
Croce, C., Koprowski, H., and Eagle, H. (1972). *Proc. Natl. Acad. Sci. U.S.A.* **69**, 1953–1956.
De Boer, E., and Loyter, A. (1971). *FEBS Lett.* **15**, 325–327.
Dubbs, D. R., and Kit, S. (1964). *Exp. Cell Res.* **33**, 19–28.
Felsted, R. L., Leavitt, R. D., and Bachur, N. R. (1975). *Biochim. Biophys. Acta* **405**, 72–81.
Littlefield, J. (1966). *Exp. Cell Res.* **41**, 190–196.
Matsumoto, I., and Osawa, T. (1972). *Biochem. Biophys. Res. Commun.* **46**, 1810–1816.
Miller, J. B., Noyes, C., Heinrikson, R., Kingdon, H. S., and Yachnin, S. (1973). *J. Exp. Med.* **138**, 939–951.
Poste, G. (1972). *Int. Rev. Cytol.* **33**, 157–252.
Poste, G. (1973). *Methods Cell Biol.* **7**, 211–249.
Poste, G. (1975). *In* "Concanavalin A" (T. K. Chowdhury and A. K. Weiss, eds.), pp. 117–152. Plenum, New York.
Poste, G., and Waterson, A. P. (1975). *In* "Negative Strand Viruses" (B. W. J. Mahy and R. D. Barry, eds.), Vol. 2, pp. 906–922. Academic Press, New York.
Poste, G., Alexander, D. J., Reeve, P., and Hewlett, G. (1974). *J. Gen. Virol.* **23**, 255–270.
Reeve, P., Hewlett, G., Watkins, H., Alexander, D. J., and Poste, G. (1974). *Nature (London)* **249**, 355–356.
Rott, R., Becht, H., Hammer, G., Klenk, H.-D., and Scholtissek, C. (1975). *In* "Negative Strand Viruses" (B. W. J. Mahy and R. D. Barry, eds.), Vol. 2, pp. 843–857. Academic Press, New York.

Sullivan, J. L., Barry, D. W., Lucas, S. J., and Albrecht, P. (1975). *J. Exp. Med.* **142**, 773–784.

Weber, T. H., Aro, H., and Nordman, C. T. (1972). *Biochim. Biophys. Acta* **263**, 94–105.

Weiss, M. C., Ephrussi, B., and Scaletta, L. J. (1968). *Proc. Natl. Acad. Sci. U.S.A.* **59**, 1132–1135.

Yachnin, S., and Svenson, R. H. (1972). *Immunology* **22**, 871–883.

Yoshida, M. C., and Ikeuchi, T. (1975). *Proc. Jpn. Acad.* **51**, 126–129.

Chapter 2

Methods of Cell Fusion with Germiston Virus

LLOYD C. OLSON

Department of Microbiology,
Indiana University School of Medicine,
Indianapolis, Indiana

I. Introduction

The phenomenon of fusion of anatomically separated membranes into a continuous structure is of considerable interest for a number of reasons. Cell fusion is a basic physiological mechanism essential to the biological activities of most cell and is involved in the processes of endocytosis, exocytosis, and cell division (for example, see Poste and Allison, 1973). Second, cell fusion has become a powerful research tool, the use of which has resulted in rapid advances of our understanding of how gene expression is regulated in eukaryotic cells (reviewed in Davidson, 1971, 1974; Handmaker, 1973). These studies have been accomplished by facilitating cell fusion with one of a number of techniques such that investigators have been able to construct hybrid cells of diverse species and origins. Finally, by fusing permissive with nonpermissive cells, it is becoming possible to demonstrate the presence of viruses in certain kinds of infected tissues wherein infectious virus is not otherwise being produced. This approach has been successful in certain chronic diseases [e.g., measles virus and subacute sclerosing panencephalitis (Barbanti-Brodano et al., 1970)] and for the rescue of tumor viruses under certain conditions (reviewed by Barski, 1970).

Cell fusion can be effected with a variety of chemical or physical agents such as lysolecithin, polyols, dimethyl sulfoxide, or lysosomes (Martin and MacDonald, 1974; Ahkong et al., 1975). However, in many instances, even though these agents are very efficient in inducing cell fusion, they also damage the cell so that the fused cells cannot be maintained in a viable state. As a result, and because their use represents a simple and convenient method, techniques using viruses have been most widely favored in cell fusion experiments. The mechanisms by which viruses induce cell fusion have been reviewed recently by Poste (1972). As noted by Poste, cell fusion may be induced by numerous kinds of viruses, all of which share the common property of possessing envelopes. To eliminate the problems associated with using infectious virus to induce cell fusion, the virus is usually inactivated before use. These techniques have been described in detail by Watkins (1971).

Recently we reported (Djinawi and Olson, 1973) that cell fusion could also be induced by two arboviruses. Wesselsbron virus is a member of the flavivirus group (group B togaviruses), and Germiston virus is classified as a bunyavirus. This report is the first in which cell fusion had been noted for members of these virus groups. In common with other viruses that effectively induce cell fusion, Wesselsbron and Germiston viruses also possess envelopes. Because the infectivity (but not fusion activity) of Germiston virus may be conveniently abolished by exposure to chloroform, we have confined most of our efforts to defining fusion conditions with this virus. In many respects Germiston virus has proved to be a useful reagent for the induction of cell fusion for the following reasons: (1) the preparation of very high-titered stocks of virus is easily accomplished; (2) infectivity and cytotoxicity can be removed by treatment with chloroform and protamine, respectively; (3) a number of different cell lines have been found to be susceptible to Germiston virus-induced fusion; (4) cell fusion occurs exceptionally rapidly (within 5 minutes in some instances); (5) the induction of fusion is extremely efficient (this may simply represent the fact that very large virus:cell ratios can be employed—of the order of $10^3:1$).

These advantages notwithstanding, one precaution must be rigorously observed. Germiston virus is infectious for man, at least by the parenteral route. The virulence and disease potentials of this agent may not be fully appreciated. The author unequivocally recommends that infectious material be handled only in a biocontainment facility, such as a laminar flow biological cabinet.

II. Preparation of Germiston Virus

Germiston virus may be obtained from the American Type Culture Collection (12301 Parklawn Drive, Rockville, Maryland 20852; Catalog No. VR–393). A permit to receive this virus must first be obtained from the U.S. Public Health Service and the United States Department of Agriculture, forms for which are available from the ATCC.

A. Stock Virus

High-titered stock preparations of Germiston virus can be prepared in various cell cultures or in suckling mouse brain. There are advantages and disadvantages to each method.

1. SUCKLING MOUSE BRAIN

Random-bred, newborn mice are used. The age of the mice is not important in respect to susceptibility to infection, but it is easier to harvest infected brain material from younger mice. Mice are inoculated with 0.05 ml of diluted virus suspension intraperitoneally with a 26-gauge needle affixed to a tuberculin syringe. The usual concentration of stock virus is of the order of 10^6–10^7 LD$_{50}$; any amount of virus greater than approximately 10^2 LD$_{50}$ will cause death in suckling mice within 48 hours. Our usual practice is to inoculate the mice early in the morning; they will generally be ready to harvest by late afternoon of the next day. Inoculated mice are checked frequently; when deaths have begun to appear, the brains of the remaining mice, most of which should be manifesting neurological abnormalities, should be harvested.

Infected brain material can be collected in either of two ways. For either, reasonable precautions with respect of sterile technique should be observed. The mice are killed with chloroform anesthesia, and immediately the brain is harvested. This can be done by dissecting away alcohol-cleansed skin over the skull, then picking away the thin cartilaginous skull. The brain may then be removed *in toto*. The brains are pooled in a sterile, preweighed screw-top tube kept in an ice bath. This method gives the largest yield. However, it is slow and the risk of contamination is increased. Consequently, we prefer collecting the infected brain material directly with a large needle and syringe; this is much faster and contamination is rare. The yield is only somewhat less, but great care must be exercised when handling the needle since it will be heavily contaminated with highly infectious material. The skin is incised transversely across the posterior cervical area, then forcibly pulled cephalad so as to expose the base of the skull. The sides of the skull are grasped firmly by a large pair of forceps in one hand while with the other hand a large needle on a plastic syringe is introduced through the foramen magnum or base of the skull into the cranium. Negative pressure (suction) is applied with the syringe while the skull is gently squeezed with forceps. The skull will be seen to collapse, and the brain material will be aspirated into the syringe. We have found it easiest to use a 13-gauge needle on a 10- or 20-ml plastic syringe. A number of brains can be quickly harvested before having to expel the syringe contents into the tube.

Once the brain material has been harvested it can either be homogenized directly or stored at $-70°C$ until ready for use. In the latter case it should be frozen quickly in a Dry Ice–alcohol bath, then stored. To prepare for homogenizing, frozen material should be thawed as quickly as possible in a 37°C water bath with constant agitation. The brain harvest should be made up as a 20% (w/v) suspension in phosphate-buffered saline, pH 7.2 supplemented with heat-inactivated (56°C for 30 minutes) fetal bovine serum (FBS) or bovine serum albumin (BSA). The final proportion of FBS should be 2%; of BSA, 0.4%. Again, virus suspensions should always be kept at 0°C, if possible, to avoid thermal inactivation of the virus. The brain suspension is homogenized at top speed for 3 minutes in an Omnimixer (Sorvall, Dupont Instruments) or equivalent, with the cup in an ice bath. The homogenate is clarified by centrifugation at 350 g for 30 minutes at 4°C; the supernatant is aspirated and stored as aliquots at $-70°C$ after quick freezing. If careful handling is maintained throughout this process, the final suspension will contain $10^8–10^9$ infectious virus particles per 0.1 ml.

2. CELL CULTURE PREPARATIONS

Working stocks of virus are much easier to prepare in monolayer cell cultures, but they also tend to yield significantly lower titers of infectious virus. Numerous different kinds of cells can be used; we have had satisfactory experience with continuous monkey kidney cell lines (Vero, BS–C–1, and LLC–MK2), human cells (HEp–2), mouse cells (L cells, macrophages), and mosquito cells (Singh's *Aedes albopictus*). Susceptibility of hamster cells (hamster kidney and BHK21 cells) to Germiston virus has also been reported (Karabotsos and Buckley, 1967). Cell cultures are handled normally except that the substitution of albumin (0.2% v/v) for serum in the maintenance medium probably increases the yield of infectious virus by at least

10-fold. Cell cultures are inoculated with virus and incubated at 37°C until the cytopathic effect involves all cells. Infected *Aedes albopictus* should also be incubated at 37°C. The virus is then most easily harvested by placing the culture bottle cell sheet at −70°C overnight, then thawing it quickly and clarifying the supernatant by centrifugation (1000 g for 30 minutes at 4°C). The supernatant is stored as aliquots at −70°C.

B. Treatment with Protamine Sulfate

We have found that precipitation of cellular proteins with protamine sulfate markedly reduces the nonspecific toxicity of Germiston virus suspensions prepared either from cell cultures or from infected mouse brain. The method used is modified from Warren *et al.* (1949). Protamine sulfate is prepared as an aqueous solution or used directly as the powder; 2 mg is added to each 1 ml of the virus suspension. The mixture is occasionally mixed and incubated at 4°C for 2 hours. A visible precipitate usually forms, which is then removed by centrifugation.

C. Treatment with Chloroform

One part of chloroform is mixed with 9 parts of virus suspension at room temperature. The mixture is then vigorously mixed periodically for 10 minutes. This is sufficient to abolish completely the infectivity of Germiston virus. The inactivated virus suspension is next centrifuged at 500 g for 15 minutes at room temperature, then the top aqueous layer containing the virus is carefully removed and is ready for use. This can either be used directly or stored at −70°C. If stored, it appears that considerable aggregation occurs, such that some fusion activity may be lost. Consequently, frozen material can be rapidly thawed, agitated vigorously, and held at 4°C for at least overnight to allow elution of virus particles from the aggregated state.

D. Assay of Viral Activity

The concentration of virus particles must be estimated before chloroform inactivation. Titrations can be carried out by plaque assays in cell cultures, such as LLC–MK2, Vero, or BHK21 cells. The most sensitive system for infectivity assay is intracerebral inoculation of suckling mice (Olson *et al.*, 1975), and we have frequently merely screened a new stock by inoculating one litter with a single dilution (10^{-7} or 10^{-8}); the end point of mouse titrations can be determined at 7 days. Germiston virus is a hemagglutinating virus, but the conditions required (pH, buffer systems, etc.) and the necessity for using goose or 1-day-old chick red cells makes this impractical unless the laboratory is already appropriately equipped. The details of hemagglutination assays of arboviruses can be found in the description by Hammon and Sather (1969); the pH optimum of Germiston virus hemagglutinin is 6.2.

Fortunately, there is an alternative method that we have found to be reliable for estimating rapidly the suitability of virus preparations for fusion experiments. Cells to be used are suspended in PBS in a concentration sufficient to yield a grossly visible suspension free of clumps. Approximately 10^5 cells per milliliter are satisfactory, but the number is not critical. To 1 ml of this preparation in a screw-top tube is added an equal volume of virus; the tube contents are mixed well and then are incubated at 37°C for up to 1 hour. At the end of 1 hour the contents are gently resuspended. The cells should appear to be visibly clumped, and this can be confirmed microscopically (Fig. 1). The minimum number of virus particles required to grossly agglutinate cells is something on the order of 10 particles per cell.

III.   Cell Fusion

A.   Conditions Influencing Cell Fusion

Some of the factors affecting the efficiency of cell fusion can be found in the recent review by Poste (1972). Germiston virus induces fusion very rapidly—we have detected fusion within 5 minutes at 37°C (Djinawi and Olson, 1973)—indicating that this is fusion from without (FFWO). The specificity of fusion associated with Germiston virus has been demonstrated in that fusion activity is not associated with normal mouse brain material, and fusion activity can be eliminated by prior incubation of the virus preparation with antiserum to Germiston virus. Fusion occurs more rapidly and is more efficient at 42°C than at lower temperatures (Gallaher and Bratt, 1972). The optimal conditions for cell fusion, however, must be determined for the particular kind of cells being used. For example, to achieve the same degree of fusion achieved for most cells, mouse L-cell fibroblasts must be incubated with Germiston virus for a longer period of time. Repeated attempts with prolonged incubation, on the other hand, were uniformly unsuccessful in inducing cell fusion in HeLa cells (Ohio strain).

B.   Techniques of Cell Fusion

Cell fusion may be induced in cells grown either as monolayers or in suspension. Similarly, heterokaryon formation between two different species

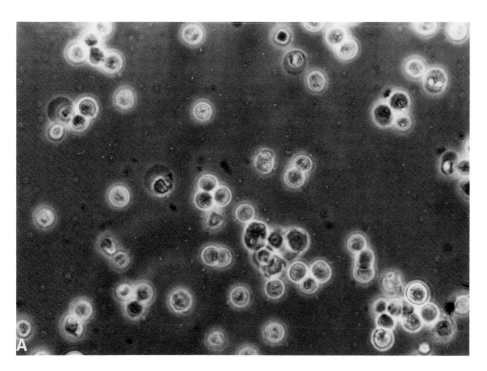

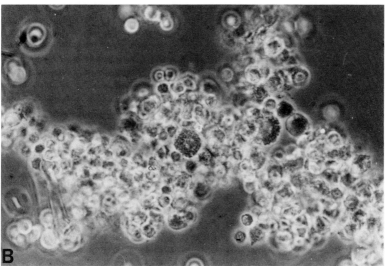

FIG. 1. Suspensions of mouse peritoneal cells incubated at 37°C for 1 hour: (a) control suspension, × 30; (b) Germiston virus, 1:100 dilution of stock preparation, added at beginning of incubation period, × 34.

of cells may be accomplished under either condition. In the suspension technique it is probably more satisfactory to grow cells separately under the conditions optimal to each. They then can be individually suspended in PBS in preparation for fusion. On the other hand, if homokaryons are to be formed, the cell cultures can be grown as monolayers on an appropriate surface and fused directly.

Whichever approach is used, it is at least theoretically important that the medium contain a minimum of serum constituents. Lipoproteins can act as nonspecific inhibitors of Germiston virus although we have not tested its effect on chloroform-inactivated virus. Virus is added to the cell preparation in the proportion desired; prior to mixing, the virus suspension and cell cultures should be cooled to 4°C. The virus is allowed to adsorb at this temperature for 1 hour, then incubated at 42°C for 2 hours. By shifting the temperature in this manner, it is possible to synchronize to some degree the fusion events. However, if maximum efficiency is not required, the process can be initiated from the start at 37°C or 42°C. With very high virus: cell ratios (> 300:1) fusion occurs very rapidly in a high proportion of cells even within a few minutes at 37°C. Figure 2 shows an early event in fusion between two mouse peritoneal macrophages. In this example, virus was mixed with suspended cells at a ratio of 30:1 and incubated for 5 minutes; it appears as though fusion has been initiated at multiple points between the villous processes of the two cells.

A curiosity encountered when very high particle: cell ratios (on the order of 1000:1) are used on monolayer cultures is that the entire culture may be fused into virtually a single syncytium (Fig. 3), a phenomenon we have termed "universal fusion." We have accomplished universal fusion in cultures as large as 25 cm². The cell boundaries disappear, but the nuclei apparently remain in their prefusion positions. The entire syncytium remains viable for some time and may be transferred intact by detaching it with EDTA-trypsin. Viability may be maintained for as long as 2 weeks, although at this point many of the nuclei have begun to exhibit karyolysis (Fig. 3B). We have not explored in any detail the characteristics of these tissues but have determined that universally fused Vero cells do not retain all the properties of normal Vero cells. Neither poliovirus nor Germiston virus replicates in syncytial Vero cells although each replicates well in normal Vero cells. The replication block is mediated by factors other than the loss of cell receptors for the virus (L. C. Olson, unpublished observations).

C.   Assessment of Fusion Efficiency

Depending on the intent of the experiment, it may or may not be important to estimate the efficiency of fusion. There are many ways to do this,

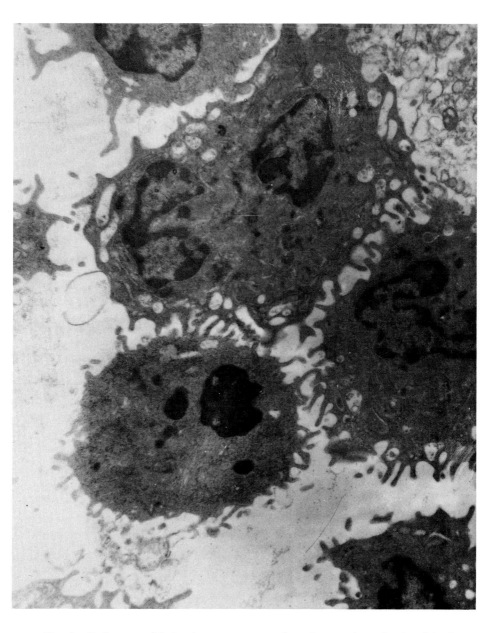

FIG. 2. Early stage of fusion between mouse peritoneal macrophages incubated for 5 minutes at 42° C with a 1:10 dilution of Germiston virus stock virus. × 16,250.

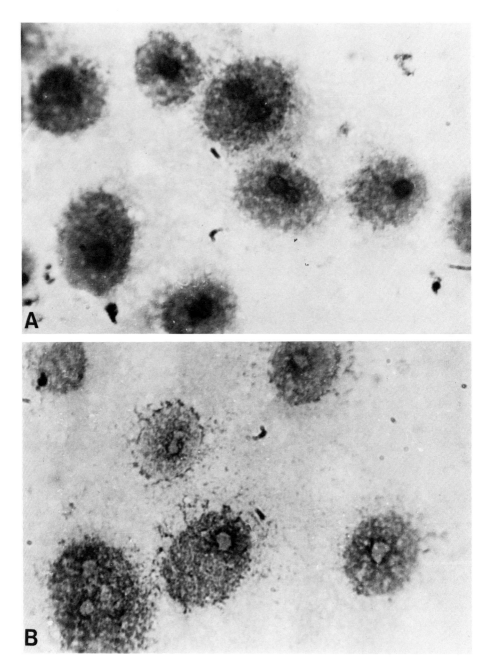

FIG. 3. Vero cell monolayers fused into a single syncytium with Germiston virus: (a) 2 hours after fusion; (b) 2 weeks after fusion. Wright's stain. × 100.

but that described by Kohn (1965) is convenient for monolayer cultures. To facilitate counting, the cells in the monolayers are dispersed with EDTA–trypsin applied for 2 minutes; after this they are stained with Wright's stain, and the nucleus: cell ratio is estimated.

IV. Comments

Cell fusion induced by Germiston offers a number of interesting features. As pointed out by Poste (1972), a valuable feature of virus-induced fusion may be that it offers a model for the study of membrane fusion in general. The rapidity and extent of fusion associated with Germiston virus may facilitate this approach.

The fraction of the Germiston virion responsible for fusion activity has not been investigated in any detail, but we assume that it resides with the envelope. We have demonstrated that fusion activity can be separated from (and is less dense than) the infectious particle after rate-zonal centrifugation on 30 to 60% (w/v) sucrose gradients. This would be consistent with the much more extensive studies by Hosaka and colleagues (Hosaka and Shimizu, 1972; Hosaka et al., 1974) wherein envelope fractions of Sendai virus have been shown to be the component of the virion possessing fusion activity.

The sequence of events leading to cell fusion can be inferred from what is known about virus–cell interactions (for general review, see Dales, 1973). Since Germiston virus has an envelope, it may be assumed that an early event is fusion of the viral envelope with the cell membrane. If there is a sufficient ratio of virus particles to cells, the chance is increased that a given particle may thus attach simultaneously to two or more cells and form a bivalent fusion bridge. This general concept has been beautifully demonstrated by the recent studies of Apostolov and Almeida (1972) on the fusion of Sendai virus with red cells.

In summary, Germiston virus is a useful tool for cell fusion, and may be unique in the extent of fusion that can be accomplished. The techniques involved in preparing the virus reagents are within the capabilities of most laboratories working with infectious agents. The infectivity may then be inactivated by simple means, eliminating the potential hazard to the laboratory worker of infectious virus. This also avoids the potential complications of using infectious virus for studies of the properties of fused cells.

ACKNOWLEDGMENTS

Dr. Neneng Djinawi participated in many of the earlier studies reported here. Part of the research was supported by a grant from the Rockefeller Foundation.

References

Ahkong, Q. F., Fisher, D., Tampion, W., and Lucy, J. A. (1975). *Nature (London)* **253**, 194–195.

Apostolov, K., and Almeida, J. D. (1972). *J. Gen. Virol.* **15**, 227–234.

Barbanti-Brodano, G., Oyanagi, S., Katz, M., and Koprowski, H. (1970). *Proc. Soc. Exp. Biol. Med.* **134**, 230–236.

Barski, G. (1970). *Int. Rev. Exp. Pathol.* **9**, 151–190.

Dales, S. (1973). *Bacteriol. Rev.* **37**, 103–135.

Davidson, R. L. (1971). *In Vitro* **6**, 411–426.

Davidson, R. L. (1974). *Annu. Rev. Genet.* **8**, 195–218.

Djinawi, N. K., and Olson, L. C. (1973). *Arch. Gesamite Virusforsch.* **43**, 144–151.

Gallaher, W. R., and Bratt, M. A. (1972). *J. Virol.* **10**, 159–161.

Hammon, W. McD. and Sather, G. E. (1969). *In* "Diagnostic Procedures for Viral and Rickettsial Infections" (E. H. Lennette and N. J. Schmidt, eds.), 4th ed., pp. 227–280. Am. Public Health Assoc. New York.

Handmaker, S. D. (1973). *Annu. Rev. Microbiol.* **27**, 189–204.

Hosaka, Y., and Shimizu, Y. K. (1972). *Virology* **49**, 627–639.

Hosaka, Y., Semba, T., and Fukai, K. (1974). *J. Gen. Virol.* **25**, 391–404.

Karabotsos, N., and Buckley, S. M. (1967). *Am. J. Trop. Med. Hyg.* **16**, 99–105.

Kohn, A. (1965). *Virology* **26**, 228–245.

Martin, F., and MacDonald, R. (1974). *Nature (London)* **252**, 161–163.

Olson, L. C., Sithisarn, P., and Djinawi, N. K. (1975). *J. Infect. Dis.* **131**, 119–128.

Poste, G. (1972). *Int. Rev. Cytol.* **33**, 157–252.

Poste, G., and Allison, A. C. (1973). *Biochim. Biophys. Acta* **300**, 421–465.

Warren, J., Weil, M. L., Russ, S. B., and Jeffries, H. (1949). *Proc. Soc. Exp. Biol. Med.* **72**, 662–664.

Watkins, J. F. (1971). *Methods Virol.* **5**, 1–32.

Chapter 3

Fusion of Mammalian Cells by Lipid Vesicles

GEORGE POSTE AND DEMETRIOS PAPAHADJOPOULOS

Department of Experimental Pathology,
Roswell Park Memorial Institute,
Buffalo, New York

I. Introduction

The fusion of different types of cultured mammalian cells to produce hybrid cells and heterokaryons is now an established research tool that is widely used in many areas of cell biology and virology. The use of viruses, Sendai virus in particular, to fuse cells was a major factor in the successful development of cell-fusion techniques in the 1960s, and viruses remain the most commonly used agents for inducing cell fusion. Although viruses are undoubtedly highly efficient agents for inducing cell fusion *in vitro*, there are certain practical difficulties associated with their use. First, the virus components responsible for inducing cell fusion are not fully characterized, and therefore the "fusion potential" of different virus preparations cannot be standardized. For example, Sendai virus preparations are quantitated in terms of hemagglutination units (HAU), yet virus stocks of equal HAU titer may differ markedly in their ability to fuse cells. In addition, the cell fusion potential of a stock virus strain may vary during continuous passage. This is particularly true for egg-grown viruses, such as Sendai, where prolonged passage of relatively undiluted virus will result in the emergence

23

and eventual predominance of incomplete (noninfective) virus with low fusion potential.

A second major difficulty is that the virus used to fuse cells may itself produce functional changes in the newly fused cells, including: alterations in cellular macromolecular synthesis (Mekler *et al.*, 1970; Yaoi and Amano, 1970; Fuchs and Kohn, 1971; Smith and Consigli, 1972; Hand and Tamm, 1973); stimulation of interferon production (Dianzani *et al.*, 1970; Clavell and Bratt, 1971) and interference with the replication of other viruses in the fused cells (ter Meulen *et al.*, 1972); chromosomal alterations (Stenman and Saksela, 1971); and changes in cell surface properties (Harris *et al.*, 1966; Okada *et al.*, 1975). These difficulties have prompted a search for alternative methods of fusing cells.

After the observations that lysolecithin was able to fuse cells (Howell and Lucy, 1969), a wide range of similar lipolytic agents have been evaluated as possible alternatives to viruses for routine use in the fusion and hybridization of mammalian cells. Even though a variety of lipophilic molecules with the capacity to fuse cells have now been identified (see Lucy, 1974), such agents appear to be of little practical value, since they are highly cytotoxic, owing mainly to their lytic action on cell membranes. The use of lysolecithin and similar lipolytic agents to induce cell fusion is accompanied by substantial cell damage, and the number of viable cells recovered is significantly lower than in cells treated with viruses (Croce *et al.*, 1971; Keay *et al.*, 1972; Papahadjopoulos *et al.*, 1973; Kataoka and Koprowski, 1975). Consequently, there is still a need for a simple procedure for inducing cell fusion that avoids the problems accompanying fusion induced by inactivated viruses or membrane lytic agents, such as lysolecithin.

Potential progress in this direction has been made with the recent demonstration that small vesicles (approximately 250–500 Å diameter) prepared from a variety of phospholipids are able to induce fusion of mammalian cells *in vitro* (Papahadjopoulos *et al.*, 1973; Martin and MacDonald, 1974). Lipid vesicles compare favorably with inactivated Sendai virus in their ability to fuse cells, yet are devoid of the cytotoxic properties associated with lysolecithin. In this chapter, we present an outline of the methods used in this laboratory to produce lipid vesicles for inducing fusion and hybridization of cultured mammalian cells.

II. Preparation of Lipid Vesicles

A general background on the preparation and properties of different types of lipid vesicles is given in Chapter 4 of this volume by Poste *et al.*

Vesicles for cell fusion experiments are prepared by ultrasonication of phospholipids dispersed in an aqueous buffer solution. The resulting product is composed predominantly of small (about 250–500 Å in diameter) vesicles surrounded by a single lipid bilayer. These are referred to as unilamellar vesicles. To prepare unilamellar vesicles for cell fusion experiments, approximately 10–20 μmoles of lipid (determined as micromoles of phosphate) are dispersed in 2–4 ml of an aqueous buffer containing 100 mM NaCl, 2 mM N-tris(hydroxymethyl)methyl-2-aminoethanesulfonic acid (TES) and 0.1 mM EDTA adjusted to pH 7.4. After shaking on a Vortex mixture for 10 minutes at 37° C, the solution is then sonicated for 1 hour in a bath-type sonicator under nitrogen. For naturally occurring phospholipids, such as phosphatidylcholine (PC) and phosphatidylserine (PS), sonication can be done at room temperature (22°–24°C), but for synthetic phospholipids with higher transition temperatures it is necessary to increase the temperature during sonication to above the transition temperature of the lipid(s) involved (see Chapter 4 of this volume). After sonication, the final preparation of unilamellar vesicles is allowed to equilibrate at room temperature for 1 hour under nitrogen before use. The vesicle preparation is then diluted in phosphate-buffered saline (PBS) before being used to treat cells at a final concentration of between 10^2 and 10^8 vesicles per cell (see Section IV). For determining the necessary dilution factor to achieve such vesicle to cell ratios, we work from the calculation that 1 μmole of sonicated lipid contains approximately 2×10^{14} vesicles (mean vesicle diameter 250 Å; 3000 lipid molecules per vesicle) (see Papahadjopoulos et al., 1973).

All the phospholipids used in the preparation of vesicles are isolated or synthesized in our laboratory. Methods for the production of various phospholipids are discussed in Chapter 4 of this volume by Poste et al.

III. Cell Fusion Techniques

Successful fusion of cultured, mammalian cells by unilamellar lipid vesicles has been achieved using similar methods to those employed in virus-induced cell fusion, the vesicles merely being substituted for virus (Papahadjopoulos et al., 1973; Martin and MacDonald, 1974).

For fusion of cells in suspension (Papahadjopoulos et al., 1973), cell populations (5×10^6 to 2×10^7 cells per milliliter) are incubated in a total volume of 1–4 ml with vesicles (typical dose 10^4 vesicles per cell added in 0.5 to 1.0 ml of PBS adjusted to pH 8.0) for 2 hours at 37°C with shaking at intervals of 5–10 minutes. The cell suspension is then pelleted by centrifugation and washed twice with prewarmed PBS; the cells are finally re-suspended in fresh culture medium and inoculated onto coverslips for sub-sequent measurement of the extent of cell fusion. The coverslip cultures are incubated for a further 16–24 hours at 37°C, after which they are fixed and stained with May–Grünwald–Giemsa. The extent of cell fusion, ex-pressed as the percentage of polykaryocytosis, is then measured by counting the number of nuclei present in multinucleate cells (polykaryocytes) and expressing this as a percentage of the total number of nuclei present in the same microscope field. At least 20 fields should be counted from each cover-slip, and ideally at least four replicate coverslips should be examined for each sample. If preferred, similar measurements can be made on fixed and stained cells attached to the surface of plastic petri dishes.

To induce fusion of cells in monolayer cultures on either glass coverslips or plastic petri dishes (Papahadjopoulos *et al.*, 1973), similar numbers of vesicles in 1.0 ml of PBS (pH 8.0) are added to near-confluent or confluent cell cultures for 2 hours at 37°C, after which the cultures are washed twice with prewarmed PBS to remove the vesicles and fresh medium is added. The cultures are then incubated at 37°C for 16–24 hours, after which the cells are fixed and stained and the extent of cell fusion is measured as described above.

In the case of vesicles containing lysolecithin incorporated into the vesicle membrane, cells are exposed to vesicles for 10 minutes rather than 2 hours. This shorter incubation time has been found necessary to limit the cytolytic activity of lysolecithin (Papahadjopoulos *et al.*, 1973).

Interspecific mammalian somatic cell hybrids were produced by fusion of thymidine kinase (TK)-deficient C1-1D mouse cells (Dubbs and Kit, 1964) and hypoxanthine guanine phosphoribosyltransferase-deficient (HGPRT) WI-18Va2 human cells (Weiss *et al.*, 1968) with subsequent selec-tion of resulting hybrid cells in HAT medium (Littlefield, 1966). Equal num-bers (5×10^6) of each cell type were suspended in 1.0 ml of serum-free Eagle's basal medium (pH 8.0) and incubated for 2 hours at 37°C with ves-icles (10^4 per cell). An aliquot of the mixed cell population was then inocula-ted onto coverslips and incubated for a further 16–24 hours, after which the coverslips were fixed and stained and the extent of cell fusion meas-ured as described above. The remainder of the mixed-cell population was inoculated into 35-mm plastic petri dishes (1×10^6 cells per dish) in 2.0 ml of HAT medium (pH 8.0). The medium was changed and replaced with fresh HAT medium (pH 8.0) 1, 3, 6, 9, and 12 days later. The culture dishes were finally fixed and stained 14 days after fusion, and the number of colonies of surviving hybrid cells were counted. The hybrid nature of cells surviving in HAT was confirmed by their expression of both mouse-specific and human-specific enzymes (lactate dehydrogenase, glucose-6-phosphate dehydrogenase, isocitrate dehydrogenase, and pyruvate kinase) as demonstrated by starch gel electrophoresis (Ruddle and Nichols, 1971; Shows, 1974).

IV. Fusion and Hybridization of Mammalian Cells by Lipid Vesicles: Comparison with Virus and Chemically Induced Fusion

The ability of vesicles of differing lipid composition to fuse mouse 3T3 and L929 cells and hamster BHK21 cells is shown in Fig. 1.

Maximum cell fusion in both monolayer and suspension cell cultures

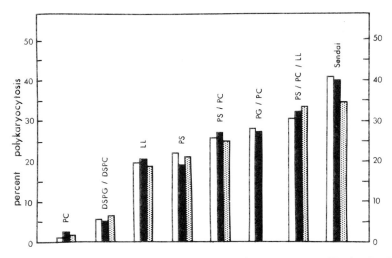

FIG. 1.   Fusion of mouse 3T3 (□), mouse L929 (■) and hamster BHK21 (□) cells by lipid vesicles, inactivated Sendai virus, and lysolecithin. Monolayer cell cultures were treated with vesicles of the indicated composition (10⁴ vesicles per cell) for 2 hours at 37°C as described in Section III, or with 2000 HAU of β-propiolactone-inactivated Sendai virus for 1 hour at 37°C, as described by Poste (1973). For fusion of cells by lysolecithin similar numbers of cells were incubated as a loose pellet with 200 μg of lysolecithin per millileter for 5 minutes at 37°C as described by Papahadjopoulos *et al.* (1973). The extent of cell fusion, expressed as the percentage of polykaryocytosis, was determined 24 hours after the various treatments. PC, egg yolk phosphatidylcholine vesicles; PS, phosphatidylserine vesicles; PS/PC, 10% phosphatidylserine in phosphatidylcholine vesicles; PG/PC, 10% phosphatidylglycerol in phosphatidylcholine vesicles; DSPG/DSPC, 10% distearoylphosphatidylglycerol in distearoylphosphatidylcholine vesicles; PS/PC/LL, 10% phosphatidylserine in phosphatidylcholine vesicles containing 3% lysolecithin; LL, free lysolecithin; and Sendai, β-propiolactone-inactivated Sendai virus.

is achieved at a dose of 10^4 vesicles per cell (Papahadjopoulos et al., 1973; Martin and MacDonald, 1974). Treatment of cells with lipid vesicles over the dose range 10^2 to 10^8 vesicles per cell for 2 hours at 37°C does not result in any significant reduction in cell viability (Papahadjopoulos et al., 1973; Martin and MacDonald, 1974). The dose of 10^4 vesicles per cell for optimum fusion is remarkably constant, and significant variation in this ratio has not been found between different cell types or between vesicles of different composition. This latter point is particularly useful since it enables direct comparison of the efficiency of vesicles of differing composition to induce fusion at a uniform vesicle-to-cell ratio. Thus, the differences in the cell fusion potential of vesicles prepared from different phospholipids shown in Fig. 1 can be attributed to the specific properties of the vesicles rather than to a failure to provide optimal conditions for fusion.

The results in Fig. 1 serve to illustrate a number of general points concerning vesicle-induced cell fusion.

First, the extent of cell fusion induced by vesicles prepared from certain phospholipid mixtures compares favorably with that induced by inactivated Sendai virus (Fig. 1).

Second, vesicles are not only more efficient than lysolecithin in fusing cells (Fig. 1), but, as mentioned above, vesicles are completely devoid of the cytotoxicity associated with the use of lysolecithin to fuse cells. Inclusion of lysolecithin directly into the membrane of vesicles (PS/PC/LL vesicles in Fig. 1) further enhances the ability of vesicles to fuse cells (Fig. 1), but this advantage is more than offset by a marked increase in the cytotoxicity of such vesicles (Papahadjopoulos et al., 1973).

Third, the presence or the absence of surface charge on vesicles and the physical state of the lipids in the vesicle membrane are of major importance in determining the fusion capacity of vesicles. Neutral vesicles (PC) are unable to fuse cells (Fig. 1), and vesicles must bear either a net negative (Papahadjopoulos et al., 1973) or a net positive (martin and MacDonald, 1974) surface charge in order to fuse cells. In addition to surface charge, the ability of vesicles to fuse cells requires that the lipids in the vesicle membrane be "fluid" at the experimental temperature, that is, above their gel-to-liquid crystalline transition temperature (T_c). As shown in Fig. 1, vesicles in which the lipids are fluid at 37°C (PS; PG/PC; PS/PC) produce significantly greater cell fusion than vesicles of comparable size composed of lipids that are "solid" at 37°C (DSPG/DSPC).

In addition to fusion of cells of the same type, lipid vesicles can also be used to fuse different cell types to produce hybrid cells (Table I). As shown in Table I, negatively charged vesicles prepared from lipids that are fluid at 37°C (PS/PC) are equally as effective as inactivated Sendai virus in producing interspecific hybrid cells by fusion of mouse and human cells. In contrast, lysolecithin appears to be of little value for hybridization of these

TABLE I

FORMATION OF HYBRID CELLS BETWEEN MOUSE CI-ID AND
HUMAN WI-18Va2 CELLS AFTER TREATMENT WITH LIPID VESICLES,
INACTIVATED SENDAI VIRUS, OR LYSOLECITHIN

Fusion agent	Initial percent polykaryocytosis[a]	Cell Viability (%)[a]	Average number of hybrid colonies per 35-mm petri dish 14 days after fusion
None (co-cultivation)	4.8	94	11.5
Vesicles (10% PS/PC)[b]	23.7	91	48.4
Sendai virus[c]	32.9	95	53.5
Lysolecithin	11.4	32	6.4

[a] Determined 24 hours after treatment with fusion agent.
[b] Fusion induced as described in Section III. PS/PC, 10% phosphatidylserine in phosphatidylcholine vesicles.
[c] Fusion induced as described in the legend to Fig. 1.

particular cell types (Table I). The very low yield of hybrid cells obtained from lysolecithin-treated mixed-cell populations probably reflects the marked cytotoxicity of this compound (Table I).

It is also of interest to note in Table I that although the extent of initial cell fusion produced by lipid vesicles is less than that produced by Sendai virus, the yield of hybrid cells from vesicle- and virus-treated cells is very similar. A possible explanation for this finding is provided by additional data on the size range of multinucleate cells found in cell populations after treatment with vesicles and with Sendai virus (Fig. 2). In vesicle-treated

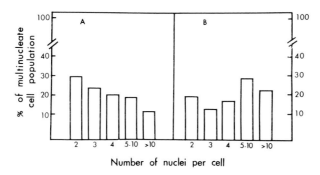

FIG. 2. Size range of multinucleate cells in mixed populations of mouse Cl-ID cells and human WI18-Va2 cells after treatment with lipid vesicles or inactivated Sendai virus. (A) At 24 hours after incubation with lipid vesicles (10% phosphatidylserine in phosphatidylcholine; 10⁴ vesicles per cell) for 2 hours at 37°C. (B) At 24 hours after incubation with 2000 HAU of β-propiolactone-inactivated Sendai virus.

cell cultures multinucleate cells with two to five nuclei predominate, whereas in cultures treated with Sendai virus a greater number of cells containing more than 5 nuclei are present (Fig. 2). Since the opportunity for formation of a hybrid cell is probably restricted to fused cells containing two or perhaps three nuclei, the higher proportion of such cells in vesicle-treated cell populations may explain why similar numbers of hybrid cells were obtained from both vesicle- and virus-treated cell populations, even though the overall extent of cell fusion is greater in virus-treated cell populations.

V. Discussion

The mechanism by which lipid vesicles induce cell fusion is not known. One possibility is that cell fusion is a by-product of fusion of vesicles with the cellular plasma membrane. If fusion were to occur simultaneously on opposite sides of a vesicle in contact with the plasma membrane of two cells, the vesicle would create a "bridge" between the two cells. Subsequent enlargement of such bridges might then lead to complete fusion of the two cells. In this sense, cell fusion induced by vesicles is considered as analogous to cell fusion induced by lipid-enveloped viruses, such as Sendai virus, where virus particles are believed to function as similar bridges between apposed cells (in this case fusion occurring between the virus envelope and the cellular plasma membrane) (Okada, 1962; Hosaka, 1970; Apostolov and Poste, 1972; Poste, 1972).

An additional possibility is that fusion of vesicles with the plasma membrane (see Chapter 4 of this volume) and the resulting incorporation of vesicle lipids into the plasma membrane might in some way render such regions of the plasma membrane more susceptible to fusion. Higher susceptibility to fusion could result from several reasons, two of which are outlined below. One possible reason is that fusion of lipid vesicles with the plasma membrane produces a "dilution" of the other plasma membrane components in favor of larger cell surface areas in which the lipid bilayer is exposed. Under these conditions, the close apposition of the two lipid bilayers from two adjacent cells would be a much more frequent event, thereby enhancing the possibility of fusion. A second possible reason, which is an extension of the first, is that the fusion of negatively charged vesicles with the plasma membrane increases the surface charge density of the lipid bilayer of the plasma membrane, thus enhancing the frequency of charge–charge interactions between lipid bilayers of adjacent cells. This second possibility arises because of the possible existence of lipid asymmetry in the plasma membrane.

Evidence for trans-bilayer asymmetry in the distribution of particular lipid species has been found in the erythrocyte (see Singer, 1974). If similar asymmetry exists in other cells, the external half of the lipid bilayer of the plasma membrane would contain the neutral lipids, and the charged acidic phospholipids would be in the inner half of the bilayer. Thus, fusion of charged vesicles with the plasma membrane might drastically alter the existing asymmetry and result in the presence of charged lipids in the outer half of the bilayer. This would be expected to increase the capacity of the membrane to make charge–charge interactions, which may be of paramount importance in order to bring the lipid bilayers of two plasma membranes into close apposition as a prerequisite to their fusion. Recent calculations based on X-ray diffraction studies have indicated that the energy required to bring two neutral lipid bilayers in close apposition (15 Å separation) is approximately 10 kT (P. Rand, personal communication). This energy barrier could not be overcome by simple Brownian movement (1.5 kT). It seems reasonable therefore that some form of charge–charge interaction would be required in order to achieve close contact (less than 15 Å) of two lipid bilayers.

The data summarized in Section IV indicate that, in order to induce cell fusion, vesicles must bear a net surface change and be composed of phospholipids that are fluid at 37°C. Neutral vesicles and charged vesicles in which the lipids are "solid" at 37°C are much less efficient in inducing cell fusion. In this respect, it is of interest to note that the latter two classes of vesicle display little or no capacity to fuse with other vesicles (Papahadjopoulos *et al.*, 1974; Kantor and Prestegard, 1975) and are unable to fuse with the plasma membrane of cells cultured *in vitro* (Poste *et al.*, this volume, Chapter 4). In contrast, charged vesicles composed of lipids that are fluid at experimental temperatures undergo rapid and extensive fusion with other vesicles (Papahadjopoulos and Poste, 1975; Papahadjopoulos *et al.*, 1974) and appear to be incorporated into cells *in vitro* by fusion with the cellular plasma membrane (Poste *et al.*, this volume, Chapter 4).

REFERENCES

Apostolov, K., and Poste, G. (1972). *Microbios* **6**, 247–261.
Clavell, L. A., and Bratt, M. A. (1971). *J. Virol.* **8**, 500–508.
Croce, C. M., Sawicki, W., Kritschevsky, D., and Koprowski, H. (1971). *Exp. Cell Res.* **67**, 427–435.
Dianzani, F., Gagnoni, S., Buckler, C. E., and Baron, S. (1970). *Proc. Soc. Exp. Biol. Med.* **133**, 324–328.
Dubbs, D. R., and Kit, S. (1964). *Exp. Cell. Res.* **33**, 19–28.
Fuchs, P., and Kohn, A. (1971). *J. Virol.* **8**, 695–700.
Hand, R., and Tamm, I. (1973). *J. Virol.* **11**, 223–231.
Harris, H., Watkins, J. F., Ford, C. E., and Schoefl, G. L. (1966). *J. Cell. Sci.* **1**, 1–30.

Hosaka, Y. (1970). *J. Gen. Virol.* **8**, 43–54.

Howell, J. I., and Lucy, J. A. (1969). *FEBS Lett.* **4**, 147–150.

Kantor, H. L., and Prestegard, J. H. (1975). *Biochemistry* **14**, 1790–1794.

Kataoka, T., and Koprowski, H. (1975). *Proc. Soc. Exp. Biol. Med.* **149**, 447–451.

Keay, L., Weiss, S. A., and Wildi, B. S. (1972). *In Vitro* **8**, 19–25.

Littlefield, J. W. (1966). *Exp. Cell Res.* **33**, 19–28.

Lucy, J. A. (1974). *FEBS Lett.* **40**, S106–S111.

Martin, F., and MacDonald, R. (1974). *Nature (London)* **252**, 161–163.

Mekler, L. B., Schlyankevich, M. A., and Shevliaghyn, V. J. (1970). *Arch. Gesamk Virusforsch.* **30**, 309–315.

Okada, Y. (1962). *Exp. Cell. Res.* **26**, 98–107.

Okada, Y., Koseki, I., Kim, J., Maeda, Y., Mashimoto, T., Kanno, Y., and Matsui, Y. (1975). *Exp. Cell Res.* **93**, 368–378.

Papahadjopoulos, D., and Poste, G. (1975). *Biophys. J.* **15**, 945–948.

Papahadjopoulos, D., Poste, G., and Schaeffer, B. E. (1973). *Biochim. Biophys. Acta* **323**, 23–24.

Papahadjopoulos, D., Poste, G., Schaeffer, B. E., and Vail, W. J. (1974). *Biochim. Biophys. Acta* **352**, 10–28.

Poste, G. (1972). *Int. Rev. Cytol.* **33**, 157–252.

Poste, G. (1973). *Methods Cell Biol.* **7**, 211–249.

Ruddle, F. H., and Nichols, E. A. (1971). *In Vitro* **7**, 120–131.

Shows, T. B. (1974). *In* "Somatic Cell Hybridization" (R. L. Davidson and F. de la Cruz, eds.), pp. 16–25. Raven Press, New York.

Singer, S. J. (1974). *Annu. Rev. Biochem.* **43**, 805–833.

Smith, G. L., and Consigli, R. A. (1972). *J. Virol.* **10**, 1091–1097.

Stenman, S., and Saksela, E. (1971). *Hereditas* **69**, 1–14.

ter Meulen, W., Koprowski, H., Iwasaki, Y., Kackell, Y. M., and Muller, D. (1972). *Lancet* **2**, 1–5.

Weiss, M. C., Ephrussi, B., and Scaletta, L. J. (1968). *Proc. Natl. Acad. Sci. U.S.A.* **59**, 1132–1135.

Chapter 4

Lipid Vesicles as Carriers for Introducing Biologically Active Materials into Cells

GEORGE POSTE,[1] DEMETRIOS PAPAHADJOPOULOS,[1]
AND WILLIAM J. VAIL[2]

[1] Department of Experimental Pathology, Roswell Park Memorial Institute, Buffalo, New York.

[2] Department of Microbiology, University of Guelph, Guelph, Ontario, Canada.

I. Introduction

Lipid vesicles have been used extensively over the past 10 years as model membranes from which to obtain information on the physicochemical organization of lipid bilayers and to study the interaction of various membrane-active drugs with lipids (Papahadjopoulos and Kimelberg, 1973; Bangham *et al.*, 1974). More recently, experiments in several laboratories have shown that mammalian cells *in vitro* and *in vivo* will incorporate large numbers of lipid vesicles without cytotoxic effects. Since a wide variety of materials can be entrapped inside vesicles and various glycoproteins and glycolipids can be inserted into the vesicle membrane, cellular uptake of vesicles of defined composition offers a potential method for modifying cellular composition and for introducing nonpermeable biologically active materials into cells.

In this chapter we outline current methods for the production of different types of lipid vesicles and discuss the mechanism by which vesicles are incorporated into cells. Recent work on the use of lipid vesicles as "carriers" to introduce biologically active materials into cells *in vitro* and *in vivo* is also reviewed, and a brief outline is presented on the feasibility of developing methods in which vesicles could be targeted to specific parts of the cell and also to particular cell types.

II. Preparation of Lipid Vesicles

A.   Multilamellar Vesicles

Multilamellar vesicles (MLV), or so-called liposomes (Bangham *et al.*, 1965), are simply the liquid–crystalline structures obtained when amphipathic lipids, such as phospholipids, are dispersed in water or aqueous salt solutions (Fig. 1). Electron microscopic and X-ray diffraction studies have established that in these structures the lipid is organized into concentric bimolecular lamellae, each separated from its neighbor by an interspersed aqueous space of variable thickness. Each bilayer lamella within the vesicle is a completely enclosed sac. Multilamellar vesicles are characteristically heterogeneous in size (0.5–10 μm in diameter), shape, and number of lamellae per vesicle.

A typical procedure for the production of MLV involves the solution of the selected lipid or lipid mixture (usually 10 μmoles of lipid) in a limited volume (1.0 ml) of chloroform in a round-bottom flask or tube to deposit

FIG. 1. Freeze-fracture electron micrograph of multilamellar lipid vesicle prepared from 10% phosphatidylserine in phosphatidylcholine by vortex shaking of the lipid mixture in 100 m*M* NaCl buffer at pH 7.4. The direction of shadowing is indicated by the arrowhead in the lower left-hand corner. Scale bar = 0.1 μm (1000 Å).

the lipid as a uniform film on the glass surface at the bottom of the vessel. The chloroform is then evaporated under vacuum. The dried lipids are next dispersed in a limited volume of water or aqueous buffer using a Vortex mixer. Some investigators also add a few glass beads to the vessel at this point to facilitate removal of the lipid film from the glass and also assist in its dispersal. Displacement of the lipid film from the wall of the vessel and resulting swelling of lipids to form MLV is indicated by an initial whitening of the film followed by its dispersal into the aqueous phase to give a suspension with a milky appearance and free of particulate matter. Detailed descriptions of the preparation of MLV are given by Bangham et al. (1965), Papahadjopoulos and Watkins (1967), and Kinsky (1975).

B. Small Unilamellar Vesicles

Ultrasonication of suspensions of MLV results in their breakdown to form small unilamellar vesicles (SUV). These are much smaller structures (250–1000 Å in diameter) comprising an internal aqueous space enclosed by a single lipid bilayer (Fig. 2). SUV populations are characteristically more homogeneous in size than MLV. The limiting size of SUV is probably defined by the constraint of the small radius of curvature on the packing of lipid molecules within the vesicle membrane. The time of sonication required for complete conversion of MLV into a homogeneous population of SUV depends on the type of lipid, ionic strength of the buffer solution, and also temperature (see Papahadjopoulos and Kimelberg, 1973; Bangham et al., 1974).

The major potential difficulty associated with the production of SUV concerns the risk of oxidation and hydrolysis of vesicle lipids during ultrasonication. Although metallic probe sonicators are certainly effective in imparting sufficient shearing energy to convert MLV to SUV, they suffer from the problem that heat transfer from the probe to the lipids may result in significant chemical alteration of the lipids. In addition, minute fragments of metal from the probe can contaminate the lipid suspension and on dissolving in the lipids may promote oxidative degradation of the lipids (Hauser, 1971). In order to avoid oxidative or hydrolytic degradation of lipids, the sonication should be done in an atmosphere of nitrogen or argon and under conditions of controlled temperature. We routinely use a bath-type sonicator rather than a probe sonicator. A 20-ml capacity tube containing the MLV suspension (usually between 2 and 20 μ moles of lipid in 1–5 ml) is flushed with nitrogen and sealed and then placed at the focal point of radiant energy in the sonication bath with the bottom of the tube 0.5 inch from the water surface in the bath. The precise resonant volume depends on the geometry of the bath and the amount of water in the bath. The bath temperature is rigorously controlled throughout the sonication process. Conversion of the MLV preparation to SUV is indicated by gradual clearing of the lipid suspension during sonication. Prolonged sonication for up to 1 hour produces a homogeneous population of SUV. In addition, effective separation of SUV from any remaining MLV can be achieved by gel filtration of the sonicated preparation on Sepharose 4B (Fig. 3) or ultracentrifugation at 100,000 g for 1 hour with recovery of SUV in the supernatant.

FIG. 2. Freeze-fracture electron micrograph of sonicated phosphatidylserine vesicles in 100 mM NaCl buffer, pH 7.4. Scale bar = 0.1 μm (1000 Å). Reproduced, with permission, from Papahadjopoulos *et al.* (1975a).

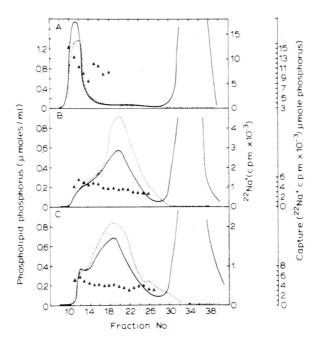

FIG. 3. Elution of phospholipid vesicles after passage through Sepharose 4B columns: (A) phosphatidylserine MLV; (B) phosphatidylserine SUV; (C) phosphatidylserine–cholesterol (equimolar ratio) SUV. Lipid, 11 μmoles in 2.0 ml, was applied to a 1.5 cm $\times$ 29 cm column which was equilibrated and eluted at room temperature by gravity. Fraction volume 1.1 ml; flow rate, 3 minutes per fraction; aqueous phase, 100 mM NaCl, 2 mM histidine; 2 mM TES, 0.1 mM EDTA (pH 7.4) containing ^{22}Na$^+$ tracer. Between 70 and 80% of the applied lipid was recovered from the column. ———, Elution profile of ^{22}Na$^+$; ..., elution profile of lipid (measured as total phosphorus); ▲, ratio of ^{22}Na$^+$ per micromole of phosphorus indicating the amount of isotope captured within vesicle. Reproduced, with permission, from Papahadjopoulos *et al.* (1972).

Two other methods described recently for the preparation of small vesicles do not involve sonication. One method involves injection of an ethanolic solution of phospholipid into a large volume of aqueous buffer (Batzri and Korn, 1973). This method can produce a homogeneous population of small vesicles in a short time. However, the possible effects of the ethanol and the large aqueous volume required present technical disadvantages in the use of this technique for capturing small water-soluble molecules. The other method involves solubilization of the phospholipid in detergent with subsequent slow removal of the detergent by dialysis. This has been used in several laboratories for the reconstitution of lipids and membrane proteins in vesicular form (Kagawa, 1972; Razin, 1972). However, the size and morphology of the vesicles obtained has not been defined or studied in detail.

C. Large Unilamellar Vesicles

We have recently described a third class of lipid vesicle, which is produced by fusion of SUV to form much larger vesicles, most of which are surrounded by a single continuous bilayer rather than multiple lipid lamellae (Papahadjopoulos *et al.*, 1975a). These structures have been designated large unilamellar vesicles (LUV). Rapid and extensive fusion of SUV of the type shown in Fig. 2 can be triggered by the addition of Ca^{2+} with resulting coalescence of vesicles to form large planar sheets of lipid that roll up to form cylinders (Fig. 4). When these "cochleate" cylinders are treated with EDTA, they undergo a further dramatic structural transformation to form large (0.2–1 μm in diameter) mostly unilamellar vesicles (Fig. 5). A schematic representation of the suggested events involved in the conversion of SUV to LUV is shown in Fig. 6.

The initial report on the formation of LUV (Papahadjopoulos *et al.*, 1975a) concerned vesicles prepared from phosphatidylserine, and it is not yet clear whether other phospholipids will form similar structures. Another technique resulting in the formation of large uni- or paucilamellar vesicles was reported several years ago (Reeves and Dowben, 1969). This was achieved with egg lecithin as the lipid component, under special conditions that exclude the presence of electrolyte or solute protein. These vesicles have not been used extensively by other investigators, and their usefulness as carriers has not been explored.

III. Vesicle Composition

Although there is no common recipe for the preparation of lipid vesicles for interaction with cells, most laboratories use a basic mixture of a phospholipid (frequently egg lecithin), cholesterol, and a charged amphiphile in varying molecular ratios, typically within the range 6:3:1 to 5:5:0.5. The precise composition of the mixture can, however, be varied considerably, and the mixture selected depends largely on the particular goal of the experiment for which vesicles are to be used. In this regard, the following comments may provide some useful guidelines.

Most phospholipids studied so far have been found to spontaneously form MLV when dispersed in aqueous solutions and can thus, in turn, be used to form SUV. These include: natural and synthetic phosphatidylcholines (PC) (Bangham *et al.*, 1965; de Gier *et al.*, Demel *et al.*, 1968); phosphatidylserine (PS) (Papahadjopoulos and Bangham, 1966); and a variety of other acidic phospholipids (Paphadjopoulos and Miller, 1967; Papahadjopoulos and Ohki, 1970). Some difficulties have been encountered with phospha-

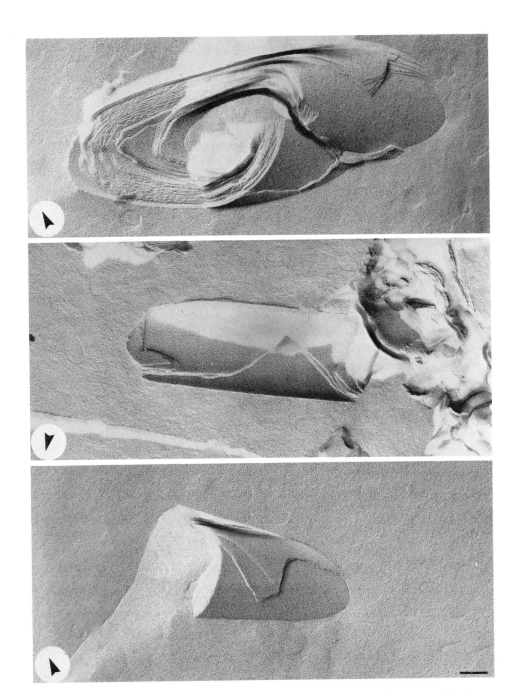

FIG. 4. Freeze-fracture electron micrographs of sonicated phosphatidylserine vesicles in 100 mM NaCl after addition of 10 mM CaCl$_2$ showing "cochleate" lipid cylinders (seen in side view). Arrowheads indicate direction of shadowing. Scale bar = 0.1 μm (1000 Å). Reproduced, with permission, from Papahadjopoulos *et al.* (1975a).

FIG. 5. Freeze-fracture electron micrograph of sonicated phosphatidylserine after addition of CaCl₂ (10 mM) and EDTA (15 mM) showing large unilamellar vesicles. Arrowhead shows direction of shadowing. Scale bar = 0.1 μm (1000 Å). Reproduced, with permission, from Papahadjopoulos *et al.* (1975a).

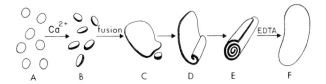

Fig. 6. Schematic representation of the effect of Ca^{2+} on sonicated phosphatidylserine vesicles leading to the formation of cochleate cylinders. (A) Sonicated vesicles in NaCl buffer before addition of Ca^{2+}, shown as spheroidal particles with an aqueous interior and a single lipid bilayer shell. (B, C, and D) Suggested intermediate steps involving the aggregation and fusion of the spheroidal vesicles into cochleate cylinders: (B) represents the step in which Ca^{2+} ruptures the vesicles, forming flat bilayer disks; (C) indicates fusion of the disks into large sheets in order to minimize hydrocarbon–water contact; (D) indicates the beginning of the folding of the flat sheet into a cylinder. (E) Cochleate cylinders formed by spiral folding of the continuous planar membrane sheets shown in (C) and (D). (F) Large unilamellar vesicles created by unfolding of the cochleate spiral membranes after incubation with EDTA. In addition to the sequence outlined in steps (B) to (E), several other processes could contribute to the formation of the cochleate cylinders shown in (E): (1) aggregation of unilameller vesicles could precede the rupture and fusion steps; (2) cochleate cylinders could be formed by the spiraling of several separate bilayer sheets together as shown in Fig. 4 (top); and (3) although the initial formation of cochleate cylinders might require membrane sheets of a "critical" size, the further growth of these cylinders could proceed by continuous fusion with unilamellar vesicles. Reproduced, with permission, from Papahadjopoulos et al. (1975a).

tidylethanolamine (PE), which tends to give suspensions that aggregate when dispersed in physiological salt solutions (Papahadjopoulos and Watkins, 1967). Stable vesicles can be obtained, however, by mixing PE with PC (Papahadjopoulos and Watkins, 1967) and by dispersal of PE in low ionic strength solutions at pH above 8.0 (Papahadjopoulos and Miller, 1967). The total lipid extracts from various body tissues or cells can also be used to form vesicles (Haxby et al., 1968; Alving et al., 1969; Kinsky et al., 1969; Knudsen et al., 1971; Papahadjopoulos et al., 1973a).

The ability of a phospholipid to form MLV when dispersed in aqueous solution is related to its transition temperature (T_c), that is, the temperature at which the hydrocarbon chains of the lipid molecule melt and convert from a gel (solid) to a liquid–crystalline (fluid) phase (see Section IV,A). Natural phospholipids are usually above this point at ambient temperature (Table I) and thus will readily form MLV. Most synthetic phospholipids with long-chain saturated acyl esters will also form similar structures if the temperature is raised close to or above their T_c (Chapman and Fluck, 1966).

The source and purity of phospholipids used in preparing vesicles provides an obvious variable in studies done in different laboratories. In our own studies, phospholipids are synthesized and purified in the laboratory and are chromatographically pure. Detailed methods have been described for the

TABLE I

THERMAL TRANSITIONS OF DIFFERENT PHOSPHOLIPIDS[a]

Lipid	Temperature for gel-to-liquid crystalline phase transition (°C)	Reference[b]
Dibehenoyl (C22) phosphatidylcholine (PC)	75	1
Distearoyl (C18) PC	58	1
Dipalmitoyl (C16) PC	41	1
Dimyristoyl (C14) PC	23	1
Dioleyl (C18:1) PC	−22	1
Egg PC	−15/−7	1
Dimyristoyl (C14) phosphatidylethanolamine (PE)	48	2
Distearoyl (C18) phosphatidylglycerol (PG)	53	3
Dipalmitoyl (C16) PG	41	4
Dimyristoyl (C14) PG	23	3
Dilauryl (C12) PG	5	5
Dipalmitoyl (C16) phosphatidic acid (PA), pH 6.5	67	4
Dipalmitoyl (C16) PA, pH 9.1	58	4
Dimyristoyl (C14) PA, pH 6.0	52	6
Dimyristoyl (C14) PA, pH 10.0	26	6
Phosphatidylserine, bovine brain (PS)	5	4
Sphingomyelin, bovine brain	40	7
Cerebroside, bovine brain	65	7

[a] Values obtained by differential scanning calorimetry in the presence of excess water except for dimyristoyl PA which was studied by fluorescence polarization techniques.

[b] 1. Ladbrooke and Chapman (1969); 2. Keough and Chapman, cited by Oldfield and Chapman (1972); 3. Kimelberg and Papahadjopoulos (1974); 4. Jacobson and Papahadjopoulos (1975); 5. Verkleij et al. (1974); 6. Trauble and Eibl (1974); 7. Oldfield and Chapman (1972).

purification of most phospholipids (see Spanner, 1973). We have routinely used, and scaled up several methods (Papahadjopoulos and Miller, 1967; Papahadjopoulos et al., 1973a) for the purification of large quantities of several phospholipids, such as egg lecithin (Hanahan et al., 1957), phosphatidylserine from bovine brain (Folch, 1942; Rouser et al., 1961), phosphatidylethanolamine from egg (Lea et al., 1955), sphingomyelin from beef brain (Spanner, 1973), dipalmitoylphosphatidylcholine (Robles and Van den Berg, 1969) and phosphatidic acid by hydrolysis of egg PC (Davidson and Long, 1958), and phosphatidylglycerol (PG) by transesterification of egg PC (Dawson, 1967).

Several commercial sources of phospholipids of reasonable purity are available for such widely used components as PC, PS, PE, and PG. It must be stressed, however, that impurities are still present in these preparations, and the extent of impurities may vary from batch to batch. The presence of divalent metal ions can change drastically the properties of acidic phospholipids (Jacobson and Papahadjopoulos, 1975).

Sterols, notably cholesterol, are also widely used as a component of both MLV and SUV. Cholesterol, which by itself does not form liquid crystals, can be mixed with phospholipids up to a 1:1 molar ratio to form vesicles. Although sterols can be omitted when vesicles are prepared from egg lecithin and some synthetic lecithins, the inclusion of cholesterol is essential for vesicle formation when phospholipids such as sphingomyelins are used. However, even in the case of the various lecithins, incorporation of cholesterol into vesicles can be of considerable value in reducing the leakage of material(s) trapped inside the vesicle (see Section IV,B).

Charged amphiphiles, such as stearylamine (positive charge) dicetylphosphate (negative charge) and other acidic phospholipids (negative charge), are frequently included as components of the vesicle membrane. The presence of charged groups on the surface of lipid bilayers increases the volume of the water space within the vesicles and thus produces a considerable enhancement of the amount of entrapped material(s) inside vesicles. It is important to emphasize, however, that the choice of charged amphiphile will be determined to a large extent by the nature of the material to be captured within the vesicle. If the material to be captured does not bear a net charge, then either negative or positive amphiphiles can be used. If the material is charged, then it may be better to use an amphiphile that displays a similar charge at the physiological pH. This minimizes the possibility of having appreciable amounts of the trapped material associated with the vesicle membrane by electrostatic binding to the bilayer. This problem is pertinent to the proposed use of vesicles as carriers to introduce drugs into cells, in which case binding of drug to the wall of the vesicle might limit its pharmacological activity. On the other hand, association of the entrapped solute by electrostatic charge interactions to a vesicle bearing opposite charge enhances considerably the amount of solute captured by the vesicle. We have actually used much a combination both for cyclic AMP (Papahadjopoulos *et al.*, 1974b) and actinomycin D (Papahadjopoulos *et al.*, 1976) without substantial loss of pharmacological activity *in vitro*. Finally, as we shall see later (Section VI,A), the presence of a charged amphiphile in the vesicle membrane appears to be important in determining the mechanism by which vesicles are incorporated into cells.

Apart from the basic vesicle components outlined above, a wide variety of other materials can be incorporated into vesicles, either in association with the vesicle membrane or trapped within the internal aqueous compartment(s) of the vesicle. Indeed, it is this particular property that makes vesicles so attractive as potential vehicles for introducing materials into

TABLE II

INCORPORATION OF MATERIAL INTO LIPID VESICLES

Material	Reference
A. Incorporation within the vesicle membrane	
1. Integral membrane proteins and glycoproteins	
Cytochrome oxidase	Hinkle et al. (1972), Jost et al. (1973)
Ca^{2+}-ATPase, Na^+-K^+-ATPase	Racker and Eytan (1973), Goldin and Tong (1974), Hilden et al. (1974), Kimelberg and Papahadjopoulos (1974)
Rhodopsin	Hong and Hubbell (1973)
S-100 protein (brain) acetylcholine receptor	Calissano et al. (1974), Michaelson and Raffery (1974)
Glycophorin	Grant and McConnell (1974) Redwood et al. (1975)
Glycophorin hydrophobic fragment	Segrest et al. (1974)
Erythrocyte band 3 membrane protein	Rothstein et al. (1975)
Myelin proteolipid apoprotein	Papahadjopoulos et al. (1975b)
2. Glycolipids Forssman antigen (GL5)	Inoue and Kinsky (1970)
Globoside (GL4)	Kataoka et al. (1973)
Galactocerebroside	Alving et al. (1974)
Monogalactosyl diglyceride	Alving et al. (1974)
Mixed gangliosides	Haywood (1975)
Bacterial lipopolysaccharides	Humphries and McConnell (1974)
B. Entrapment within the internal aqueous space (SUV) or spaces (MLV) of the vesicle	
1. Cations Na^+, K^+, Rb^+, Ca^{2+}	Bangham et al. (1965), Demel et al. (1972), Papahadjopoulos et al. (1972), Spielvogel and Norman (1975)
2. Anions Cl^-, CrO_4^-, TcO_4^-, SO_4^-	Papahadjopoulos and Watkins (1967), Papahadjopoulos et al. (1972), McDougall et al. (1974), Spielvogel and Norman (1975)

(cont.)

TABLE II (cont.)

Material		Reference
3. Sugars and simple	Glucose	Haxby et al. (1968)
nonelectrolytes	Sucrose	Papahadjopoulos et al. (1972)
	Erythritol, Mannitol	Reeves and Dowben (1970)
4. Nucleotides	cAMP	Papahadjopoulos et al. (1974b)
	cGMP	D. Papahadjopoulos and
		G. Poste (unpublished)
	Polynucleotides	Straub et al. (1974)
5. Enzymes	Lysozyme	Sessa and Weissmann (1970)
	Amyloglucosidase	Gregoriadis and Ryman (1972)
	Hexokinase, glucose-6-	Kataoka et al. (1973)
	phosphate dehydrogenase,	
	β-galactosidase	
	Dextranase	Colley and Ryman (1974)
	Horseradish peroxidase	Magee et al. (1974),
		Weissmann et al. (1975)
6. Antimetabolites	Actinomycin D	Gregoriadis (1973),
		Rahman et al. (1974),
		Papahadjopoulos et al.
		(1976)
	Bleomycin	Gregoriadis and
		Neerunjun (1975b)
	Cytosine arabinoside	D. Papahadjopoulos
		(unpublished)
	5-Fluorouracil	Segal et al. (1975)
	Methotrexate	Colley and Ryman (1975)
7. Miscellaneous	Albumin	Gregoriadis and Ryman (1972)
	Anti-α-glucosidase	de Barsy et al. (1975)
	Diphtheria toxoid	Allison and Gregoriadis (1974)
	EDTA	Rahman et al. (1973)
	Pepstatin	Dean (1975)

cells. The precise nature of the additional components incorporated into vesicles will be dictated by the experimental requirements of each individual investigator. Table II summarizes *some* of the components that have been successfully incorporated into lipid vesicles so far. Given the diversity of the components listed in Table II, it would appear that the range of materials that could be incorporated into vesicles may prove to be restricted more by the imagination of the investigator than by technical limitations.

As shown in Table II, a number of materials can be inserted directly into the vesicle membrane. The successful incorporation of glycolipids, proteins, and glycoproteins isolated from cellular membranes into vesicles (Table II) is of particular relevance to studies discussed later (Section VII) in which the use of vesicles as agents to alter plasma membrane composition in living cells will be discussed. Of importance is that glycolipids and glycoproteins can be bound hydrophobically within the vesicle membrane and retain their antigenic properties (Kataoka *et al*., 1973; Humphries and McConnell, 1974; Joseph *et al*., 1974), suggesting that they are oriented within the lipid bilayer of the vesicle in a similar fashion to their organization within natural membranes.

In addition to the hydrophobic insertion of components within the vesicle membrane, various materials can be entrapped within the internal aqueous space(s) of the vesicle. If they are present in the aqueous phase during the initial formation of the vesicles, theoretically any aqueous-soluble molecule or macromolecule can be trapped within a vesicle. It is clear, however, that SUV are in fact too small to permit entrapment of significant amounts of large macromolecules. Furthermore, the sonication step required for the production of SUV may also inactivate certain materials, notably nucleic acids. In this respect, the recently developed LUV may be a useful alternative to MLV for entrapping macromolecules for subsequent delivery to cells. LUV not only have a significantly larger internal aqueous space than SUV, but can also be produced without the need for sonication. Preliminary work in our laboratories have shown that polynucleotides can be successfully incorporated into LUV by inclusion in the aqueous solution surrounding a cochleate lipid cylinder before the final treatment of these structures with EDTA to produce LUV.

IV. Physicochemical Properties of Lipid Vesicles

A. Thermotropic Phase Transitions in Lipid Membranes: "Fluid" and "Solid" Vesicles

Data presented in Section V indicate that the physical state of lipids within the vesicle membrane is of major importance in determining the mechanism by which vesicles enter the cell. It is considered pertinent therefore to briefly outline here the physicochemical basis for changes in the physical state of lipids in vesicle membranes and how this can be used experimentally to produce so-called "solid" and "fluid" vesicles.

As mentioned earlier, phospholipids undergo a first-order thermal phase transition from a solid to a liquid–crystalline phase. The temperature at which this transition occurs (T_c) differs for individual phospholipid species (Table I). Below the T_c the lipids are in a solidlike state, and their hydrocarbon chains are relatively ordered (Fig. 7). Above the T_c the chains are

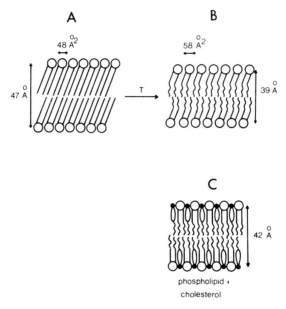

FIG. 7. Schematic representation of molecular orientation in phospholipid bilayers below (A) and above (B) the phase transition. The head groups of the phospholipid molecules arranged as a bilayer are shown by the open circles, and the fatty acyl chains by solid lines. Below the phase transition, the motion of the acyl chains is inhibited as indicated by the straight lines. Above the phase transition (B) the acyl chains have greater motional freedom, and this is indicated by the curved lines. The effect of the introduction of cholesterol into a fluid bilayer is shown in (C). Cholesterol is represented by a small dark circle (hydroxyl), a larger circular area (the four-ring steroid skeleton), and a curved line (hydrocarbon tail). The dimensions for the thickness of the bilayer in (A) and (B) refer to dipalmitoyl phosphatidylcholine bilayers and in (C) to a phosphatidylcholine–cholesterol bilayer. Full references to the original work done in several laboratories from which these figures are constructed can be obtained from the review by Papahadjopoulos and Kimelberg (1973).

more disordered, having greater motional freedom, and are commonly referred to as being in a "fluid" state (Fig. 7). The characteristic T_c for each phospholipid species is defined mostly by the configuration of the acyl chain and also by the degree of hydration and the chemistry of the polar headgroup (Phillips, 1972). The presence of cis-double bonds, branching and decreasing chain length all tend to reduce the T_c. The existence of characteristic T_c values for each phospholipid dictates that, by appropriate selection of either a single lipid or a mixture of lipids, it is possible to prepare vesicles that are either "fluid" or "solid" at the required experimental temperature. As discussed in Sections V and VII, this property may prove to be of substantial experimental value in developing methods for targeting lipids to particular intracellular sites.

B. Permeability of Lipid Vesicles

Several methods have been developed for monitoring the permeability of vesicles to material(s) trapped inside them. The most straightforward approach involves measurement of the rate of efflux of radiolabeled material from vesicles (see Bangham *et al.*, 1965; Papahadjopoulos *et al.*, 1972). The radiolabeled marker is usually added to the aqueous salt solution during initial formation of the vesicle and thus becomes entrapped within the internal aqueous space(s) of the vesicle. Vesicles with entrapped marker are then separated from untrapped marker material either by dialysis or by passage through a gel-filtration column (usually G-50). The efflux of the trapped marker can then be followed by sequential dialysis of the vesicles against an appropriate aqueous buffer. This method is particularly suited for studying the efflux of slowly diffusing solutes (Papahadjopoulos *et al.*, 1972).

Measurements of the efflux of nonradiolabeled entrapped materials from vesicles can also be followed in the same way, assuming of course that an appropriate assay for the released material is available. For example, Haxby *et al.* (1968) measured the efflux of entrapped glucose from vesicles by a spectrophotometric assay involving enzyme-induced reduction of NAD using the glucose released from the vesicles as a substrate for the enzyme reaction.

Another method that is particularly suitable for studying the efflux of fast-diffusing solutes, such as water, glycerol, and urea, has been devised by Bangham *et al.* (1967), based on changes in the optical density of vesicle suspensions occurring with alterations in the osmolarity of the external aqueous environment.

Vesicles prepared from *unmodified* phospholipids show an extreme impermeability to cations, though anion diffusion rates are in general higher (Papahadjopoulos and Kimelberg, 1973). In contrast to their low permeability to simple ions, vesicles exhibit considerable permeability to water, with permeability coefficients ranging from 5 to 100×10^{-4} cm/sec (see Papahadjopoulos and Kimelberg, 1973). These values are well within the range of water permeability in natural membranes. The coincidence of osmotic and isotopic water fluxes found for various vesicle populations also indicates that there are no aqueous pores within the vesicle membrane, a conclusion consistent with the high electrical resistance displayed by such membranes.

The fatty acyl chain of the phospholipid molecule plays an important role in determining the efflux of material(s) from inside vesicles. Several studies using a range of entrapped markers have shown that vesicle permeability generally increases with increasing unsaturation and decreasing chain length of the phospholipid (see Papahadjopoulos & Kimelberg, 1973).

TABLE III

EFFECT OF PROTEINS ON NA[+] PERMEABILITY OF UNILAMELLAR PHOSPHATIDYLSERINE (PS) VESICLES WITH AND WITHOUT CHOLESTEROL[a]

Protein[b]	Concentration (mg/ml)	$^{22}Na^+$ Self-diffusion rate (% per hour)		Increase in permeability		Ratio[d]
		PS (°C)	PS + cholesterol (°C)[c]	PS	PS + cholesterol[c]	
—	—	0.06 (36)	0.02 (36)	—	—	—
—	—	0.02 (24)	0.01 (24)	—	—	—
Cytochrome c	10	49.9 (36)	0.4 (36)	780	20	39.
A₁ basic myelin protein	0.6	40.6 (24)	0.9 (24)	2030	90	22.5
Hemoglobin	0.18	36.0 (36)	1.8 (36)	600	60	10
N₂ myelin proteolipid protein	0.36	15.3 (24)	4.9 (24)	765	490	1.5

[a] Data were taken from Papahadjopoulos et al. (1973b).
[b] Proteins were added at the indicated concentrations to vesicle preparations containing approximately 1 μmole of lipid per milliliter.
[c] Cholesterol was present in equimolar amounts with PS.
[d] Ratio of the increase in permeability induced by the indicated proteins in PS vesicles to that in vesicles containing PS and cholesterol.

The final general feature of vesicle permeability that is relevant to experiments on vesicle–cell interactions is that the binding of proteins to the wall of the vesicle may substantially increase the rate of efflux of entrapped material from vesicles (Table III). The ability of proteins to increase the permeability of vesicle membrane appears to be correlated with the degree of penetration of the protein into the lipid bilayer (Papahadjopoulos and Kimelberg, 1973). Cell culture media abound with serum components and released cellular proteins that are capable of increasing the leakage of entrapped material from vesicles. Consequently, for experiments in which vesicles are to be used to transfer entrapped material(s) into cells, it is important to establish that the *in vitro* culture environment (or body fluids where vesicles are to be used *in vivo*) does not induce rapid leakage of material out of the vesicles, so that little material is finally introduced into cells. The severity of this problem increases where prolonged incubation of vesicles with cells is required.

Some protection against rapid efflux of entrapped material from vesicles can, however, be achieved by inclusion of cholesterol in vesicle membranes (Papahadjopoulos *et al.*, 1973b). Numerous studies (Jain, 1975) have shown that the incorporation of cholesterol into lipid vesicles substantially reduces their permeability to trapped material, even in the presence of proteins (Table III). This effect of cholesterol in reducing membrane permeability is attributed to its so-called "condensing" effect on the packing of lipid molecules within the bilayer. Incorporation of cholesterol into lipid membranes has been shown to restrict the molecular motion of phospholipid molecules in the region of the first 8–10 carbon atoms of the acyl chain from the lipid–water interface, leaving the rest of the hydrocarbon chain relatively free (Fig. 7). This effect is accompanied by a reduction in the area occupied by each phospholipid molecule, which leads to a more perpendicular orientation of lipid molecules and a resulting increase in the thickness of the membrane (Fig. 7). The condensing effect of cholesterol on vesicle membranes results in a significant reduction in their permeability to anions, cations, water, and a range of nonelectrolyte solutes (Papahadjopoulos and Kimelberg, 1973; Jain, 1975).

V. Incorporation of Lipid Vesicles by Cells *in Vitro* and *in Vivo*

A. *In Vitro*

Over the past 4 years, studies by several investigators have established that a variety of cells cultured *in vitro* can incorporate vesicles of widely

TABLE IV

UPTAKE OF UNILAMELLAR (SUV) AND MULTILAMELLAR LIPID VESICLES (MLV) OF DIFFERING COMPOSITION BY CELLS CULTURED *in Vitro*

Cell type	Species	Type of vesicle	Vesicle composition[a]	Reference
ML	Mouse	MLV	Sphingomyelin/chol/dicetylphosphate	Magee and Miller (1972)
Peritoneal macrophages	Mouse	MLV	PC/chol/PA	Gregoriadis and Buckland (1973)
Lung fibroblasts (MRC-5)	Human	MLV	PC/chol/PA	Gregoriadis and Buckland (1973)
YAC leukemia	Mouse	SUV	PC/chol	Inbar and Shinitzky (1974)
HeLa	Human	SUV	Sphingomyelin/chol/stearylamine	Magee et al. (1974)
Spleen cells (unspecified)	Mouse	SUV	PC	Ozato and Huang (1974)
3T3, SV3T3, and L929	Mouse	SUV and MLV	Sphingomyelin/chol, PC/chol/stearylamine; PS/DSPC/DPPC	Papahadjopoulos et al. (1974b,c, 1975c)
Erythrocytes	Human	SUV	PS/PC/chol	Papahadjopoulos et al. (1974b)
L929	Mouse	SUV and MLV	Sphingomyelin/chol, PC/chol/stearylamine, PC/chol/dicetrylphosphate, PE/chol/stearylamine	Straub et al. (1974)
Ehrlich ascites tumor	Mouse	SUV	PC/chol	Alderson and Green (1975a)
Lymphocytes	Bovine	SUV	PC/chol	Alderson and Green (1975b)
AKR leukemia	Mouse	MLV	PC/chol/PA	Gregoriadis and Neerunjun (1975b)
Peritoneal macrophages	Mouse	SUV	PC; PC/chol	Johnson (1975)
Lung fibroblasts (V-79)	Hamster	SUV	PC; dioleyl PC	Pagano and Huang (1975)
Phagocytes (unspecified)	Dogfish	MLV	PC/chol/dicetylphosphate	Weissmann et al. (1975)
Lung fibroblasts (DC3F)	Hamster	SUV	PS/PC/chol	Papahadjopoulos et al. (1976)

[a] Abbreviations: chol, cholesterol; DSPC, distearoylphosphatidylcholine; DPPC, dipalmitoylphosphatidylcholine; PA, phosphatidic acid; PC, phosphatidylcholine; PE, phosphatidylethanolamine; PS, phosphatidylserine.

differing lipid composition without cytotoxic effects (Table IV). Most of the studies listed in Table IV were concerned with testing the feasibility of using lipid vesicles as carriers to introduce materials into cells, and relatively little attention has been given to the kinetics of vesicle–cell interaction and the mechanism(s) by which vesicles might be incorporated into cells. It is clear, however, that the use of vesicles as carrier vehicles to introduce materials into cells demands at least some basic appreciation of the mechanism(s) by which these structures are incorporated into cells.

Cultured mammalian cells display a remarkable appetite for lipid vesicles. Incorporation of up to 1×10^6 vesicles per cell is easily achieved and does not impair cell viability or the ability of cells to proliferate (Papahadjopoulos et al., 1974b, c, 1975c, 1976; Pagano and Huang, 1975; Poste and Papahdjopoulos, 1976). Although different cell types may vary significantly in the number of vesicles that they can incorporate, the general kinetics of vesicle uptake at 37°C appear to be similar in different cells, with rapid uptake of vesicles occurring within the first 2 hours, after which uptake levels off, reaching a plateau after 3–8 hours depending on the cell type (Fig. 8). It has been noted, however, that the uptake of SUV is linear over a broad vesicle concentration range (Fig. 9), although high levels of vesicle incorporation ($> 5 \times 10^6$ vesicles per cell) are accompanied by a reduction in cell viability and cell growth rate. It appears that the presence or the absence of a net negative or positive charge, or preparation of SUV from lipids that are "solid" or "fluid" at 37°C, has little effect on the overall extent of vesicle incorporation (Papahadjopoulos et al., 1974c, 1975c). However, both vesicle surface charge and the physical state of vesicle lipids are of major importance in determining the mechanism by which vesicles are incorporated into cells (see below).

Information on the kinetics of MLV uptake by cells in vitro does not appear to be available, even though vesicles of this type have been used in several studies (Table IV).

Observations on the cellular uptake of SUV containing a radiolabeled component in the vesicle membrane and a different radiolabeled component trapped inside the vesicle have established that both components are incorporated into cells in identical amounts at 37°C, indicating that the vesicles are taken up as intact structures (Papahadjopoulos et al., 1974b, c). Further indication that SUV are actually incorporated into cells (as opposed to simple binding to the cell surface without true intracellular incorporation) is provided by the recovery of vesicle components from subcellular fractions (Magee et al., 1974; Papahadjopoulos et al., 1974c; Huang and Pagano, 1975) and by the demonstration of vesicle components within cells by electron histochemistry (Magee et al., 1974) and electron radioautography (Huang and Pagano, 1975). Finally the demonstration that biologically

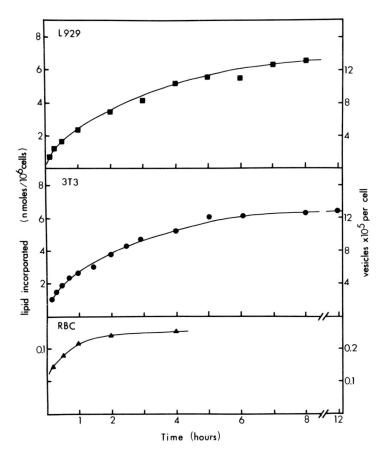

FIG. 8. Incorporation of small unilamellar vesicles by mouse L929 and 3T3 cells and human erythrocyte ghosts (RBC) at 37°C. L929 cells and erythrocyte ghosts were incubated with vesicles composed of phosphatidylserine, phosphatidylcholine, and cholesterol (1:9:8 molar ratio plus trace amounts of ^{3}H-labeled dipalmitoylphosphatidylcholine) and 3T3 cells with vesicles composed of phosphatidylcholine, cholesterol, and stearylamine (4:3:1 molar ratio and ^{14}C-labeled cholesterol as a tracer). L929 cells, 3T3 cells, and erythrocyte ghosts were exposed, respectively to 50, 50, and 1.3 nmoles of phospholipid per 10[6] cells. Vesicle uptake was determined at the indicated intervals by measuring the amount of cell-associated radioactivity derived from vesicles. Reproduced, with permission, from Papahadjopoulos et al. (1974c).

active materials trapped inside vesicles not only can be recovered from cells, but also can produce specific changes in cellular metabolism and cell behavior (see Section VI) provides convincing evidence that vesicles are actually incorporated into cells, not merely adsorbed to the cell surface.

Since vesicles are incorporated into cells as intact structures, it is not unreasonable to conclude that their uptake must occur either by endocytosis or by fusion with the cellular plasma membrane, or both.

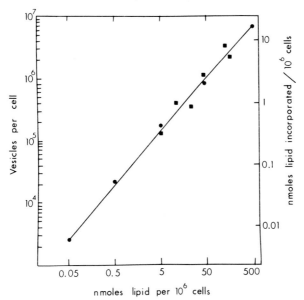

FIG. 9.   Incorporation of small unilamellar vesicles by mouse 3T3 (●) and L929 cells (■) exposed to different concentrations of vesicles. 3T3 cells were incubated for 4 hours at 37°C with vesicles composed of phosphatidylcholine, cholesterol, and stearylamine (4:3:1 molar ratio with trace amounts of ³H-labeled dipalmitoylphosphatidylcholine). L929 cells were incubated for 4 hours at 37°C with vesicles composed of phosphatidylserine, phosphatidylcholine, and cholesterol (1:9:8 molar ratio with trace amounts of ³H-labeled dipalmitoylphosphatidylcholine). Vesicle uptake was determined by measurement of cell-associated radio-activity. Reproduced, with permission, from Papahadjopoulos *et al.* (1974b).

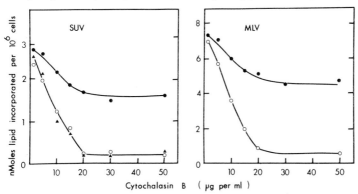

FIG. 10.   The effect of cytochalasin B on the uptake of unilamellar (SUV) and multilamellar (MLV) vesicles of differing composition by mouse 3T3 cells. ●, Vesicles prepared from 10% phosphatidylserine in phosphatidylcholine (PS/PC); ○, vesicles prepared from phosphatidylserine, dipalmitoylphosphatidylcholine, and distearoylphosphatidylcholine (1:4.5:4.5 moles) (PS/DPPC/DSPC); and▲, vesicles prepared from phosphatidylcholine. Monolayer cell cultures grown in 60-mm petri dishes were exposed to vesicles (approximately 200 nmoles of lipid per 1 × 10⁶ cells) containing trace amounts of ³H-labeled dipalmitoylphosphatidylcholine, and the amount of cell-associated radioactivity was determined after incubation for 2 hours at 37°C.

Cytochalasin B (CB) has been shown to inhibit the endocytotic uptake of materials in a wide range of cell types (Allison and Davies, 1974). This drug also inhibits the uptake of MLV and SUV by cells *in vitro* (Fig. 10), although the extent of inhibition varies significantly for vesicles of differing composition (Papahadjopoulos *et al.*, 1975c; Poste and Papahadjopoulos, 1976). As shown in Fig. 10, CB causes marked inhibition of cellular uptake of negatively charged MLV and SUV prepared from lipids that are "solid" at 37°C and from neutral lipids, even though these lipids are "fluid" at 37°C. In contrast, cellular uptake of negatively charged vesicles of similar size prepared from lipids that are "fluid" at 37°C is inhibited to a much lesser extent by CB (Fig. 10). Similarly, pretreatment of cells with inhibitors of glycolysis and respiration to block endocytosis (see Karnovsky, 1962) results in complete inhibition of the uptake of fluid neutral and solid charged

TABLE V

EFFECT OF METABOLIC INHIBITORS ON UPTAKE OF SUV OF DIFFERING
COMPOSITION BY 3T3 CELLS

		Vesicle uptake[b]	
Vesicle composition	Treatment[a]	Nmoles of lipid, 1×10^6 cells	% of control
PS/PC[c]	Control	2.45	100
(fluid charged)	Sodium azide	2.18	89
	2-desoxyglucose azide plus	2.03	83
	Desoxyglucose	1.57	64
PS/DSPC/DPPC[d]	Control	2.32	100
(solid charged)	Sodium azide	1.98	85
	2-Desoxyglucose azide plus	1.83	79
	Desoxyglucose	0.28	12
PC (fluid neutral)	Control	2.12	100
	Sodium azide	1.74	82
	2-Desoxyglucose azide plus	1.65	78
	Desoxyglucose	0.19	9

[a] Cells were incubated with inhibitors for 30 minutes at 37°C before exposure to vesicles: sodium azide, $5 \times 10^{-3} M$; 2-desoxyglucose, $5 \times 10^{-2} M$.

[b] Vesicles were labeled with dipalmitoylphosphatidylcholine-^{3}H and incubated with cells (approximately 160 nmoles of lipid per 10^6 cells) at 37° for 2 hours, and vesicle uptake was determined by measurement of cell-associated radioactivity.

[c] 10% phosphatidylserine (PS) in phosphatidylcholine (PC).

[d] Phosphatidylserine (PS), 10%, in equimolar mixture of dipalmitoyl (DPPC) and distearoylphosphatidylcholine (DSPC); T_c 43°C.

vesicles (Table V), but uptake of fluid charged vesicles is again inhibited by only 30–40% compared with untreated control cells (Table V). Poste and Papahadjopoulos (1976) have interpreted these results as indicating that neutral vesicles and solid negatively charged vesicles are incorporated into cells primarily by endocytosis whereas at least 60% of fluid negatively charged vesicles enter cells by a nonendocytotic mechanism. These investigators proposed that the nonendocytotic pathway for vesicle uptake involves fusion of vesicles with the cellular plasma membrane. Support for this possibility has been obtained from experiments on the effect of temperature on vesicle uptake and from ultrastructural observations.

Cellular uptake of SUV and MLV is temperature-sensitive, but the kinetics of inhibition at lower temperatures differ depending on the composition of the vesicle. The uptake of fluid negatively charged SUV and MLV displays a distinct change in slope at 16°–18°C (Fig. 11), but a similar transition is not found with fluid neutral vesicles or solid negatively charged vesicles of comparable size (Fig. 11). Processes requiring membrane fusion have been found in other studies to display a similar distinct transition at about 18°C (Lagunoff and Wan, 1974) while endocytosis exhibits a simple linear temperature dependence (Steinman et al., 1974). Poste and Papahadjopoulos (1976) therefore interpreted the data shown in Fig. 11 as further evidence in support of their proposal that solid negatively charged and fluid neutral vesicles are incorporated into cells by endocytosis, while fluid negatively charged vesicles are incorporated predominantly by a nonendocytotic route, involving fusion of vesicles with the cellular plasma membrane. Electron micrographic evidence for fusion of fluid charged SUV (Martin and Mac-

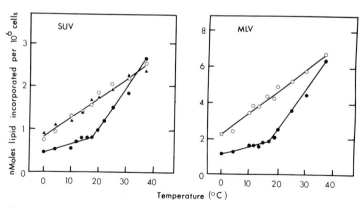

FIG. 11. The effect of temperature on the uptake of unilamellar (SUV) and multilamellar vesicles (MLV) of differing composition by 3T3 cells. Experimental conditions and vesicle composition are identical to those described in the legend to Fig. 10. ●, PS/PC vesicles; ○, PC/DSPC/DPPC vesicles; and ▲, PC vesicles.

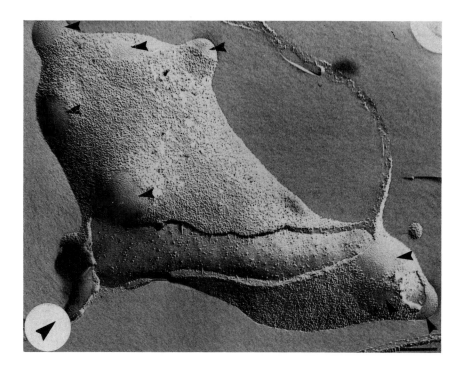

FIG. 12. Freeze-fracture electron micrograph of human erythrocyte ghost after incubation with multilamellar lipid vesicles (10% phosphatidylserine in phosphatidylcholine; 1.3 nmoles of lipid per 1×10^6 cells) for 30 minutes at 37°C showing cell surface "patches" devoid of intramembranous particles (arrows). The dimensions and surface topography of these patches are similar to those of vesicles, suggesting that they may represent points at which vesicles have fused with the cell surface. Similar patches were not found on untreated control ghosts. Arrowheads indicate direction of shadowing. Reproduced, with permission, from Papahadjopoulos *et al.* (1974b).

Donald, 1974) and MLV (Fig. 12) with cells *in vitro* lends further support to this proposal.

The apparent inability of solid negatively charged and fluid neutral vesicles to fuse with cell surface is consistent with the observation that these classes of vesicle are unable to induce fusion of cells (Papahadjopoulos *et al.*, 1973a), a process that possibly requires fusion of vesicles with the cell surface (Papahadjopoulos *et al.*, 1973a; Martin and MacDonald, 1974). Vesicles of this type also show a very low capacity to fuse with other vesicles (Papahadjopoulos *et al.*, 1974a; Kantor and Prestegard, 1975). In contrast, fluid charged vesicles not only induce substantial cell fusion (Papahadjopoulos *et al.*, 1973a; Martin and MacDonald, 1974), but also undergo rapid and extensive fusion with other vesicles (Papahadjopoulos *et al.*, 1974a; Papahadjopoulos and Poste, 1975).

Pagano and Huang (1975) have claimed that neutral lecithin vesicles (SUV) do in fact fuse with the cellular plasma membrane. However, the results presented by these investigators are not entirely convincing. Pagano and Huang found that pretreatment of hamster V79 cells with inhibitors of *either* glycolysis *or* respiration did not impair their ability to subsequently incorporate lecithin vesicles. From this they concluded that the vesicles were not entering the cell by endocytosis and that fusion of vesicles with the plasma membrane must therefore be involved. However, no direct evidence to support this mechanism was presented. More important, they did not establish whether their treatment of cells with metabolic inhibitors had in fact blocked endocytosis. This is by no means certain since they employed only single metabolic inhibitors while in many cell types addition of two inhibitors to block both respiration and glycolysis is required to achieve effective inhibition of endocytosis (see Karnovsky, 1962; Steinman *et al.*, 1974). Thus, Poste and Papahadjopoulos (1976), like Pagano and Huang, found that uptake of neutral lecithin vesicles was not impaired after treatment of cells with single metabolic inhibitors of either respiration *or* glycolysis (Table V). However, measurements on the uptake of ^{14}C-labeled sucrose by such cells (a specific test for endocytotic activity; see Wagner *et al.*, 1971), revealed that endocytosis was still occurring and that exposure of cells to inhibitors of both glycolysis and respiration was necessary to achieve complete inhibition of endocytosis. Under these conditions, uptake of neutral lecithin vesicles was completely inhibited (Table V), indicating that they are incorporated exclusively by endocytosis.

Poste and Papahadjopoulos (1976) were unable to define any differences in the cellular uptake pathways for MLV and SUV of identical composition, indicating that the physical state of lipids in the vesicle membrane and the presence or the absence of surface charge were more important than vesicle size in determining whether the predominant pathway for incorporation into cells was by endocytosis or by fusion with the cellular plasma membrane. However, Weissmann *et al.* (1975) have claimed that MLV are *always* endocytosed and do not fuse with the cell surface but this generalization does not withstand detailed scrutiny. Weissmann *et al.* examined *only* fluid charged MLV and did not investigate the cellular uptake of MLV of differing lipid composition. Their proposal that fluid-charged MLV were endocytosed was based on ultrastructural identification of MLV (or MLV fragments) within the lysosomes of treated cells. This does not provide unequivocal evidence that the *entire* vesicle has been endocytosed. An alternative possibility is that the outer bilayer of a MLV can fuse with the plasma membrane

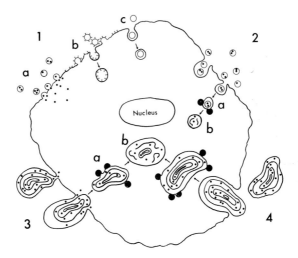

FIG. 13.   Schematic representation of the possible mechanisms involved in the uptake of small unilamellar vesicles (SUV) and multilameller vesicles (MLV) of differing composition by cells *in vitro*. MLV and SUV have not been drawn to scale with respect to each other or in relation to the cell.

1. Uptake of charged SUV prepared from lipids that are fluid at experimental temperatures. Fusion of such vesicles with the cellular plasma membrane (1a) would permit release of material entrapped within vesicles (●) directly into the cytoplasm. In addition, fusion of vesicles will result in the incorporation of the vesicle membrane into the cellular plasma membrane (1b). Vesicle-derived components could either remain in the plasma membrane or be internalized by endocytosis (1b). A certain proportion of fluid charged SUV appear to be endocytosed directly without fusing with the cell (1c), and these vesicles will enter the pathway outlined in 2 below.

2. Uptake of fluid neutral SUV or solid charged SUV. These vesicles are endocytosed, after which lysosomes (stippled structures) fuse with the endocytotic vacuole containing the vesicle (2a). The action of lysosomal lipases on the vesicle membrane will probably result in vesicle breakdown and release of any material trapped inside the vesicle (2b). The resistance of the released material to lysosomal hydrolases and the permeability properties of the lysosomal membrane will then influence whether the material will be able to diffuse out of the lysosome and reach other intracellular sites in an active form.

3. Uptake of charged fluid MLV. Fusion of the outer lamella of the vesicle with the cellular plasma membrane would result in the introduction of the bulk of the vesicle directly into the cytoplasm, where it would probably undergo fusion with lysosomes (3a) followed by eventual breakdown of the vesicle with release of its contents (3b). A certain proportion of MLV of this composition are directly endocytosed and would thus enter pathway 4 described below.

4. Uptake of fluid neutral MLV and solid charged MLV. Endocytosis of vesicles results in their subsequent incorporation into the intracellular digestive apparatus with eventual breakdown of vesicles by lysosomal enzymes as described above for pathway 2.

with resulting introduction of the remaining "bulk" of the vesicle into the cytoplasm, where it would then be free to undergo subsequent fusion with lysosomes (see Fig. 13, pathway 3). Ultrastructural proof that endocytosis of *entire* vesicles had occurred would require the use of MLV in which the outermost lamella was labeled with an electron dense marker, such as ferritin, and the subsequent demonstration that the labeled lamella was still present in the vesicles found in lysosomes. In addition, the data presented by Weissman *et al.* on the effect of cytochalasin B on cellular uptake of fluid charged MLV also argues against endocytosis being the only uptake pathway for vesicles of this type. Weissmann *et al.* reported that treatment of cells with cytochalasin B produced only a 40% reduction in their activity to incorporate fluid charged MLV, an identical figure to that shown in Fig. 10 for fluid-charged (PS/PC) MLV. However, as shown in Fig. 10, cytochalasin B produces much greater inhibition of solid-charged (PS/DSPC/DPPC) MLV. The limited effect of this drug on cellular uptake of fluid charged MLV suggests that the major portion (about 60%) of such vesicles are entering by a nonendocytotic mechanism, most probably involving fusion with the plasma membrane.

The major pathways for the incorporation of vesicles of differing composition by cells *in vitro* suggested by the evidence reviewed in this section are shown in Fig. 13. As shown in Fig. 13, by manipulating the lipid composition of vesicles it may be possible to achieve some degree of "targeting" of vesicles (and their contents) to specific regions of the cell (see Section VII).

B. *In Vivo*

Little is known concerning the mechanism(s) of vesicle incorporation into cells *in vivo*. There is no obvious reason to conclude that the *in vivo* uptake mechanism(s) should differ from those operating *in vitro*. In this respect, it is of interest to note that Rahman and Wright (1975) observed that neutral lecithin/cholesterol MLV were endocytosed by liver cells *in vivo*.

Numerous studies over the past three years have shown that MLV and SUV are well tolerated by experimental animals after administration by a variety of routes: intravenously (McDougall *et al.*, 1974; Segal *et al.*, 1974; Colley and Ryman, 1975; Juliano and Stamp, 1975), intraperitoneally (Gregoriadis and Neerunjun, 1975b), subcutaneously (Allison and Gregoriadis, 1974), intramuscularly (Allison and Gregoriadis, 1974), and by intratesticular injection (Segal *et al.*, 1975).

VI.   Use of Lipid Vesicles to Introduce Materials into Cells
in Vitro and *in Vivo*

A.   *In Vitro*

The first demonstration that vesicles could be used as carrier vehicles to introduce biologically active materials into cells *in vitro* was made by Magee and Miller in 1972, who showed that treatment of ML cells with MLV containing IgG with a high neutralizing titer to Coxsackie virus A-21 protected the cells against subsequent infection by this virus. The IgG-containing vesicles were 3000–10,000 times more effective in protecting cells than an equivalent titer of free antibody. Since direct incubation of IgG-containing vesicles with isolated virions also resulted in significant virus neutralization, it would appear that the IgG was associated with the surface of the vesicle rather than trapped inside. Although Magee and Miller did not formally localize the distribution of vesicles within the treated cells, the observed inhibition to virus penetration clearly suggests that effective amounts of virus-neutralizing IgG remained at the cell surface. Since the IgG could not be removed from the cell surface by extensive washing (in contrast to free IgG), it is tempting to speculate that the vesicles had in fact fused with the cellular plasma membrane, thus incorporating IgG directly into the cell surface. In view of the comments in the previous section on the importance of surface charge in enabling vesicles to fuse with cells, it is of interest that Magee and Miller found it necessary to add a charged amphiphile (stearylamine) to vesicles in order to achieve significant association of IgG with the cell surface.

Evidence that vesicles might also be useful for introducing material directly into the intracellular environment was next obtained by Gregoriadis and Buckland (1937). They showed that MLV could serve as carriers to introduce invertase into invertase-deficient mouse macrophages and human fibroblasts (MRC-5) cultured *in vitro*. Cells were first prelabeled with sucrose–^{14}C, which could not be metabolized because of the invertase deficiency and instead accumulated within lysosomes. Successful vesicle-mediated transfer of invertase into these cells was demonstrated by the fact that the treated cells acquired the ability to hydrolyze sucrose–^{14}C to glucose–^{14}C and fructose–^{14}C. This study did not, however, include a control experiment to test whether simple addition of free invertase to the culture medium might also result in uptake of the enzyme. This is by no means an extreme possibility. Several studies using fibroblast cultures derived from patients with specific lysosomal enzyme deficiencies have shown that simple addition of the appropriate deficient enzyme to the culture medium will result in its incorporation into lysosomes in a functional state with accompanying temporary remission of the deficient state (for references, see Lagunoff *et al.*, 1973; Brady, 1975). However, a recent report by Weissmann *et al.* (1975) has shown that the uptake of horseradish peroxidase trapped inside MLV by dogfish phagocytes is up to 100 times greater than in cells exposed to the free enzyme.

The feasibility of using SUV, as opposed to MLV, as carriers to enhance the incorporation of nonpermeable materials into cultured cells was demonstrated by Papahadjopoulos *et al.* (1974b,c). They showed that incorporation of fluid charged SUV containing entrapped cyclic AMP (cAMP) by mouse 3T3 cells and virally transformed 3T3 cells (SV3T3) resulted in significant inhibition of cell growth at nucleotide concentrations where free cAMP or dibutyryl cAMP together with theophyline had little or no effect on cell proliferation. This work also demonstrated the potential importance of vesicle composition in determining whether material entrapped within the vesicle can be released intracellularly in an active form. The marked inhibition of cell growth produced by vesicle-derived cAMP obtained using fluid charged vesicles as carriers was not found when cAMP entrapped within solid charged vesicles, even though cells incorporated similar amounts of both types of vesicle.

Papahadjopoulos *et al.* (1974b) proposed that the differing cellular responses to cAMP delivered via solid and fluid vesicles were due to the fact that solid and fluid vesicles were incorporated into cells by different mechanisms, with resulting differences in the intracellular site(s) at which cAMP was eventually released. They proposed that fluid charged vesicles were able to fuse with the cellular plasma membrane and thus release the entrapped cAMP directly into the cytoplasm (see Fig. 13, pathway 1a). In contrast, it was proposed that solid charged vesicles were unable to fuse with the plasma membrane and were incorporated by endocytosis with release of the en- cAMP into the lysosomal apparatus (Fig. 13, pathway 2). The lack of effect of cAMP on cell growth when introduced into lysosomes was suggested as resulting from failure of cAMP to be released from the lysosome in an active form at sufficient concentrations. The limited permeability of the lysosomal membrane (Gordon and Cohn, 1973) and the likelihood of inactivation of cAMP by lysosomal phosphodiesterases (Tappel, 1969) would both favor this situation.

Unilamellar vesicles have also been used successfully to introduce antitumor drugs into tumor cells *in vitro*. Resistance of tumor cells to such drugs as actinomycin D, methotrexate, nitrogen mustard, daunomycin, and the vinca alkaloids is considered to be due to changes in plasma membrane permeability that result in reduced drug uptake by resistant cells compared with their sensitive counterparts (Kessel, 1974). Support for this hypothesis has been based largely on demonstrated differences in drug uptake by sensitive and resistant cells (see Kessel, 1974) and on the finding that enhancement of drug uptake by detergent-induced modification of membrane permeability results in increased killing of tumor cells (Yamada and Iwanami, 1962; Riehm and Biedler, 1972). The latter method suffers, however, from the shortcoming that detergent treatment *per se* causes significant cytotoxicity and thus complicates interpretation of cellular response to the drug. The use of drug-containing vesicles to enhance drug uptake into low permeability drug-resistant variants offers a potential solution to this problem (Papahadjopoulos *et al.*, 1976). Vesicle-mediated introduction of drugs directly into tumor cells not only produces significant enhancement of drug uptake, but is also devoid of complicating side effects on cell viability.

Papahadjopoulos *et al.* (1976) have used SUV as a carrier vehicle to enhance the uptake of actinomycin D (AD) into an AD-resistant hamster cell line (DC3F-ADX), which is resistant as a result of a decreased capacity to transport AD across the plasma membrane. This cell line is able to grow in media supplemented with concentrations of AD (10 μg/ml) that are highly toxic for the sensitive parent cell line (DC3F). Incubation of resistant DC3F-ADX cells with AD-containing fluid charged SUV produced a 5-fold increase in intracellular drug concentration over that achieved by exposure to identical concentrations of free drug and produced significant inhibition of cellular RNA synthesis and cell growth at drug concentrations that had little or no effect when added as free drug to the culture medium (Papahadjopoulos *et al.*, 1976).

B. *In Vivo*

In 1932, J. Y. Johnson filed a successful patent application for a procedure in which "pharmaceutical preparations for injection into the muscular system or subcutaneously can be prepared by combining medicaments with liquids, such as fats or fatty oils, if necessary together with waxes or wax-like substance, with water, or other liquids, and a dispersing agent whereby a system, hereinafter called 'depot' is found capable of holding any desired dose of medicament but releasing it over any desired space of time . . . without the slightest detriment to the organism." Lipid vesicles might well fall within this specification. Indeed, one of the detailed specifications given in the patent application reads like a contemporary recipe for liposomes. A "depot" to achieve gradual release of strophantin within the body is listed as: "an emulsion prepared from 25 parts of lecithin; 20 parts of water; 1.5 of cholestin; 0.03 parts of strophantin; and 0.5 parts of Nipasol (p-hydroxybenzoic acid normal propyl ester)." As Bangham *et al.* (1974) pointed out, it seems likely that any success achieved by such preparations can probably now be ascribed to the fact that at least some of the drug was sequestered within a lipid smectic mesophase, i.e., a multilamellar vesicle.

Although little use seems to have been made of the above application, interest in the use of lipid vesicles as drug carriers *in vivo* has recently emerged (or perhaps more correctly re-emerged) as an area of research interest (Gregoriadis, 1973).

The sequestration of therapeutic agents or other biologically active materials within vesicles for delivery to cells *in vivo* offers a number of advantages over the more straightforward administration of the free drug. Many foreign substances will provoke host immune responses, and their repeated injection will be accompanied by risk of dangerous allergic reactions. Entrapment of materials within immunologically inert lipid vesicles avoids this problem entirely (Allison and Gregoriadis, 1974). Capture of materials within vesicles also protects them against rapid breakdown by enzymes present in the circulation and various tissue fluids.

One area of current interest concerns the possible value of lipid vesicles as carriers for drug delivery in cancer chemotherapy. Rahman *et al.* (1974) have shown that administration of MLV containing actinomycin D(AD) to mice with Ehrlich ascites turmors significantly increased their survival time compared with untreated animals and animals treated with free AD. Similar results have been obtained by Gregoriadis and Neerunjun (1975b), who used MLV containing AD to enhance the survival of mice inoculated with AKR leukemia cells.

Rahman *et al.* (1973) have also used MLV as carriers for introducing chelating agents into cells *in vivo*. The polyaminopolycarboxylic acid chelating agent EDTA is used therapeutically in the treatment of heavy-metal poisoning, but its effectiveness is restricted by its failure to enter cells. Rahman *et al.* (1973) encapsulated EDTA within MLV and showed that administration of such vesicles to mice resulted in their uptake by cells in a range of organs and effective removal of intracellular deposits of toxic metals, such as plutonium, lead, and inorganic mercury.

Straub *et al.* (1974) have reported recently that intravenous injection of MLV and SUV containing the synthetic polynucleotide, poly(1):poly(c), into mice produced a significant increase in interferon production over that obtained with free poly(1):poly(c). Control animals inoculated with lipid vesicles without polynucleotides failed to produce interferon.

Inbar and Shinitzky (1974) have claimed that lipid vesicles can be used to alter the plasma membrane composition of tumor cells with resulting reduction of their *in vivo* tumorigenicity. Inbar and Shinitzky incubated YAC mouse leukemia cells with PC/cholesterol SUV *in vitro* at 4°C and found that this resulted in a significant increase in the cholesterol content of the plasma membrane in these cells. Such cholesterol-enriched tumor cells were claimed to show a reduced capacity to form tumors when injected into mice. Recent work in the present authors' laboratory has confirmed that vesicles can indeed be used to enrich the cholesterol content of tumor cells, but we have so far failed to observe any change in the tumorigenicity of such cholesterol-enriched tumor cells (Poste, *et al.* 1976).

VII.   Targeting of Lipid Vesicles to Specific Intracellular Compartments

The data presented earlier in Section V, and summarized in Fig. 13, suggest that by manipulating the lipid composition of vesicles it should be possible to direct vesicles to specific regions of the cell. By producing vesicles that will be incorporated by endocytosis, the vesicles and their contents can be targeted to the lysosomal apparatus. In contrast, vesicles whose composition dictates that they can fuse with the plasma membrane can be utilized to introduce novel components into the plasma membrane and to achieve release of entrapped material direct into the cytoplasm.

Targeting of vesicles to the lysosomal apparatus offers new experimental opportunities for introducing material into lysosomes. Vesicles could thus be used to introduce a variety of defined substrate materials directly into lysosomes to study lysosomal enzyme activity and the kinetics of substrate degradation in different cell types (cf. Dean, 1975). Vesicle-mediated introduction of drugs into lysosomes could be similarly exploited to modify the properties of the lysosomal membrane. Also of interest is the possibility of using enzyme-containing vesicles in the treatment of lysosomal storage deficiency diseases (Gregoriadis and Buckland, 1973; Weissmann *et al.*, 1975). However, as mentioned earlier lipid vesicles may not offer any unique experimental opportunities for this particular situation since cells appear to be able to incorporate free enzymes from the culture medium. The possibility remains, however, that vesicles might still be a better candidate for therapeutic applications since the lipid membrane of the vesicle would at least protect the enzyme from enzymic degradation within the circulation and the various tissue fluids. Also, if techniques for "homing" vesicles to specific cell types can eventually be developed (Section VIII) then enzyme-containing vesicles with appropriate homing abilities would have obvious advantages over simple injection of the free enzyme.

De Duve and his colleagues have suggested that the endocytotic uptake of various biologically active materials bound to biodegradable carriers with subsequent incorporation into lysosomes could provide a general pathway for introducing nonpermeable materials into cells (de Duve *et al.*, 1974). This concept, or so-called lysosomotropism, proposes that biologically active materials bound to various carriers would be taken up into lysosomes after which the active material would be released by the action of lysosomal enzymes on the carrier portion. The biologically active material would then be free to act either within the lysosome or diffuse out to other parts of the cell.

Endocytosis of lipid vesicles containing biologically active materials clearly falls within the definition of a lysosomotrophic agent, since the vesicle membrane can be readily degraded to release the vesicle contents into the lysosome. Certain reservations must be expressed, however, as to whether lysosomotropism represents the best route for introduction of materials that are required to act at an extralysosomal intracellular site. The risk of enzymic degradation of the introduced material within the lysosome, inactivation of material by the low pH of the intralysosomal environment, and the limited permeability of the lysosomal membrane all suggest that it may be more profitable to target materials destined for extralysosomal sites by means of a nonendocytotic pathway. Fusion of vesicles with the plasma membranes might provide such a pathway, since this would result in the introduction of vesicle contents directly into the cytoplasm (Fig. 13, pathway 1).

In addition to its value as an experimental tool for the delivery of non-permeable materials directly into the cytoplasm, the fusion of lipid vesicles with the cellular plasma membrane also offers a potential method for direct introduction of new components into the plasma membrane of cells *in vitro* and *in vivo* (Fig. 13, pathway 1). The ability to incorporate isolated hydrophobic plasma membrane proteins and glycoproteins into vesicle membranes (Table II) raises the obvious experimental possibility that vesicles could be used to transfer plasma membrane components between different cell types. Research on the feasibility of this approach is currently in progress in this laboratory.

VIII. "Homing" of Vesicles to Specific Cell Types

In addition to targeting of vesicles to specific regions of individual cells, the value of lipid vesicles as an investigative tool in cell biology would be further enhanced if techniques could be devised whereby vesicles would "home" to specific target cells. Such specificity would be of enormous value for *in vivo* studies where drug-containing vesicles would be directed to specific target organs and tissues.

An encouraging initial approach toward this goal has been made by Gregoriadis and Neerunjun (1975a). They incorporated cell-specific immunoglobulins into the membranes of MLV and showed that this produced a substantial degree of homing of vesicles to the cell type against which the immunoglobulin had been raised. Similar success could perhaps be achieved using vesicles that contain other natural or synthetic ligands capable of reacting with specific determinants on the surfaces of different cells.

IX. Conclusions

Lipid vesicles offer a new and potentially powerful technique for achieving cellular uptake of materials that would not ordinarily be taken up by cells or would be incorporated at insufficient concentrations to affect cellular function. The availability of a well established technology for the production and characterization of different types of vesicle provides a firm foundation for the further development of studies on the interaction of vesicles with cells *in vitro* and *in vivo*. Vesicles can be constructed to meet a diversity of experimental requirements by altering such characteristics as surface charge, permeability to ions, and the physical state of the lipids. Vesicles can also be prepared over a wide size range (250 Å to 10 μm diameter) and homogeneous subpopulations of differing size isolated by density gradient centrifugation and gel-filtration techniques. This size range enables materials of widely differing molecular weight to be incorporated into vesicles, either in association with the vesicle membrane or trapped within the internal aqueous space(s) of the vesicle.

Uptake of lipid vesicles of defined composition by cells *in vitro* and *in vivo* offers a new experimental approach for modifying cellular composition and for introducing materials into cells. By manipulating the lipid composition of vesicles, it is possible to achieve some degree of targeting of vesicles and their contents to different regions of the cell. The incorporation of vesicles into cells both by endocytosis and by fusion with the cellular plasma membrane enables vesicle-associated materials to be targeted to the lysosomal apparatus, the cytoplasm, and the plasma membrane. Finally, by inserting appropriate "recognition" ligands into the membranes of vesicles, it may prove to be possible to achieve homing of vesicles to specific cell types.

REFERENCES

Alderson, J. C. E., and Green, C. (1975a). *FEBS Lett.* **52**, 208–211.
Alderson, J. C. E., and Green, C. (1975b). *FEBS Letts.* (in press).
Allison, A. C., and Davies, P. (1974). *In* "Transport at the Cellular Level" (M. A. Sleigh and D. H. Jennings, eds.), pp. 419–446. Cambridge Univ. Press, London and New York.

Allison, A. C., and Gregoriadis, G. (1974), *Nature (London)* **252**, 252.

Alving, C. R., Kinsky, S. C., Haxby, J. A., and Kinsky, C. B. (1969), *Biochemistry* **8**, 1582–1587.

Alving, C. R., Fowble, J. W., and Joseph, K. C. (1974). *Immunochemistry* **11**, 475–481.

Bangham, A. D., Standish, M. M., and Watkins, J. C. (1965). *J. Mol. Biol.* **13**, 238–252.

Bangham, A. D., de Gier, J., and Greville, G. D. (1967). *Chem. Phys. Lipids* **1**, 225–236.

Bangham, A. D., Hill, M. W., and Miller, N. G. A. (1974). *Methods Membrane Biol.* **1**, 1–68.

Batzri, S., and Korn, E. D. (1973). *Biochim. Biophys. Acta.* **298**, 1015–1019.

Brady, R. O. (1975). *Ann. Intern. Med.* **82**, 257–261.

Calissano, P., Alema, S., and Fasella, P. (1974). *Biochemistry* **13**, 4553–4560.

Chapman, D., and Fluck, D. J. (1966). *J. Cell Biol.* **30**, 1–14.

Colley, C. M., and Ryman, B. E. (1974). *Biochem. Soc. Trans.* **2**, 871–872.

Colley, C. M., and Ryman, B. E. (1975). *Biochem. Soc. Trans.* **3**, 157–159.

Davidson, F. M., and Long, C. (1958). *Biochem. J.* **69**, 458–467.

Dawson, R. M. C. (1967). *Biochem. J.* **102**, 205–210.

Dean, R. T. (1975). *Nature (London)* **257**, 414–416.

de Barsy, T., Devos, P., and Van Hoof, F. (1975). *Biochem. Soc. Trans.* **3**, 159–160.

de Duve, C., de Barsy, T., Poole, P., Trouet, A., Tulkens, P., and Van Hoof, F. (1974). *Biochem. Pharmacol.* **23**, 2495–2531.

de Gier, J., Manderslot, J. G., and van Deenen, L. L. M. (1968). *Biochim. Biophys. Acta.* **150**, 666–674.

Demel, R. A., Kinsky, S. C., Kinsky, C. B., and Van Deenen, L. L. M. (1968). *Biochim. Biophys. Acta* **150**, 655–663.

Demel, R. A., Bruckdorfer, K. R., and Van Deenen, L. L. M. (1972). *Biochim. Biophys. Acta* **255**, 321–336.

Folch, J. (1942). *J. Biol. Chem.* **146**, 35–43.

Goldin, S. M., and Tong, S. W. (1974). *J. Biol. Chem.* **249**, 5907–5915.

Gordon, S., and Cohn, Z. A. (1973). *Int. Rev. Cytol.* **36**, 171–214.

Grant, C. W. M., and McConnell, H. M. (1974). *Proc. Natl. Acad. Sci. U.S.A.* **71**, 4653–4657.

Gregoriadis, G. (1973). *FEBS Lett.* **36**, 292–296.

Gregoriadis, G., and Buckland, R. A. (1973), *Nature (London)* **244**, 170–172.

Gregoriadis, G., and Neerunjun, E. D. (1975a). *Biochem. Biophys. Res. Commun.* **65**, 537–544.

Gregoriadis, G., and Neerunjun, E. D. (1975b). *Res. Commun. Chem. Pathol. Pharmacol.* **10**, 351–362.

Gregoriadis, G. and Ryman, B. E. (1972). *Eur. J. Biochem.* **24**, 485–490.

Hanahan, D. J., Dittmer, J. C., and Warashina, E. (1957). *J. Biol. Chem.* **228**, 685–694.

Hauser, H. O. (1971). *Biochem. Biophys. Res. Commun.* **45**, 1049–1052.

Haxby, J. A., Kinsky, C. B., and Kinsky, S. C. (1968). *Proc. Natl. Acad. Sci. U.S.A.* **61**, 300–304.

Haywood, A. M. (1975). *J. Gen. Virol.* **29**, 63–68.

Hilden, S., Rhee, H. M., and Kokin, L. E. (1974). *J. Biol. Chem.* **249**, 7432–7440.

Hinkle, P. C., Kim, J. J., and Racker, E. (1972). *J. Biol. Chem.* **247**, 1338–1339.

Hong, K., and Hubbell, W. L. (1973). *Biochemistry* **12**, 4517–4522.

Huang, L., and Pagano, R. E. (1975). *J. Cell Biol.* **38**, 38–48.

Humphries, G. K., and McConnell, H. M. (1974). *Proc. Natl. Acad. Sci. U.S.A.* **71**, 1691–1695.

Inbar, M., and Shinitzky, M. (1974). *Proc. Natl. Acad. Sci. U.S.A.* **71**, 2128–2130.

Inoue, K., and Kinsky, S. C. (1970). *Biochemistry* **9**, 4767–4776.

Jacobson, K., and Papahadjopoulos, D. (1975). *Biochemistry* **14**, 152–158.

Jain, M. K. (1975). *Curr. Top. Membr. Transp.* **6**, 1–57.

Johnson, J. Y. (1934). British Patent 417,715 (on behalf of I.G. Farbenindustrie Aktiengesellschaft, accepted Oct. 1, 1934).

Johnson, S. M. (1975). *Biochem. Soc. Trans.* **3**, 160–162.

Joseph, K. C., Alving, C. R., and Wistar, R. (1974). *J. Immunol.* **112**, 1949.

Jost, P. C., Capaldi, R. A., Vanderkooi, G., and Griffith, O. H. (1973). *J. Supramol. Struct.* **1**, 269–280.

Juliano, R. L., and Stamp, D. (1975). *Biochem. Biophys. Res. Commun.* **63**, 651–658.

Kagawa, Y. (1972). *Biochim. Biophys. Acta* **265**, 297–338.

Kantor, H. L., and Prestegard, J. H. (1975). *Biochemistry* **14**, 1790–1794.

Karnovsky, M. L. (1962). *Physiol. Rev.* **42**, 143–168.

Kataoka, T., Williamson, J. R., and Kinsky, S. C. (1973). *Biochim. Biophys. Acta* **298**, 158–169.

Kessel, D. (1974). *Biochem. Pharmacol., Suppl.* **2**, 47–50.

Kimelberg, H. K., and Papahadjopoulos, D. (1974). *J. Biol. Chem.* **249**, 1071–1080.

Kinsky, S. C. (1975), In "Methods in Enzymology" (S. Fleischer, L. Packer, and R. Estabrook, eds.), Vol. 32, Part B, pp. 501–513. Academic Press, New York.

Kinsky, S. C., Haxby, J. A., Zopf, D. A., Alving, C. R., and Kinsky, C. B. (1969), *Biochemistry* **8**, 4149–4155.

Knudsen, K. C., Bing, D. H., and Kater, L. (1971). *J. Immunol.* **106**, 258–296.

Ladbrooke, B. D., and Chapman, D. (1969). *Chem. Phys. Lipids* **3**, 304–309.

Lagunoff, D., and Wan, H. (1974). *J. Cell Biol.* **61**, 809–811.

Lagunoff, D., Nicol, D. M., and Pritzl, P. (1973). *Lab. Invest.* **29**, 449–453.

Lea, C. H., Rhodes, D. N., and Stoll, R. D. (1955). *Biochem. J.* **60**, 353–359.

McDougall, I. R., Dunnick, J. K., McNamee, M. G., and Kriss, J. P. (1974). *Proc. Natl. Acad. Sci. U.S.A.* **71**, 3487–3491.

Magee, W. E., and Miller, O. V. (1972). *Nature (London)* **235**, 349–350.

Magee, W. E., Goff, C. W., Schoknecht, J., Smith, D. M., and Cherian, K. (1974). *J. Cell Biol.* **63**, 492–504.

Martin, F., and MacDonald, R. (1974). *Nature (London)* **252**, 161–163.

Michaelson, D. M., and Raftery, M. A. (1974). *Proc. Natl. Acad. Sci. U.S.A.* **71**, 4768–4772.

Oldfield, E., and Chapman, D. (1972). *FEBS Lett.* **21**, 303–306.

Ozato, K., and Huang, L. (1974). *In* "Carnegie Institution of Washington Annual Report to the Director," pp. 83–84. Carnegie Institution, Baltimore, Maryland.

Pagano, R. E., and Huang, L. (1975). *J. Cell Biol.* **67**, 49–60.

Papahadjopoulos, D., and Bangham, A. D. (1966). *Biochim. Biophys. Acta* **126**, 185–188.

Papahadjopoulos, D., and Kimelberg, H. K. (1973). *Prog. Surf. Sci.* **4**, 141–232.

Papahadjopoulos, D., and Miller, N. (1967). *Biochim. Biophys. Acta* **135**, 624–638.

Papahadjopoulos, D., and Ohki, S. (1970). *In* "Liquid Crystals and Ordered Fluids" (J. F. Johnson and R. S. Porter, eds.), pp. 13–32. Plenum, New York.

Papahadjopoulos, D., and Poste, G. (1975). *Biophys. J.* **15**, 945–948.

Papahadjopoulos, D., and Watkins, J. C. (1967). *Biochim. Biophys. Acta* **135**, 639–652.

Papahadjopoulos, D., Nir, S., and Ohki, S. (1972). *Biochim. Biophys. Acta* **266**, 561–564.

Papahadjopoulos, D., Poste, G., and Schaeffer, B. E. (1973a). *Biochim. Biophys. Acta* **323**, 23–42.

Papahadjopoulos, D., Cowden, M., and Kimelberg, H. K. (1973b). *Biochim. Biophys. Acta* **330**, 8–26.

Papahadjopoulos, D., Poste, G., Schaeffer, B. E., and Vail, W. J. (1974a). *Biochim. Biophys. Acta* **352**, 10–28.

Papahadjopoulos, D., Poste, G., and Mayhew, E. (1974b). *Biochim. Biophys. Acta* **363**, 404–418.

Papahadjopoulos, D., Mayhew, E., Poste, G., Smith, S., and Vail, W. J. (1974c). *Nature (London)* **252**, 163–166.

Papahadjopoulos, D., Vail, W. J., Jacobson, K., and Poste, G. (1975a). *Biochim. Biophys. Acta* **394**, 483–491.

Papahadjopoulos, D., Vail, W. J., and Moscarello, M. (1975b). *J. Membr. Biol.* **22**, 143–164.

Papahadjopoulos, D., Poste, G., and Mayhew, E. (1975c). *Biochem. Soc. Trans.* **3**, 606–608.
Papahadjopoulos, D., Poste, G., Vail, W. J., and Biedler, J. B. (1976). *Cancer Res.* in press.
Phillips, M. C. (1972). *Prog. Surface Memb. Sci.* **5**, 139–221.
Poste, G., and Papahadjopoulos, D. (1976). *Proc. Natl. Acad. Sci. U.S.A.* **73**, 1603–1607.
Poste, G., Bernacki, R., Papahadjopoulos, D., Jacobson, K., and Porter, C. (1976). In preparation.
Racker, E., and Eytan, E. (1973). *Biochem. Biophys. Res. Commun.* **55**, 174–178.
Rahman, Y.-E., and Wright, B. J. (1975). *J. Cell Biol.* **65**, 112–122.
Rahman, Y.-E., Rosenthal, M. W., and Cerny, E. A. (1973). *Science* **180**, 300–302.
Rahman, Y.-E., Cerny, E. A., Tollaksen, S. L., Wright, B. J., Nance, S. L., and Thomson, J. F. (1974). *Proc. Soc. Exp. Biol. Med.* **146**, 1173–1176.
Razin, S. (1972). *Biochim. Biophys. Acta* **265**, 241–296.
Redwood, W. R., Jansons, V. K., and Patel, B. C. (1975). *Biochim. Biophys. Acta* **406**, 347–361.
Reeves, J. P., and Dowben, R. M. (1969). *J. Cell. Physiol.* **73**, 49–57.
Reeves, J. P., and Dowben, R. M. (1970). *J. Membr. Biol.* **3**, 123–134.
Riehm, H., and Biedler, J. M. (1972). *Cancer Res.* **32**, 1195–1200.
Robles, E. C., and Van den Berg, D. (1969). *Biochim. Biophys. Acta* **187**, 520–526.
Rothstein, A., Cabantchik, Z. I., Balshin, M., and Juliano, R. (1975). *Biochem. Biophys. Res. Commun.* **64**, 144–150.
Rouser, G., Bauman, A. J., Kritchevsky, G., Heller, P., and O'Brien, J. S. (1961). *J. Am. Oil Chem. Soc.* **38**, 544–549.
Segal, A. W., Wills, E. J., Richmond, J. E., Slavin, G., Black, C. D. V., and Gregoriadis, G. (1974). *Br. J. Exp. Pathol.* **55**, 320–327.
Segal, A. W., Gregoriadis, G., and Black, C. D. V. (1975). *Clin. Sci. Mol. Med.* **49**, 99–106.
Segrest, J. P., Gulik-Krzywicki, T., and Sardet, C. (1974). *Proc. Natl. Acad. Sci. U.S.A.* **71**, 3294.
Sessa, G., and Weissmann, G. (1970). *J. Biol. Chem.* **245**, 3295–3301.
Spanner, S. (1973), *In* "Form and Function in Phospholipids" (G. B. Ansel, J. N. Hawthorn, and R. M. C. Dawson, eds.), pp. 43–65. Elsevier, Amsterdam.
Spielvogel, A. M., and Norman, A. W. (1975). *Arch. Biochem. Biophys.* **167**, 335–344.
Steinman, R. M., Silver, J. M., and Cohn, Z. A. (1974). *J. Cell Biol.* **63**, 949–969.
Straub, S. X., Garry, R. F., and Magee, W. E. (1974). *Infect. Immun.* **10**, 783–792.
Tappel, A. L. (1969), *In* "Lysosomes in Biology and Pathology" (J. T. Dingle and H. B. Fell, eds.), Vol. 2, pp. 207–242. North-Holland Publ., Amsterdam.
Trauble, H., and Eibl, H. (1974). *Proc. Natl. Acad. Sci. U.S.A.* **71**, 214–219.
Verkleij, A. J., De Kruyff, B., Ververgaert, P. H. J., Tocanne, J. F., and Van Deenen, L. L. M. (1974). *Biochim. Biophys. Acta* **339**, 432–437.
Wagner, R., Rosenberg, M., and Estensen, R. (1971). *J. Cell Biol.* **50**, 804–817.
Weissmann, G., Bloomgarden, D., Kaplan, R., Cohen, C., Hoffstein, C., Collins, T., Gottlieb, A., and Nagle, D. (1975). *Proc. Natl. Acad. Sci. U.S.A.* **72**, 88–92.
Yamada, T., and Iwanami, Y. (1962). *Gann* **52**, 225–233.

Chapter 5

Use of Erythrocyte Ghosts for Injection of Substances into Animal Cells by Cell Fusion

MITSURU FURUSAWA,[1] MASARU YAMAIZUMI,[2]
TOSHIKAZU NISHIMURA,[1,3] TSUYOSHI UCHIDA,[2] AND
YOSHIO OKADA[2]

I. Introduction

HVJ (Sendai virus)-mediated cell fusion of cultured mammalian cells and mammalian erythrocytes results in transfer of hemoglobin to the cytoplasm of the cultured cells (Furusawa et al., 1974). We considered that if erythrocytes containing a foreign substance were used in this system instead of intact erythrocytes, we could introduce the substance into the cells. As the first trial, Furusawa et al. (1974) used fluorescein isothiocyanate (FITC) as a marker substance and prepared FITC-containing erythrocyte ghosts. By cell fusion of the ghosts and cultured cells, introduction of FITC into the

[1]Laboratory of Embryology, Department of Biology, Faculty of Science, Osaka City University, Osaka, Japan.

[2]Department of Animal Virology, Research Institute for Microbial Disease, Osaka University, Osaka, Japan.

[3]*Present address*: Department of Anatomy, Aichi Medical University, Nagakute-cho, Aichi-ken, Japan.

cells was attained although the injection frequency was low. Subsequently, we improved the technique for replacing contents of erythrocytes and succeeded in raising the fusion frequency (Nishimura et al., 1976).

II. Method of Injection

The injection procedure consists of two steps: the first step is to introduce foreign substance into erythrocyte ghosts, and the second is to fuse the ghosts containing the substance with target cells using HVJ.

A.   Introduction of Immunoglobulin G (IgG) into Erythrocytes

Human erythrocytes can be used for this purpose irrespective of their blood group. Blood was collected in a syringe containing isotonic sodium citrate to prevent coagulation. The erythrocytes (HRBCs) were washed 3 times with phosphate-buffered saline (PBS: 137 mM NaCl, 2.7 mM KCl, 8.1 mM Na$_2$HPO$_4$, 1.5 mM KH$_2$PO$_4$, 4 mM MgCl$_2$; pH 7.2) to remove leukocytes and once with reverse PBS (137 mM KCl, 2.7 mM NaCl, 8.1 mM Na$_2$HPO$_4$, 1.5 mM KH$_2$PO$_4$, 4 mM MgCl$_2$; pH 7.2) with centrifugation at 450 g for 5 minutes. Packed HRBCs were used. The HRBC suspension is storable at 4°C for several days. Blood supplied by a blood bank is also usable.

For introduction of IgG into HRBC ghosts, the gradual hemolysis method of Seeman (1967) was used with some modifications. The procedure is schematically shown in Fig. 1. Packed HRBCs, 0.2 ml, and 1.8 ml of various concentrations of rabbit IgG dissolved in reverse PBS were put into a small dialysis tubing (1.6 cm in diameter) and dialyzed, with vigorous stirring of the outer solution, against 480 ml of sixfold diluted reverse PBS containing 4 mM MgCl$_2$ in a 500-ml beaker for about 30 minutes at room temperature, if necessary at 4°C. During this dialysis the content of the tubing was occasionally mixed by shaking the tubing. As hemolysis proceeded, the contents became transparent, and ghosts sedimented. During this procedure, outflow of hemoglobin and penetration of IgG into ghosts took place. After the absence of intact, nonlysed HRBCs was confirmed under phase contrast microscope, the sample was dialyzed against 500 ml of isotonic PBS for 30 minutes under the same conditions. In the second dialyzing step, ruptured membrane resealed, and finally IgG was trapped in the ghosts. The ghosts were then washed several times with PBS with centrifugation at 500 g for 5 minutes and suspended in balanced salt solution, BSS$^+$ (140 mM NaCl,

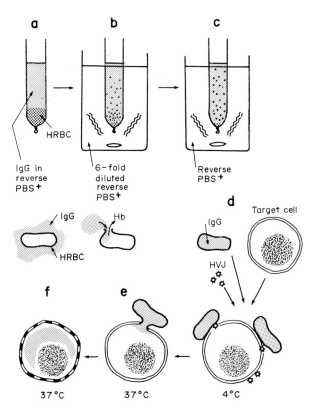

FIG. 1. Diagrammatic representation of the steps in replacing hemoglobin of erytho-cytes with IgG and injecting the IgG into target cells by cell fusion. (a) IgG and HRBCs (human erythrocytes) are mixed in a dialysis tubing under isotonic conditions; HRBC (shown in the middle row) are intact. (b) During dialysis against hypotonic phosphate-buffered saline (PBS $^+$), the outflow of Hb (hemoglobin) and penetration of IgG into HRBC ghosts occur. (c) The ruptured HRBC membrane is resealed and IgG is trapped in the ghost during dialysis against isotonic PBS$^+$. (d) mixing of ghosts containing IgG, target cells, and HVJ (Sendai virus) causes agglutination. (e, f) HRBC ghosts and target cells fuse at 37°C, IgG is introduced into the cytoplasm of the cell, and molecules of HRBC membrane and target cell are intermixed.

54 mM KCl, 0.34 mM Na$_2$HPO$_4$, 0.44 mM KH$_2$PO$_4$, 2 mM CaCl$_2$, buffered with 10 mM Tris-HCl at pH 7.6). The ghosts thus prepared retain the con-cave shape of intact erythrocytes. The ghost suspension can be stored in a refrigerator at 4°C for few days without leakage of IgG.

An experiment was done on the amount of IgG incorporated into ghosts using ^{125}I-labeled IgG. As shown in Table I, the IgG trapped in ghosts increased linearly with increase in concentration of IgG in the tubing, and the concentration in the ghosts was about one-third of the original con-

TABLE I

AMOUNT OF IgG TRAPPED IN ERYTHROCYTE GHOSTS

Total amount of IgG added in a dialysis tubing (μg/2 ml)	960	480	240	120
Total amount of IgG trapped in erythrocyte ghosts (μg/0.2 ml)	31.1	15.9	7.4	3.8

[a] Reverse phosphate-buffered saline, 1.8 ml containing various amounts of ^{125}I-labeled IgG (1500 cpm/μg), and 0.2 ml of packed human erythrocytes (HRBCs) were mixed and dialyzed. The amount of IgG trapped in the ghosts was calculated from radioactivity of trapped ^{125}I-labeled IgG in the total ghosts of 0.2 ml.

centration in the tubing. After staining of acetone-fixed ghosts with FITC-labeled antirabbit IgG, each ghost showed nearly equal intensity of fluorescence, indicating that each ghost contained nearly an equal amount of IgG.

B. Intracellular Injection of IgG

1. INJECTION OF IgG INTO TARGET CELLS

Injection of IgG is achieved by cell fusion of target cells with the ghosts containing IgG. Cell fusion was performed according to the method of Okada and Murayama (1966), an outline of which is schematically illustrated in Fig. 1.

Cells to be injected were suspended in BSS$^+$ at a concentration of 2.5×10^6 cells/ml and kept in ice until used. Concentrated HVJ was inactivated by UV irradiation by the method of Okada and Tadokoro (1962) and diluted with cold BSS (CaCl$_2$-free BSS$^+$) to 1000 HAU/ml (hemagglutinating unit: Salk, 1944). In a round-bottom test tube 0.25 ml of the cell suspension, 0.25 ml of ghosts containing IgG suspended (20% v/v) in cold BSS$^+$, and 0.5 ml of the inactivated HVJ were mixed together. The mixture was kept at 4°C for 20 minutes for cell agglutination, for 5 minutes in a water bath at 37°C, and then shaken gently in the bath for 15 minutes to allow cell fusion between ghosts and cells. Incubations longer than 20 minutes did not cause an increase in efficiency of cell fusion. The sample was transferred to a centrifuge tube and centrifuged at 40 g for 5 minutes and the supernatant, which contained free ghosts and HVJ, was discarded. The sedimented cells were retained for subsequent culture or experiments. A small proportion of them was used to check fusion efficiency. Hemoglobin which had been introduced into target cells was detected by benzidine staining (LoBue et al., 1963) to determine the fusion efficiency. On the average, the fusion or injection efficiency was about 40%, but this value was variable in each trial. During incubation, cell fusion between target cells also occurred. About 70% of the total cells fused with ghosts were mononucleated ones which survive and divide.

To avoid cell fusion between target cells plated cells may be used. Cells, 1×10^5 or fewer, were inoculated into a plastic dish, 60 mm in diameter. The dish was placed in an incubator until the cells settled onto the substratum. After removal of the culture medium, the dish was washed with cold BSS. Inactivated HVJ, 2 ml (1000 HAU), was poured into the dish and kept at 4°C for 5–10 minutes to allow the viruses to be adsorbed to the surface of the plated cells. Next, the culture medium was removed carefully with a pipette, and 2 ml of suspension (20% v/v) of ghosts containing IgG were added in the dish. During incubation for 20 minutes at 4°C, the ghosts were adsorbed onto the surface of the cells, and the aggregates formed rosettes. The culture was then incubated at 37°C for 20 minutes to allow cell fusion of the ghosts with cells. Most of the adsorbed ghosts remained separated from the cell surface, and only a small number of them could fuse with cells. The medium and free ghosts were carefully removed with a pipette, 5 ml of warm culture medium were added, and the culture was continued. The fusion efficiency or efficiency of IgG injection, which were examined by benzidine staining, was lower than in the case of suspension cells.

Besides human erythrocytes, dog, guinea pig, and cow erythrocytes can be used for the injection.

2. BEHAVIOR OF CELLS RECEIVING INJECTIONS

Viability percentage of mononucleated cells receiving injection was about 100% when HVJ of 1000 HAU was used. Titers greater than 1000 HAU caused a decrease in the viability of cells, although the fusion frequency became higher. The cells fused with ghosts can divide at a normal rate, but they show a delay of 1 day in the start of cell division in comparison with nonfused cells. In the process of injection the cell membrane is inevitably inserted into the cell membrane of the target cell, but during subsequent cell divisions these transplanted molecules must undergo progressive geometric dilution.

III. Advantages and Limitations

This injection method is simple, and we can inject foreign material within a short time into a large number of cells. It should be emphasized that the injection procedure itself is not harmful to the cell and that it is applicable to almost all types of animal cells. tRNA, viral RNA, and variously charged proteins can be injected into cells (M. Yamaizumi and M. Furusawa, unpublished data). Moreover, Loyter *et al.* (1975) reported injection of highly structural materials, such as T2 phages and latex particles 1000 Å in diameter, by a similar method.

The erythrocyte membrane transplanted into the target cells may be a good marker to identify cells that have received an injection, because erythrocyte membrane-specific antigens can be detected by staining of the cell surface with fluorescein-labeled antibody (Furusawa and Adachi, 1968). The dose of trapped material in a single ghost can be also controlled. Therefore, it should be possible to select cells injected with a given quantity of a given substance, using a cell separator/sorter that can sort and collect cells according to the intensity of fluorescence or cell volume.

There is another injection technique using artificial vesicles, liposomes (Magee, 1972; Papahadjopoulos *et al.*, 1974). However, it is uncertain whether the contents of the liposomes are introduced into the cell by direct injection or endocytosis. By our method, substances can be injected directly into the cytoplasm. The procedure of trapping substances into ghosts does not involve sonication as in the liposome method, and virtually no proteolytic enzymes, DNase, or RNase are present in the ghosts; so destruction of macromolecules is minimal.

Disadvantages of our method are (1) the unwanted introduction of erythrocyte membrane, residual hemoglobin, viral envelopes, and inactivated viral RNA, which can have undesirable effects on cells, (2) the possibility of denaturation of materials by exposing them to hypotonic condition, and (3) wastefulness of the materials to be injected.

IV.   Possible Applications

This method may be useful in studies in almost all fields in cell biology: (1) molecular biology at the cellular level; e.g., introduction of regulating substances into cells to evoke an expression of masked genes; (2) cellular genetics; e.g., introduction of genes or chromosomal fragments into cells to alleviate genetic defects; (3) cellular physiology; e.g., introduction of biologically active substances into cells to study the mechanism of intracellular regulation; (4) virology; e.g., introduction of a viral genome into a cell that has no receptor for the virus in order to study the host–virus interrelationship.

V. An Example of the Use of This Method: Neutralization of Diphtheria Toxin in Living Cells

We show here an application of this method by analyzing the mode of action of diphtheria toxin. Diphtheria toxin is a single polypeptide which inhibits peptide chain elongation in translation (Honjo *et al.*, 1968; Gill *et al.*, 1969; Gill and Pappenheimer, 1971). The toxin consists of fragments A and B. The inhibitory activity lies in fragment A with a molecular weight of 24,000 (Gill and Pappenheimer, 1971; Drazin *et al.*, 1971), while fragment B has a site for attachment to the cell surface receptors (Uchida *et al.*, 1972). Antifragment A was prepared in rabbits. Contrary to expectation, excess antifragment A [6 "AU"/ml (flocculating antitoxin unit); 1 AU can precipitate fragment A from 1 Lf (flocculation titer against standard antitoxin; 1 Lf is equal to 2 μg of toxin)] added in culture medium could not protect FL cells at all from the attack of the toxin (0.2 Lf/ml); the survival ratio was < 0.1%. We examined whether the susceptible cells become resistant to the toxin when antifragment A, which is not effective when added to the medium, is introduced into these cells. FL cells, which originated from human amniotic membrane, were used as targets. Injection of antifragment A into FL cells was performed by the injection method. The FL cells, some of which had received injections of antifragment A, were exposed to 0.2 Lf/ml of toxin for 8 hours and then cultured in toxin-free medium for 7 days (3×10^4 cells per dish) to allow formation of colonies. The colony number obtained increased with increasing doses of injected antifragment A. At the highest dose (6 "AU"/ml), the colony number was 80.5 times greater than that of the controls (the number being adjusted for fusion efficiency). The results indicate that the antibody retains its function in living FL cells and neutralizes fragment A and that toxin molecules split in the cell membrane or in cytoplasm into a form such that the fragment A is exposed and can react with antifragment A. We estimate that 450 or fewer toxin molecules, or fragment A, enter the cytoplasm of a single FL cell. As shown here, this method could be very useful for studying functions of macromolecules in a living cell.

ACKNOWLEDGMENTS

The authors are indebted to Dr. M. Kotani for reading the manuscript. The work was partially supported by a Grant-in-Aid for Scientific Research from the Ministry of Education, Science and Culture, Japan, and for Cancer Research from the Ministry of Education, Science and Culture, Japan.

References

Drazin, R., Kamdel, J., and Collier, R. J. (1971). *J. Biol. Chem.* **246**, 1504–1510.

Furusawa, M., and Adachi, H. (1968). *Exp. Cell Res.* **50**, 497–504.

Furusawa, M., Nishimura, T., Yamaizumi, M., and Okada, Y. (1974). *Nature (London)* **249**, 449–450.

Gill, D. M., and Pappenheimer, A. M., Jr. (1971). *J. Biol. Chem.* **246**, 1492–1495.

Gill, D. M., Pappenheimer, A. M., Jr., Brown, R., and Kurnick, J. T. (1969). *J. Exp. Med.* **129**, 1–21.

Honjo, T., Nishizuka, Y., Hayaishi, O., and Kato, I. (1968). *J. Biol. Chem.* **243**, 3533–3535.

LoBue, J., Dornfest, B. S., Gordon, A. S., Hurst, J., and Quastler, H. (1963). *Proc. Soc. Exp. Biol. Med.* **112**, 1058–1062.

Loyter, A., Zakai, J., and Kulda, R. G. (1975). *J. Cell Biol.* **66**, 292–304.

Magee, W. E. (1972). *Nature (London)* **235**, 339–341.

Nishimura, T., Furusawa, M., Yamaizumi, M., and Okada, Y. (1976). *Cell Struct. Function* (in press).

Okada, Y., and Murayama, F. (1966). *Exp. Cell Res.* **44**, 527–551.

Okada, Y., and Tadokoro, J. (1962). *Exp. Cell Res.* **26**, 98–128.

Papahadjopoulos, D., Poste, G., and Mayhew, E. (1974). *Biochim. Biophys. Acta* **363**, 404–418.

Salk, J. E. (1944). *J. Immunol.* **49**, 87–98.

Seeman, P. (1967). *J. Cell Biol.* **32**, 55–70.

Uchida, T., Pappenheimer, A. M., Jr., and Harper, A. A. (1972). *Science* **175**, 901–903.

Chapter 6

The Production of Binucleate Mammalian Cell Populations

LOIS H. DICKERMAN AND ROBERT D. GOLDMAN

*Department of Biology, Case Western Reserve University, Cleveland, Ohio; and
Department of Biological Sciences, Carnegie-Mellon University, Pittsburgh, Pennsylvania*

I. Introduction

Binucleate cells are potentially useful agents for the study of a wide variety of problems in cell biology. Questions of nuclear-cytoplasmic balance, cell cycle control, mitotic synchrony, and gene dosage effects are but a few examples of topics that can be investigated by means of binucleate cells (Kelly and Sambrook, 1973; Rao and Johnson, 1974; Fournier and Pardee, 1975). Another advantage of binucleate mammalian cells is the frequency with which such cells become tetraploid (Defendi and Stoker, 1973; Hoehn et al., 1973; Snow, 1973). The production of binucleate cells is, therefore, a stepping stone to the formation of tetraploidy, another nuclear state of potential usefulness in answering genetic and physiological questions.

Binucleate mammalian cells can be induced by treatment with the fungal metabolite, cytochalasin B. Although the mechanisms for the drug's action are not clearly understood, a major consequence in mitotic cells of exposure to cytochalasin B is the inhibition of the completion of cytokinesis (Carter, 1967, 1972; Schroeder, 1970). This inhibitory effect on cell division can be combined experimentally with mitotic synchrony techniques to produce nearly uniform populations of binucleate mammalian cells in minimum time periods (Dickerman and Goldman, 1973).

II. Methods

BHK21/C13 fibroblasts were cultured as previously described (Goldman, 1971) in BHK21 medium (Grand Island Biological Co., Grand Island, New York) supplemented with 10% calf serum, 10% tryptose phosphate broth, and 100 units of penicillin and of streptomycin per milliliter. Synchronized cultures of mitotic cells were produced by a modification of the mitotic shake-off technique (Terasima and Tolmach, 1963; Scharff and Robbins, 1966). Twelve to 18 hours prior to collection of mitotic cells, fresh cultures of BHK21/C13 cells were initiated in 150-ml screw-capped flasks (Falcon Plastics, Bioquest, Oxnard, California) at a concentration of 5×10^4 cells/ cm^2. This cell density was selected to ensure maximum mitotic activity. After a 12–18-hour growth period, the normal medium was replaced by medium containing 0.5 μg of colchicine per milliliter (Sigma Chemicals, St. Louis, Missouri), and the cultures were incubated at 37°C for 2.5–3 hours to accumulate a substantial number of metaphase-arrested cells. After this time interval, the medium containing colchicine was removed, and 5 ml of medium containing 1 μg of cytochalasin B (ICI, Ltd., London) per milliliter and 0.01% DMSO were added to each flask of cells. Mitotic cells were suspended by vigorous shaking, and the resulting cell suspension was collected in sterile centrifuge tubes. After a brief 3–5-minute spin at 1000 rpm, the pellet of mitotic cells was washed, pelleted, and resuspended two more times in small volumes (approximately 5 ml) of fresh medium containing 1 μg of cytochalasin B per milliliter and 0.01% DMSO. After the final wash and resuspension, the cells were plated in small culture dishes, so that karyokinesis could be completed in the presence of cytochalasin B. This washing procedure served two purposes. One, colchicine was diluted sufficiently to remove the mitotic block, and, second, the metaphase cells were immediately exposed to the agent which inhibits cytokinesis, cytochalasin B. Such immediate exposure to cytochalasin B at the time of mitotic shake-off produced the maximum number of binucleate cells in the synchronized population. Approximately 5×10^4 cells from each culture flask were obtained by this procedure. After a 3–4-hour interval, sufficient time to allow all cells to proceed from mitosis into G_1, the medium containing cytochalasin B was replaced with normal growth medium.

III. Results

Within 90 minutes after mitotic shake-off and removal of colchicine, 90–95% of the cells are in the final stages of mitosis. In the presence of 1 μg of cytochalasin B per milliliter, however, cytokinesis is not completed (Fig. 1– 10). Cleavage furrow formation occurs, but the furrow regresses before final separation of daughter cells is achieved. The end result is a population of cells that are 90–95% binucleate, the mononucleate cells probably representing nonmitotic contaminants in the original shake-off procedure (Dickerman and Goldman, 1973). In the presence of cytochalasin B, these binucleate cells do not demonstrate locomotory activity or membrane ruffling. Such motile activities are rapidly resumed, however, when normal medium is added to the binucleate cells.

The concentration of cytochalasin B used in this method is important for observations of the effects on cytokinesis. With BHK21/C13 cells, 1 μg of cytochalasin B per milliliter is the optimum concentration to inhibit daughter cell separation without preventing the spreading of the postmitotic cells on the substrate. At higher concentrations of cytochalasin B (5.0–10.0 μg/ml) the postmitotic cells remain rounded, with a mass of cytoplasm obscuring the nuclear region, so that detection of the number of binucleate cells in the population is impossible. In our hands, the number of binucleate BHK21 cells is not substantially increased with higher concentrations of cytochalasin B, although Fournier and Pardee (1975) have reported an increase of binucleate BHK21/C13 cells in 5 and 10 μg of the drug per milliliter in asynchronous populations.

IV. Discussion

If asynchronous populations of BHK21/C13 fibroblasts are exposed to cytochalasin B for 24 hours or longer, 40–65% of the cells become binucleate (Defendi and Stoker, 1973; Kelly and Sambrook, 1973). By combining the use of cytochalasin B with mitotic synchrony techniques, it is possible to obtain up to 95% binucleates in one-quarter of the time. This combined technique is important not only from the standpoint of experimental efficiency, but also because it minimizes the exposure period of the cells to cytochalasin B, which has several deleterious effects on mammalian cells (e.g., Cohn et al., 1972; Zigmond and Hirsch, 1972). An additional advantage to the protocol is that it produces a binucleate population of cells in approximately the same stage of the cell cycle, which can be useful for experimental purposes.

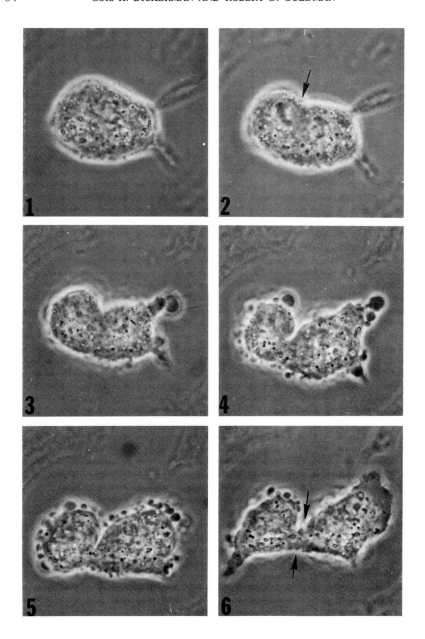

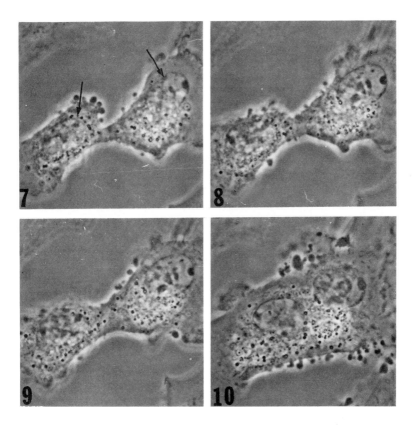

FIGS. 1–10.   A series of micrographs of a living BHK21 cell in mitosis. The medium in which the cell is immersed contains 2 μg of cytochalasin B per milliliter. Figure 1 was taken during anaphase. Figure 2 depicts late telophase and early cleavage furrow formation (arrow). Figures 3–5 show the inward progression of the cleavage furrow. The bridge of continuous cytoplasm between the two cells (arrows) remains, as the presumptive daughter cells begin to spread on the substrate (Figs. 6 and 7). Both nuclei can be seen in Fig. 7 (arrows), indicating that karyokinesis has been completed. Figures 8 and 9 show the regression of the cleavage furrow, which leads to an increase in thickness of the bridge connecting the two parts of the cell. Figure 10 (taken 1 hour after Fig. 1) shows that the cleavage furrow has completely regressed and that the single cell is binucleate. Phase contrast × 1000.

Although we have not determined the frequency with which binucleate cells give rise to tetraploid cells, the level appears to be quite high (Defendi and Stoker, 1973; Hoehn et al., 1973). Undoubtedly the rate is dependent upon the nuclear synchrony in binucleate cells at the next round of mitosis. Therefore, we anticipate that our experimental protocol would result in a significantly greater proportion of tetraploids than the level arising from an asynchronous population. Detection of tetraploid cells would also be substantially easier in populations of cells that are initially binucleate.

V. Summary

Populations of BHK21/C13 fibroblasts that are 90–95% binucleate can be obtained within a short time period by combining the use of cytochalasin B with mitotic synchrony techniques. Such binucleate cell populations are potentially useful for investigating a wide variety of genetic and physiological phenomena, and can be used as a means to produce tetraploid cell lines.

ACKNOWLEDGMENTS

This work was supported by grants from the National Science Foundation and the National Cancer Institute.

REFERENCES

Carter, S. B. (1967). *Nature (London)* 213, 261–264.
Carter, S. B. (1972). *Endeavour* 31, 77–82.
Cohn, R. H., Banerjee, S. D., Shelton, E. R., and Bernfield, M. R. (1972). *Proc. Natl. Acad. Sci. U.S.A.* 69, 2865–2869.
Defendi, V., and Stoker, M. G. P. (1973). *Nature (London), New Biol.* 242, 24–26.
Dickerman, L., and Goldman, R. D. (1973). *Exp. Cell Res.* 83, 433–436.
Fournier, R. E., and Pardee, A. B. (1975). *Proc. Natl. Acad. Sci. U.S.A.* 72, 869–873.
Goldman, R. D. (1971). *J. Cell Biol.* 51, 752–762.
Goldman, R. D. (1972). *J. Cell Biol.* 52, 246–254.
Hoehn, H., Sprague, C. A., and Martin, G. M. (1973). *Exp. Cell Res.* 76, 170–174.
Kelly, F., and Sambrook, J. (1973). *Nature (London) New Biol.* 242, 217–219.
Krishan, A. (1972). *J. Cell Biol.* 54, 657–664.
Rao, P. N., and Johnson, R. T. (1974). *In* "Control of Proliferation in Animal Cells" (B. Clarkson and R. Baserga, eds.), pp. 785–800. Cold Spring Harbor Lab., Cold Spring Harbor, New York.
Scharff, M. D., and Robbins, E. (1966). *Science* 151, 992–995.
Schroeder, T. E. (1970). *Z. Zellforsch. Mikrosk. Anat.* 109, 431–449.
Snow, M. H. L. (1973). *Nature (London)* 244, 513–515.
Terasima, T., and Tolmach, L. J. (1963). *Exp. Cell Res.* 30, 344–362.
Zigmond, S. H., and Hirsch, J. G. (1972). *Science* 176, 1432–1434.

Chapter 7

Enucleation of Mammalian Cells in Suspension

MICHAEL H. WIGLER, ALFRED I. NEUGUT, AND
I. BERNARD WEINSTEIN

*College of Physicians and Surgeons, Departments of Microbiology,
Pathology, and Medicine, and Institute of Cancer Research,
Columbia University, New York, New York*

I. Introduction

In 1972 Prescott *et al.* reported an efficient, relatively simple technique for the enucleation of large numbers of cultured mammalian cells. The method entails plating cells on a glass coverslip or some other surface and centrifuging the surface, cell-side down in medium containing the mold metabolite cytochalasin B. The centrifugation creates a shearing force between the nucleus, which is suspended in cytoplasm, and the cytoplasm, which is anchored by way of the plasma membrane to a surface. In the presence of cytochalasin B, the intracellular cohesive forces are disrupted and allow physical separation of the nucleus and cytoplasm to occur.

Prescott's method or variations of it have found widespread application. The use of enucleated cells, or cytoplasts, thus prepared is justified by the findings that cytoplasts retain many of the morphological and behavioral characteristics of whole cells (Goldman *et al.*, 1973, 1975; Pollack *et al.*, 1974; Shay *et al.*, 1974). Cytoplasts have been used for studying the sites of replication and development of arboviruses (Kos *et al.*, 1975), rhabdoviruses (Sokol and Koprowski, 1975), poliovirus (Pollack and Goldman, 1973),

RNA tumor viruses (Varmus et al., 1974; Donner et al., 1974), and other viruses (Follett et al., 1975). Cytoplasts have also found use in the study of the cellular responses to interferon (Young et al., 1975), glucocorticoids (Ivarie et al., 1975), and cAMP (Miller and Ruddle, 1974); in the study of the metabolism of cellular constituents such as mRNA (Aloni et al., 1975) and NAD (Rechsteiner, 1974); and in the localization of certain cellular functions (Croce and Koprowski, 1973). An entirely different set of experiments becomes possible with the use of cell fusion and the introduction of heterologous nuclei or cytoplasm into whole cells or cytoplasts (Croce and Koprowski, 1973; Veomett et al., 1974; Wright and Hayflick, 1975; Ege et al., 1974).

The enucleation method of Prescott nevertheless has certain drawbacks. These include the fact that the yield is limited by the surface area that can be exposed to centrifugal forces, and the method cannot be used on cells that adhere poorly or not at all to solid surfaces. We have therefore explored the possibility that cells can be enucleated while in suspension by isopycnic centrifugation in the presence of cytochalasin B, and have developed a procedure of high efficiency and yield that can, theoretically, be applied to almost any eukaryotic cell lacking a cell wall (Wigler and Weinstein, 1975).

II.   Enucleation in Isopycnic Suspension

A protocol for enucleation of mouse L cells was published previously (Wigler and Weinstein, 1975). The method is generally applicable to a variety of cell types and is briefly outlined below.

Discontinuous density gradients are poured in ultracentrifuge tubes from solutions of Ficoll (Pharmacia Fine Chemicals, New Jersey) in minimal essential medium (MEM) for suspension cultures (GIBCO) containing cytochalasin B (Aldrich Chemicals) at $10 \mu g/ml$. The gradients are incubated at 37°C for 6–18 hours prior to use. Just prior to enucleation, mouse L cells which have been grown as monolayer cultures are detached from their substrate by incubation with 0.5 mM EDTA in phosphate-buffered saline. Cells are collected, counted, pelleted at low speed at room temperature, resuspended in a small volume of MEM containing Ficoll and cytochalasin B, and layered onto the preformed gradients. The gradients are then centrifuged for 60 minutes in a Beckman ultracentrifuge maintained at 31°C. After centrifugation, cytoplasts or karyoplasts may be collected from the appropriate banding interfaces with a syringe and cannula, diluted into fresh medium without cytochalasin B or Ficoll, pelleted by low speed centrifugation, resuspended in fresh medium, and plated at the desired concentration for further studies.

Isopycnic centrifugation of living cells is made possible by the use of highly water-soluble macromolecules, such as albumin, dextrans, and Ficoll, which give solutions of high density but low osmolarity. Ficoll is a synthetic polymer with an average molecular weight of 400,000. It is relatively non-toxic and inexpensive and can be autoclaved.

The gradient is designed to span the buoyant density of the cells to be enucleated. We routinely prepare discontinuous gradients only because it is difficult to prepare sterile continuous gradients. Sharp density interfaces, however, will cause clumping of cells, which will greatly reduce enucleation efficiency. It is therefore important to preincubate gradients at 37°C for 6–18 hours to allow diffusion at density interfaces.

The buoyant density of cells depends on the cell type and the osmolarity of the cellular environment. To obtain reproducible results it is necessary to prepare gradients of reproducible density and osmolarity. To this end, Ficoll is prepared as a stock solution of 50% w/w in distilled water (dissolved by continual stirring for 24 hours), autoclaved, and stored indefinitely at $-20°C$. A second stock ("25%") is prepared by diluting the 50% stock 1:1 by volume with twice concentrated (2X) MEM for suspension cultures supplemented with antibiotics and 20 μg of cytochalasin B per milliliter. Cytochalasin B is stored as a stock solution of 2 mg/ml in dimethyl sulfoxide at 4°C. MEM for suspension cultures is used because its low divalent cation concentration inhibits cell aggregation. From the 25% Ficoll stock, other solutions of lower density are prepared by mixing with 1X medium containing 10 μg/ml of cytochalasin B. Prior to making the gradients, all solutions are preequilibrated with a CO_2/air mixture to a pH of 7.0–7.6. The refractive index of the Ficoll solutions provides a reliable standard for batch-to-batch comparisons.

The gradient used for enucleating L cells in the Spinco SW 41 rotor contains (listed in order of addition to the centrifuge tube): 2 ml of 25%, 2 ml of 17%, 0.5 ml of 16%, 0.5 ml of 15%, and 2 ml of 12.5% Ficoll with refractive indices of 1.381, 1.365, 1.363, 1.361, and 1.355, respectively. After pre-incubation at 37°C in a 5% CO_2 atmosphere, and just prior to enucleation, this gradient is overlayered with 3 ml of cells evenly suspended in 12.5% Ficoll. Finally, 3 ml of 1X medium are overlayered. All the layers of the gradient contain 10 μg/ml cytochalasin B, and all centrifuge tubes are sterilized by UV irradiation prior to use.

L-cell cytoplasts band between 15% and 17% Ficoll. Other cells may have a higher or lower buoyant density and require gradients of different design.

Cells growing as monolayers may be detached with 0.25% trypsin in balanced salt solution. However, it was found necessary to detach L cells with EDTA since detachment with trypsin resulted in loss of viable cyto-plasts. If the cells to be enucleated have been grown in suspension or are obtained as suspensions from an animal, this is not a consideration. It is important, however, that the cells in Ficoll suspension be free of cell clumps when layered onto the gradient, as cell clumping is the most common cause of poor results in this procedure. We customarily load $1-6 \times 10^7$ cells per gradient. Loading with greater than 6×10^7 cells will result in decreased enucleation efficiency.

The centrifugal force on cells at the banding interface determines the efficiency of enucleation. With L cells, centrifugation at 25,000 to 30,000 rpm in a Spinco 41 rotor (resulting in g forces of about 100,000) for 60 min produces high efficiency of enucleation with minimum cell loss due to frag-mentation. Some cell lines, for example NRK and Muntjac, require higher g forces. Other cell lines may not be enucleated by this method because the forces required for enucleation cause extensive cell fragmentation. Enuclea-tion requires both centrifugation and the presence of cytochalasin B. Furthermore, it is a temperature-dependent process, as has been noted previously (Prescott *et al.*, 1972). Gradients, rotor, and centrifuge must be prewarmed to at least 31°C, and the centrifugation step should also be done at 31°C.

After centrifugation there should be visible cellular material with the nuclei, or karyoplasts, in the pellet, cytoplasts in the intermediate density range, and cell debris near the top of the gradient.

III.   Results

Yields of cytoplasts obtained with the above procedure range from 50% to 70% of input cells. Some loss is undoubtedly due to complete cell fragmenta-tion. After plating, most of the cytoplasts adhere to plastic culture dishes within 30 minutes, and, depending on the cell type, undergo partial to complete morphological recovery within 2–4 hours. Some of the cytoplasts do not, however, recover morphologically but remain as adherent refractile bodies when observed with the phase microscope. The cytoplasts that do recover generally assume the morphology of their parent cell (see Fig. 1), and exhibit thinly spread membrane edges typical of ruffled membranes. Indirect evidence strongly suggests that cytoplasts retain the capacity for locomotion since they can be observed to redistribute themselves with time from focal areas of high cell concentration to surrounding areas of low cell concentration. The morphological longevity of cytoplasts depends on their state of growth prior to enucleation as well as on their origin. L-cell cyto-

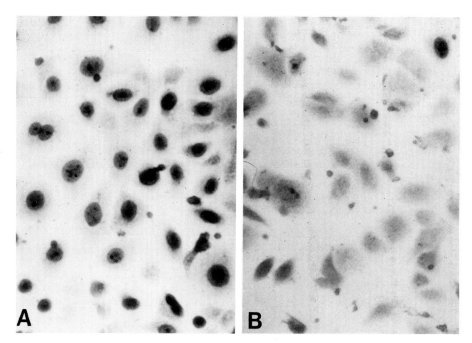

FIG. 1.   Cytoplasts from a rat liver cell line, K-16. (A) Cells maintained in suspension in a Ficoll solution containing cytochalasin B, which were not subjected to centrifugal forces, were replated into plastic petri dishes and stained 4 hours later with Giemsa. (B) Cells treated as above, but which were subject to 100,000 g forces, were replated into plastic petri dishes and stained 4 hours later with Giemsa. Note the presence of a few enucleated cells in group A and one nucleated cell in the upper right corner in group B. See text for details. × 150.

plasts remain morphologically intact for about 24 hours. Prior to plating, the volumes of the cytoplasts obtained by the Ficoll method and by the method of Prescott were similar and were approximately 50% of the volumes of intact cells, when analyzed in suspension in a Coulter Counter (Coulter Electronics, Hialeah, Florida).

The efficiency of enucleation in suspension is a function of cell type and centrifugal force. With L cells, greater than 99.9% pure cytoplasts can be obtained routinely. Enucleation efficiency can be accurately determined by DNA-specific stains such as Feulgen. Giemsa staining is not always adequate for this purpose because cytoplasts that have not recovered morphologically will stain deeply with Giemsa and frequently be indistinguishable from karyoplasts or rounded nucleated cells. Remaining intact cells may also be determined from the cloning efficiency of a cytoplast preparation.

Studies on macromolecular synthesis by L-cell cytoplasts have been reported (Wigler and Weinstein, 1975). Incorporation of radioactive precursors into RNA and DNA generally confirm the high efficiency of enucleation. Protein synthesis, as determined by cycloheximide-inhibitable incorporation of radiolabeled leucine into acid-precipitable counts, is initially 50% of the level observed in intact cells when normalized for total cell protein. This value is the same value we obtained for the identical L cell line when enucleated by Prescott's method. Cytoplasts continue to synthesize protein at a diminishing rate for at least 24 hours.

Although enucleation in suspension may not be applicable to all cell types, we have successfully applied it to several cell lines, including mouse L cells, NRK (normal rat kidney fibroblasts), Vero (African Green Monkey fibroblasts), a Muntjac fibroblast line (Wurster and Benirschke, 1970) and K-16 (normal rat epithelial liver cell line). In addition, we have found it possible to "enucleate" a population of L cells consisting of greater than 90% colchicine-arrested mitotic cells using the same conditions as for enucleating an asynchronous population of L cells.

IV. Discussion

Cytoplasts obtained by enucleation in isopycnic suspension have not been as well characterized as cytoplasts obtained by enucleation of adherent cells; but, to the extent that they have been characterized, they appear to be comparable. They retain the morphology of intact cells, can incorporate amino acids into protein and are probably capable of locomotion. The cell volumes of cytoplasts obtained by the two methods are also comparable. These are not surprising results, since the mechanism of enucleation in suspension is similar to the mechanism of enucleation of cells adhering to a surface: differential forces acting on nucleus and cytoplasm cause separation in the presence of cytochalasin B. Enucleation of adherent cells is, however, a minimal perturbation method, requiring fewer manipulations.

The method of enucleation in suspension compares favorably with previous methods in several other respects. The total yield of cytoplasts is greater than by conventional techniques, although there are several published accounts of ways to increase the yields of cells enucleated while adherent (Ivarie et al., 1975; Croce and Koprowski, 1973). The efficiency of the Ficoll method is at least as great, being limited only by that centrifugal force which might cause extensive cell fragmentation. Finally, although not all cells may be enucleated in suspension, our method complements previous methods in that cells which do not adhere to surfaces may be enucleated. An example of the latter is the enucleation of colchicine-arrested mitotic cells.

In the selection of a method for enucleation, consideration should be given to the fact that enucleation in suspension is, at this time, a less well characterized and somewhat more complex technique than previous methods.

ACKNOWLEDGMENTS

This work was supported by National Cancer Institute Contract No. 72-3234, Research Grant CA 02332, and U.S. Public Health Service Medical Scientist Training Grant GM 02042-07.

REFERENCES

Aloni, Y., Shani, M., and Reuveni, Y. (1975). *Proc. Natl. Acad. Sci. U.S.A.* **72**, 2587.
Croce, C. M., and Koprowski, H. (1973). *Virology* **51**, 227.
Donner, L., Turek, L., and Svoboda, J. (1974). *Int. J. Cancer* **14**, 657.
Ege, T., Krondahl, U., and Ringertz, N. R. (1974). *Exp. Cell Res.* **88**, 428.
Follett, E. A. C., Pringle, C. R., and Pennington, T. H. (1975). *J. Gen. Virol.* **26**, 183.
Goldman, R. D., Pollack, R., and Hopkins, N. H. (1973). *Proc. Natl. Acad. Sci. U.S.A.* **70**, 750.
Goldman, R. D., Pollack, R., Chang, C. M., and Bushnell, A. (1975). *Exp. Cell Res.* **93**, 175.
Ivarie, R. D., Fan, W. J. W., and Tomkins, G. M. (1975). *J. Cell. Physiol.* **85**, 357.
Kos, K. A., Osborne, B. A., and Goldsby, R. A. (1975). *J. Virol.* **15**, 913.
Miller, R. A., and Ruddle, F. H. (1974). *J. Cell Biol.* **63**, 295.
Pollack, R., and Goldman, R. (1973). *Science* **179**, 915.
Pollack, R., Goldman, R. D., Conlon, J., and Chang, C. (1974). *Cell* **3**, 51.
Prescott, D. M., Myerson, D., and Wallace, J. (1972). *Exp. Cell Res.* **71**, 480.
Rechsteiner, M. (1974). *J. Cell. Physiol.* **84**, 481.
Shay, J. W., Porter, K. R., and Prescott, D. M. (1974). *Proc. Natl. Acad. Sci. U.S.A.* **71**, 3059.
Sokol, F. and Koprowski, H. (1975). *Proc. Natl. Acad. Sci. U.S.A.* **72**, 933.
Varmus, H. E., Guntaka, R. V., Fan, W. J. W., Heasley, S., and Bishop, J. M. (1974). *Proc. Natl. Acad. Sci. U.S.A.* **71**, 3874.
Veomett, G., Prescott, D. M., Shay, J., and Porter, K. R. (1974). *Proc. Natl. Acad. Sci. U.S.A.* **71**, 1999.
Wigler, M. H., and Weinstein, I. B. (1975). *Biochem. Biophys. Res. Commun.* **63**, 669.
Wright, W. E., and Hayflick, L. (1975). *Proc. Natl. Acad. Sci. U.S.A.* **72**, 1812.
Wurster, D. H., and Benirschke, K. (1970). *Science* **168**, 1364.
Young, C. S. H., Pringle, C. R., and Follett, E. A. C. (1975). *J. Virol.* **15**, 428.

Chapter 8

Cell Culture with Synthetic Capillaries

PAUL SCHRATTER

Amicon Corporation, Lexington, Massachusetts

I. Introduction

There appear to be three primary factors critical for the control of mass cell culture: adequate supply of nutrients to each cell, removal of metabolic by-products, and relative constancy as well as suitability of the environment. The latter requires not only that such aspects as temperature and pH remain within acceptable limits, but also that the medium in contact with the tissue should not change drastically. Changes in the pericellular environment may alter behavioral control of cells needed for the formation of tissue masses (Knazek, 1974). There are also indications that optimal tissue growth is a function of cell density, mutual contact (Knazek *et al.*, 1974) and compression of tissues between close surfaces (Rose, 1967). Contact inhibition is diminished by perfusion, as shown by Kruse and Miedema (1965). It can be concluded that an optimized culture environment should enhance multilayer culture. It should also promote the maintenance of cells with specific, differentiated functions.

Perfusion by synthetic capillaries imitates the *in vivo* state by interposing a semipermeable membrane between circulating culture medium and the cell mass. Closely spaced hollow fibers create a protective cellular microenvironment. The tissue is relatively shielded from sudden changes in chemical composition of the medium and from hydraulic effects. Choice of the membrane barrier's diffusivity for macromolecules provides an additional degree of control for protection of the cell environment, for recovery of desired species, and for the design of special experiments. Also, as suggested by Nettesheim and Makinodan (1967), cells protected by a membrane barrier are not vulnerable to immune rejection.

The close proximity of the medium stream in bundled capillaries to the surrounding tissue is important for another reason. Penetration of soluble nutrients (including O_2) by diffusion decreases rapidly with distance and is thus limited to relatively thin boundary layers (Knisely et al., 1969; Tromwell, 1958; Thomlinson and Gray, 1955). The density and volume of tissue that may be cultured in any perfusion system is controlled not only by length of the diffusion path for nutrients, but also by molecular diffusivity and exclusion limitations of specific apparatus. This would be particularly true for cells with high nutrient needs.

II. General Characteristics

Artificial capillaries are hollow fibers of nontoxic materials compatible with anchorage-dependent cells, having selectively permeable thin walls—typically 25–75 μm thick—and internal diameters of about 200 μm. The membrane wall permits relatively unhindered passage of small species such as electrolytes, salts, and dissolved gases.

Capillaries made by Amicon Corporation consist of a spongelike body with a very thin (0.5–1 μm), dense layer on the lumen side. Retention of this "skin" for macromolecules may be selected in steps from about 10,000 to 100,000 MW—about 20 to 120 Å. From that layer outward, the pores increase in size to 5–10 μm at the perimeter, presenting a large, reticulated surface for cell attachment (Fig. 1). Made of environmentally resistant polymers, several types of the above capillaries are steam-autoclavable.

Capillaries made by Dow Chemical Corp. and offered by Flow Laboratories or Bio-Rad Laboratories are homogeneous structures of cellulosic materials with smooth exterior as well as interior surfaces and macromolecule retention for species below 30,000 MW. Those sources provide gas-permeable silicone polycarbonate capillaries admixed with the cellulosic capillaries. They are intended to supplement transfer of oxygen and carbon dioxide between medium and tissue, as discussed by Knazek et al. (1972).

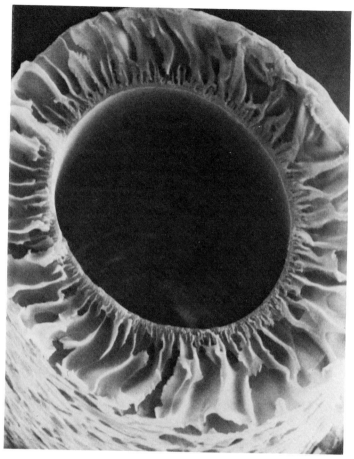

FIG. 1. Anisotropic synthetic capillary. Spongelike body opens progressively outward from selectively permeable lumen surface.

III. Transport Aspects

When culture medium flows through the capillary interior, solute molecules diffuse radially through the stream and through the fluid imbibed in the walls, in both directions, under the driving force of radial and axial concentration gradients (Knisely *et al.*, 1969). Transport rates are inversely proportional to a fractional power of solute molecular weight, so that small molecules permeate relatively rapidly and large species more sluggishly, reaching a zero rate somewhere above the stated exclusion limit of the membrane. Since rated retention levels are only approximations and reten-

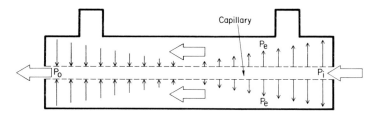

FIG. 2. Schematic representation of secondary convective flow in synthetic capillary unit, due to differential hydrostatic pressures (Starling effect) at inlet (P_i), outlet (P_0) and in extracapillary space (P_e). $P_i > P_e$; $P_o < P_e$.

tion is determined by such factors as molecular shape, concentration, time, and transmembrane movement of fluid, actual diffusion characteristics differ for each set of conditions.

The transmembrane diffusion pattern is affected by depletion of nutrient in the medium stream as it moves through the capillary lumen, and by the increasing concentration of cell excreta in the same stream between inlet and outlet. The effect of these factors would be diminished at high circulation rates of the medium.

An additional and sometimes more important driving force of transmembrane exchange is the hydrostatic pressure gradient created by pumping culture medium through the capillaries. While the average pressure differential between intra- and extracapillary regions is zero, the intracapillary pressure is higher than that of the closed extracapillary space at the inlet and lower at the outlet. Known as the Starling effect and discussed by White *et al.* (1973) as well as Russ (1976), the pressure gradient is believed to cause a secondary, codirectional flow in the extracapillary region, at about 2% of intracapillary flow rates. This convective flow component—ultrafiltration— enhances the exchange of macromolecules in the system (Fig. 2).

The above applies when the extracapillary space is closed. Transport of large molecules from the nutrient may be increased by controlled "bleeding" of the extracapillary fluid, creating a net positive pressure gradient. This can be used to harvest or remove nonpermeating species from the extracapillary side.

IV.   Capillary Perfusion Systems

A typical system as described by Knazek *et al.* (1972), consists of one or more culture units, reservoir and pump for the circulating medium, as well as a means of charging the medium stream with oxygen and carbon dioxide

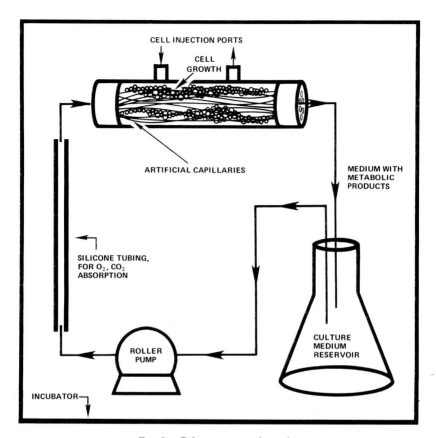

FIG. 3. Culture system schematic.

(Fig. 3). The latter was achieved by Knazek (1974), with the use upstream of silicone tubing through which O_2 and CO_2 from the incubator atmosphere (5% CO_2 in air) diffuse readily. These gases are dissolved in the medium before it enters the culture unit, and diffuse across the capillary membranes sufficiently to support a small cell mass. Larger systems will require higher-capacity provisions for oxygenation.

Each unit (Fig. 4) incorporates a bundle of capillaries whose ends are encapsulated so that the medium can only enter their interiors. The extra-capillary space is accessible through ports for cell injection and recovery. Culture fluid is circulated by a peristaltic pump at a rate between 1 and 10 ml per minute for a unit of about 60 cm^2 lumen area.

After the system is sterilized by autoclaving or chemical methods (which must be followed by thorough detoxification), and supplied with circulating

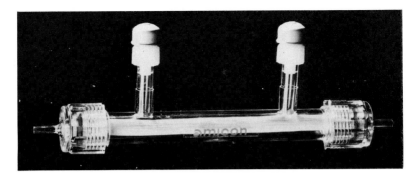

FIG. 4. Artificial capillary cell culture unit (Amicon Corporation).

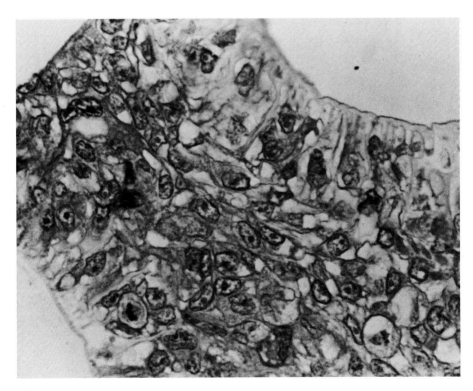

FIG. 5. Breast cancer cell culture. Sectors of synthetic capillaries at lower left and upper right. Courtesy of Dr. R. A. Knazek, National Cancer Institute, N.I.H.

medium, the cell suspension or explant material is injected into the extra-capillary region. Progress may be monitored by glucose consumption, production of lactate and cell-specific peptides or protein products, or visually through the transparent walls of the culture unit. The medium—typically 125–200 ml for a small unit—is changed periodically, depending on by-product accretion, usually every 2–7 days. Both the medium and the culture space are accessible for addition or extraction during the run.

Cells tend to attach temselves to the exterior surfaces of the capillaries and, under the proper conditions, in the interstices between adjacent capillaries. In some cases, they have grown to fill all available space (Knazek, 1974). On Amicon's anisotropic fibers, whose walls consist largely of inter-connected chambers with dimensions increasing toward the outer surface, cells have been shown to grow into the wall as well (Fig. 5).

When the desired growth level has been reached, tissue may be harvested by chemical detachment (e.g., interior circulation of trypsin) and flushing. For histological study of the culture, the entire capillary bundle can be encapsulated in agarose or by other suitable techniques, removed from the outer shell, and sectioned.

V. Applications and Experience

Synthetic capillary units are now being used in many ways. These include: recovery of products, such as hormones or other valuable substances (e.g., immunoglobulins); study of cell function and metabolism under conditions of controlled nutrient supply; experimentation involving nonpermeating macromolecules; culture of cells difficult to maintain by conventional techniques; propagation and recovery of virus and production of viral vaccines; investigation of the effects of therapeutic agents on tissue; study of cell interactions in culture units operated in series; development of artificial organs.

Artificial capillary units have been used to culture a wide variety of normal and neoplastic cells, primary tissue explants, and cell lines. Knazek *et al.* reported in 1972 that an inoculum of 1.5×10^6 human choriocarcinoma cells into each of several units doubled its human chorionic gonadotropin production every 1.2 days and after 4 weeks produced hCG at a rate equal to 145 times the initial number of cells in monolayer, all in a volume of 3 cm^3. Knazek later (1974) illustrated the formation of solid tissue masses and logarithmic increase in prolactic secretion by rat pituitary tumor cells. There are indications of satisfactory growth of various other tissue such as Chinese hamster ovary, monkey kidney, chick and mouse embryo fibroblasts, Chang liver and WI38 cells; also, repeated harvest of 10^9 virus-transformed hamster embryo fibroblasts from a single unit with 60 cm^2 lumen surface area about 9 days after inoculation with 5×10^5 cells. Pancreatic beta cells have been maintained and have produced insulin for long periods (Chick *et al.*, 1975). Cells of hepatic origin cultured on artificial capillaries have performed complex functions of bilirubin uptake, conjugation, and excretion (Wolf and Munkelt, 1975).

VI. Future Development

The development and application of membrane perfusion systems is now at an early stage. Current emphasis is on exploring their potential and evaluating them comparatively. Imaginative uses are likely to lead to configurations for special purposes such as continuous-harvest systems, organ culture devices, or operation with cells inside the fibers. Continuous expansion of membrane technology will support progress in this field through availability of materials with different retention characteristics or physical features and by chemical surface modifications with desirable biospecific properties. As applications for industrial processes are defined, large-scale systems can be developed, combining tissue culture results in the laboratory with existing experience in production of large hollow fiber apparatus for other purposes.

In summary, synthetic capillary systems provide the cell biologist with a new tool for investigation of cell physiology and for efficient tissue culture in a vascular environment closer to that of the natural body than provided by previously available techniques.

ACKNOWLEDGMENTS

The advice of Drs. W. F. Bowers, E. N. Lightfoot, T. Q. Garvey, III, and T. W. Mix is gratefully acknowledged.

REFERENCES

Chick, W. L., Like, A. A., and Lauris, V. (1975). *Science* **187**, 847–848.
Knazek, R. A. (1974). *Fed. Proc. Fed. Am. Soc. Exp. Biol.* **33**, 1978–1981.
Knazek, R. A., Gullino, P. M., Kohler, P. O., and Dedrick, R. I. (1972). *Science* **178**, 65–67.
Knazek, R. A., Kohler, P. O., and Gullino, P. M. (1974). *Exp. Cell Res.* **84**, 251–254.
Knisely, M. H., Reneau, D. D., and Bruley, D. F. (1969). *Angiology* **20**, 1–56,
Kruse, P. F., and Miedema, E. (1965). *J. Cell Biol.* **27**, 273.

Nettesheim, P., and Makinodan, T. (1967). *In* "Methods in Developmental Biology," p. 471. Crowell-Collier, New York.

Rose, G. G. (1967). *J. Cell Biol.* **32**, 108.

Russ, M. B. (1976). M. S. Thesis, University of Delaware, Newark.

Thomlinson, R. H., and Gray, L. M. (1955). *Br. J. Cancer* **16**, 841.

Tromwell, O. A. (1958). *Exp. Cell Res.* **16**, 118–147.

White, A., Handler, P., and Smith, E. L. (1973). *In* "Principles of Biochemistry," 5th ed., p. 905. McGraw-Hill, New York.

Wolf, C. F. W., and Munkelt, B. E. (1975). *Trans. Am. Soc. Artif. Intern. Organs* **21**, 16–27.

Chapter 9

Magnesium Acetate Treatment of Glass Roller Bottles to Facilitate Cell Attachment

JOHN J. MONAHAN

Department of Cell Biology,
Baylor College of Medicine,
Houston, Texas

I. Introduction

The need for increasing quantities of tissue culture cells, viruses, and antigens has lead to the widespread use of cylindrical roller bottles for the growth of certain cell cultures. There are advantages to growing such cultures in suspension, and there is now a wealth of experience and equipment available for this purpose (Telling and Radlett, 1970). In some situations it is actually an advantage to grow cells attached to a solid substrate, such as a roller bottle, for example, when cell cycle synchronization studies are being carried out in which frequent changes of the medium are required, when cell life span or contact inhibition studies are being carried out, or when it is desired to grow normal diploid cells that are anchorage dependent. In these situations, a large-scale preparation of cells usually either involves growing cells either in suspension culture containing glass beads (Robineaux *et al.*, 1970), on multisurface glass plates (Weiss and Schleicher, 1968), in gas-permeable plastic bags (Munder *et al.*, 1971), in roller bottles, or in bottles containing a spiral plastic sheet (House *et al.*, 1972). A review of the advantages and disadvantages of the above culturing procedures for anchorage-dependent cells has been presented in detail elsewhere (Maroudas, 1973).

Growth of cells in roller bottles leads to good oxygenation and exposure of cells to media, helps to maximize space utilization, and leads to significant time savings in planting, feeding, and harvesting of cells. The technique itself is not new. Originally described by Gey in 1933, it is based upon the idea of utilizing the entire wall of a rotating cylindrical culture bottle as an anchorage surface for cell attachment and growth. Because the vessels are continuously rolled at an optimum speed for cell attachment and growth, minimal volumes of media are required for an abundant cell yield. The reduced handling of the cultures tends also to minimize the possibilities of contamination.

An abundant cell yield, however, is strongly dependent upon a rapid attachment of cells to the roller bottle surface. Unlike the situation in Blake bottles, or when cells are cultured on multisurface glass plates where the cell has ample opportunity to attach to the surface, unattached cells in roller bottles are in a dynamic state and are brought only briefly in contact with the bottle surface by gravity. For this reason processes that effect cell anchorage become very important in determining the feasibility of culture growth using this technique. Cells that would otherwise grow well in monolayer cultures may attach poorly to a roller bottle surface and thus give low yields with this cell culture method.

Polystyrene roller bottles are now available commercially which have had their surface treated by electrical corona discharge. This results in a plastic surface that leads to excellent attachment of many anchorage-dependent cell types leading to good yields of cells. However, such "disposable" vessels can be expensive for large-scale preparations. Often it is more convenient to use reusable glass roller bottles. Such bottles are usually made of borosilicate glass. Both a low-cost mold-blown or a more expensive hand-made, ripple-free surface type of bottle are available. The hand-made bottles are fabricated to precision tolerances and have better optical viewing characteristics. However, both types of bottles appear to act similarly as far as cell attachment is concerned.

The reuse of these bottles clearly necessitates that they be thoroughly washed and sterilized each time they are used. The use of radioactive precursors in metabolic studies of cells grown in roller bottles bring about an additional difficulty in that the bottles have to be washed with chromic acid (or one of the commercial radioactivity decontaminants presently available) in order to remove completely residual radioactivity in the bottle. It is commonly observed that glassware treated in this way no longer behaves optimally for the attachment of cells to the glass surface. New roller bottles used for the first time often provide more uniform cell cultures with a higher yield than old extensively washed bottles. The reason for this is unclear, although it almost undoubtedly involves processes that affect the glass surface. By leaving the glass surface in contact with 1 mM magnesium acetate for a few hours followed by removal of excess magnesium acetate and auto-claving, it is possible to obtain a glass surface that is very effective in binding cell suspension to roller bottles (Monahan and Hall, 1974). This simple procedure is described below.

II.   Design of Equipment and Procedures for Using Roller Bottles

A precision roller bottle apparatus can be obtained from a number of manufacturers. Some are modular in design in that they can accommodate additional bottle as the need arises. The speed of bottle rotation should be precisely regulated and be continuously monitored with an electrical tacho-meter. One should remember that the speed of rotation of a roller bottle is dependent upon the diameter of the bottle itself. If different-sized roller bottles are being used, it will be necessary to determine the optimum tacho-meter setting for each type. For at least one type of bottle, the bottle caps (which should be left slightly open for gas diffusion) have a habit of un-winding themselves off as the bottle is rotated. This rather frustrating occurrence inevitably leads to a contaminated culture; it can be very simply prevented by attaching a piece of adhesive tape from the cap to the bottle neck.

The procedure described below has been tried with a number of different cell cultures, these include L-929, KB, 3T3, DON-C, HTC, and BHK21 cells. Essentially the same observations were made in each case. The cells were grown in Minimum Essential Medium for monolayer cultures (GIBCO) supplemented with 5% fetal calf serum, potassium penicillin G (75 units/ml), streptomycin sulfate (50 ug/ml), and amphotericin B (2.5 ug/ml). The bottle was rotated at 0.2 rpm at 37°C in a 5% CO_2/air incubator. Speeds below 0.7 rpm did not appear to in any way affect cell attachment, however, it may be advantageous to determine the optimum rotation speed for other cell types.

The amount of medium that should be added to the bottle will often dep-end upon the seeding density of the initial inoculum of cells. For some per-manent cell lines, this may not be too critical. A relatively small inoculum (0.5×10^6 cells), for example, of L-929 cells, can be added to a 849 cm², 28 cm long, 11 cm in diameter roller bottle that is filled with medium such that one-quarter of the cylindrical glass wall is bathed in medium at any one time. Without changing the medium it is possible to obtain a confluent monolayer of cells ($\sim 10^5$ cells/cm²) in 4 days. For other cells, particularly nontransformed cultures, a higher initial inoculum of cells is required in a smaller volume of medium. The amount of medium, however, should not be allowed to drop below that required to cover about one-eight of the glass surface at any one time. Overcrowding of cells in smaller volumes of medium leads to ridges of clumped cells forming on the glass.

Small volumes of medium in the roller bottle of course necessitate that the medium be changed more frequently. For a large-scale preparation of cell cultures using roller bottles, it is advisable to make an initial determination of the optimum seeding density of cells required per bottle per one or more changes of medium. Any reduction in handling of bottles not only saves in time and effort but also helps minimize contamination.

Care is required in washing glass roller bottles. If they have been previously used for radioactive studies, they should be washed with chromic acid or one of the commercial radioactive decontaminants presently available. Grease films remaining upon the glass interior lead to a nonuniform cell monolayer, often characterized by patches of cells which fall off the glass surface and form cell aggregates in the media. It is equally important to ensure that, after adequate detergent wash of the bottle, all the residual detergent has been removed. Generally, we wash each bottle with running tap water overnight to ensure the complete removal of detergent (and chromate ions, if chromic acid was used). An incompletely cleaned bottle is often characterized by the tissue culture medium turning strongly acidic (or in some cases for detergents, basic) when a few milliliters of medium are added to a bottle after autoclaving.

III.   Treatment of Roller Bottles to Enhance Cell Attachment

If the above simple precautions are taken, reasonable attachment of many cells to the bottle surface can be obtained. However, from time to time one observes bottles in which cell attachment and growth are poor in varying degrees. This is particularly so for bottles treated to remove residual radioactivity. Figure 1 shows a typical incomplete attachment of L-929 cells to a roller bottle 24 hours after the addition of 1×10^7 cells to a 849 cm^2 bottle washed as described above. Cells that fail to attach to the glass tend to form clumped aggregates in the medium. These aggregates appear to trap further mitotic cells poorly attached to the glass, leading to larger aggregates of cells. As these aggregates grow, cells in the interior of the aggregate die, leading to poor growth of the whole culture. In cases where the incorporation of precursors in one roller bottle culture is being compared to the next, such monolayers are very unsatisfactory.

This problem can be overcome very simply by filling each bottle with a solution of 1 mM magnesium acetate after the overnight washing with tap water, and leaving this solution to stand at room temperature for a few

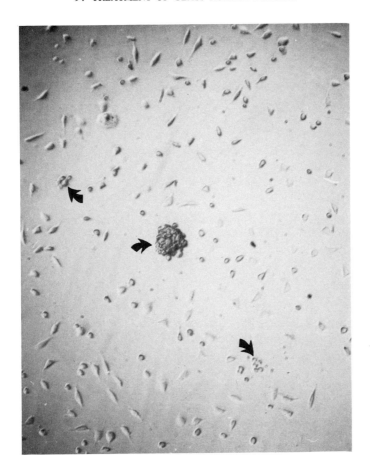

FIG. 1. Incomplete attachment of L-929 cells to a roller bottle which was not pretreated with 1 mM magnesium acetate. Cells on the curved glass surface are out of focus to show clumps of cells. Arrows indicate cells that are unattached to the glass.

hours. The bottle is then emptied (washed once with glass-distilled water) and autoclaved and dried in the normal manner. Figure 2 shows the complete attachment of an identical preparation of L-929 cells seen in Figure 1 to the glass surface. No clumps of unattached cells are seen. A uniform cell monolayer on the glass was obtained. The procedure leads to a rapid and uniform attachment of almost all the cells to the roller bottle, thereby allowing a realistic comparison to be made in metabolic studies between cultures in different roller bottles. No adverse effect of treating the glass with magnesium acetate was observed upon the incorporation of ^{32}P-labeled H_3PO_4, ^{3}H-labeled uridine, ^{3}H-labeled leucine into L-929 cells.

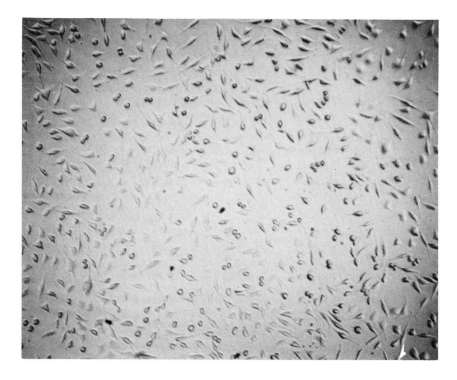

Fig. 2. Complete attachment of L-929 cells to a roller bottle pretreated with 1 m*M* magnesium acetate. Almost all cells are completely attached to the glass surface and are well spread over the entire area.

IV. Discussion

The mechanisms involved in the binding of cells to a surface is largely unknown at present. The parameter known as the "surface energy" (Zisman, 1964) does define a wettability character, which is useful in determining whether a surface is suitable for cell attachment. A surface energy of greater then 40 dynes/cm appears to be necessary for anchorage of most cells (Maroudas, 1973). For this reason polystyrene with a surface energy of ~ 33 dynes/cm is usually specially treated to increase its surface energy to ~ 56 dynes/cm to make it a good substrate for cell attachment (Maroudas, 1973). However, most glasses have a very high surface energy (~ 700 dynes/cm) well in excess of that required for cell attachment, indicating that other parameters may be involved since cells will not always bind to glass.

Borosilicate glass does act as a weak ion exchanger (Jones, 1971; Holland, 1972). The surface of tissue culture cells appears to be negatively charged since the cells tend to adhere particularly well to a positively charged matrix (van Wezel, 1967). It is possible that the procedures used to wash glassware "leach out" essential cations used by cells to attach themselves to the glass surface. Any difficulty cells may have in attaching to such glass surfaces will be particularly apparent in the case of roller bottle cultures, because a rapid attachment of the cells to the glass surface is required. The markedly improved attachment of cells to the magnesium acetate-treated glass would not appear to be simply a question of a lack of Mg^{2+} ions in the culture medium itself, since increasing the concentration of $MgSO_4$ in the medium from 200 mg to 250 mg per liter did not improve the attachment of cells to the glass surface of untreated bottles.

It is possible that traces of magnesium acetate remaining on the glass surface during autoclaving and drying interact with the glass in a way that cannot be attained simply by addition to Mg^{2+} ions to the glass in solution at 37°C. Since this method was found to be very satisfactory, the feasibility of substituting other cations has not yet been explored.

REFERENCES

Gey, G. O. (1933). *Amer. J. Cancer* **17**, 752–756.
Holland, L. (1972). *In* "The Properties of Glass Surfaces," pp. 122–192. Chapman & Hall, London.
House, W. M., Searer, M., and Maroudas, N. G. (1972). *Exp. Cell Res.* **71**, 293–296.
Jones, G. O. (1971). "Glass," p. 97. Chapman & Hall, London.
Maroudas, N. G. (1973). *In* "New Techniques in Biophysics and Cell Biology" (R. H. Pain and B. J. Smith, eds.), Vol. 1, pp. 67–86. Wiley, New York.
Monahan, J. J., and Hall, R. H. (1974). *Prep. Biochem.* **4**, 353–358.
Munder, P. G., Modolell, M., and Wallach, D. F. H. (1971). *FEBS Lett.* **15**, 191–196.
Robineaux, R., Lorans, G., and Beaure d'Angères, C. (1970). *Eur. J. Clin. Biol. Res.* **15**, 1066–1071.
Telling, R. C., and Radlett, P. J., (1970). *Adv. Appl. Microbiol.* **13**, 91–116.
van Wezel, A. L., (1967). *Nature (London)* **216**, 64.
Weiss, R. E., and Schleicher, G. B. (1968). *Biotechnol. Bioeng.* **10**, 601–611.
Zisman, W. A. (1964). *Adv. Chem. Ser.* **43**, 20–32.

Chapter 10

Use of Plastics for Suspension Culture Vessels

ROLAND A. COOK, FREDERICK T. COUNTER, AND
JOYCE K. MCCOLLEY

Eli Lilly and Company, Indianapolis, Indiana

I. Introduction

The field of suspended culture of mammalian cells is now over 20 years old, but continues to receive impetus from technological advances. Mass suspension culture systems have been devised for the production of virus vaccines and diagnostic reagents; large-scale cultures of lymphoblastoid, or hematopoietic cells—which grow naturally in suspension—have been utilized for the production of antilymphocyte serum and serve as a source of material for analysis of cell membranes and subcellular components. The recent use of DEAE-Sephadex beads (Horng and McLimans, 1975; van Wezel, 1967) and porous silica spherules (Bizzini *et al.*, 1973) as microcarriers may make possible the growth in suspension of many types of functional, differentiated cells, which are normally anchorage dependent.

Our laboratory has been involved in the adaptation of cell lines derived from normal tissues to growth in continuous suspension culture. Plastic culture vessels were found to facilitate this work, as we have previously reported (Cook *et al.*, 1974a,b; Cook and Counter, 1974). We believe that other workers will appreciate the ease with which plastic materials can be used to construct suspension culture vessels to a given experimental design. In this report we describe our experiences with the use of these vessels and provide detailed information on their construction.

II. Use of Plastics for Monolayer Cultures

The use of plastics for monolayer culture of mammalian cells has been well documented. Recent refinements in plastics technology have led to the development of plastic surfaces that are uniform enough for microtitration and microcloning techniques (Fuccillo *et al.*, 1969; Robb, 1970). A method of treating plastic petri dishes for use in cloning chick embryo cells has been described (Trager and Rubin, 1966); however, processes used by commercial firms to modify plastic surfaces to enhance cell adhesion have not been published.

The advances mentioned above have involved alteration of the basic "non-stick" qualities shared by most plastics. For certain operations, adhesion of cells to vessels is undesirable and unaltered plastics can be used to advantage. Polystyrene tubes have been used to eliminate cell adhesion in radiation experiments (Walker *et al.*, 1962). Industrial firms have developed a large market for plastic-coated cookware based on the ease with which these utensils can be cleaned.

III. Adaptation of Cell Lines to Growth in Suspension

A. Problems with Glass Suspension Culture Vessels

Initial studies were performed with the LLC-MK$_2$ cell line—a Lilly Laboratory cell line derived from normal rhesus monkey kidney in R. N. Hull's laboratory (Hull *et al.*, 1956). It was necessary to maintain suspensions of viable cells for several weeks. During this maintenance period, cells were observed to grow to increasing densities on the walls of glass flasks, with occasional sloughing of clumps of cells back into the culture medium. Calcium has been reported to cause clumping of cells in suspension (Eagle,

TABLE I
ADHESION OF CELLS TO GLASS[a]

Day	Culture volume (ml)	Cells in suspension $\times$ 10^6	Cells on flask wall[b]	Total cells
–	366	183	–	183×10^6
1	360	180	+	Unknown
2	354	89	+ + +	Unknown
3	479	144	+[c]	Unknown
6	467	42	70×10^{6d}	112×10^6

[a]From Cook et al., 1974a.
[b]Relative density of cells in ring on flask wall at meniscus of medium.
[c]Increase in media volume resulted in numerous large chunks of cells sloughing off ring into medium.
[d]Cells on wall of flask removed by trypsinization and counted separately.

1959); however, no calcium was present in our medium formulation (Cook et al., 1974a). Actual growth of suspended cells could not be determined because of these problems. Results shown in Table I were typical of problems encountered with glass vessels. The LLC-MK$_2$ cells used for this culture had previously been grown in bottles and were not adapted to suspension growth. At the end of 6 days, only 37.5% of surviving cells were in suspension; the remainder had adhered to the flask wall. Capstick et al. (1962) reported similar difficulties during the early stages of adaptation of the BHK21/13 (baby hamster kidney) cell line to growth in suspension. Conversely, investigators working with HeLa and L cell cultures reported minimal tendencies of these cells to adhere to glass in spinner culture (McLimans et al., 1957). The use of methyl cellulose in culture media (0.12%) and silicone treatment of glass flasks was found to retard but not prevent adhesion of cells to the flask walls.

B. Use of Plastic Suspension Culture Vessels

It seemed possible that cells would be less likely to adhere to plastic surfaces. Polypropylene Erlenmeyer 125-ml flasks (Kimble) were obtained and equipped with magnetic stirring bars suspended from fishing swivels as described by Cherry and Hull (1956). Cultures of LLC-MK$_2$ cells did not adhere to the walls of these flasks. Suspended cells grew singly, or in small clumps of 3–4 cells and could be accurately counted.

Subsequently, PK-15 cells—which are of normal pig kidney origin—were obtained from the American Type Culture Collection. These cells had previously been grown as monolayers and initially grew as large clumps in

TABLE II

ADHESION OF PK-15 SUSPENSION CULTURES TO NEW AND
PREVIOUSLY USED POLYCARBONATE STIRRER FLASKS AFTER 20
HOURS SETTLING AT DIFFERENT TEMPERATURES[a,c]

Temperature (°C)	New flasks	Previously used flasks[b]
36	−	+
25	−	+
4	−	−

[a] Symbols: +, Cells adhered to flask bottoms—did not detach after shaking. −, Cells readily resuspended.

[b] Previously used flasks are flasks that have been used for several weeks and appear to have surface characteristically altered by this use.

[c] From Cook et al., 1974a.

suspension. It was necessary to disperse these clumps several times during the adaptation process. These cells also did not adhere to plastic stirrer flasks. In more recent studies, however, PK-15 cells were found to adhere to polycarbonate stirrer flasks, which had been used as suspension culture vessels for several months, but not to new polycarbonate flasks (Table II).

IV. Toxicity Testing of Plastic Laboratory Ware

The polypropylene Erlenmeyer flasks which we initially used as suspension culture vessels were not treated or washed prior to use. No evidences of cell toxicity were observed. To determine whether other varieties of plastic might also be suitable, eleven different types were obtained from commercial suppliers and modified for use as suspension culture vessels (Table III). Some of the materials were not autoclavable, so all were sterilized with β-propiolactone and rinsed with medium prior to inoculation with suspensions of LLC-MK$_2$ cells. Glass flasks were used for control cultures. No evidence of gross toxicity could be observed in cultures after 3 days of incubation at 36°C. Cells did not adhere to any of the plastic containers; heavy rings of cells had grown at the meniscus on the glass control flasks.

It should be mentioned here that all plastics contain a variety of compounds, such as plasticizers, fillers, and stabilizers. Also, formulas for a given type of plastic may differ from one manufacturer to the next. It is possible that a given formulation might be toxic to cultured cells; however, many of our plastic culture vessels have been in continuous use for several months and toxicity problems have not yet been encountered.

TABLE III

Plastics Obtained for Toxicity Testing[a]

Type of plastic	Item	Supplier	Catalog No.
Polypropylene	Bottle, 4-ounce	Kimble	53250
Polycarbonate	Erlenmeyer flask, 125-ml	Matheson	24093-25
Teflon FEP	Bottle, 4-ounce	Matheson	4356-10
Teflon TFE	Evaporating dish, 100-ml	Star	S-3450
Linear polyethylene (LPE)	Bottle, 4-ounce	Matheson	4371-20
Conventional poly- ethylene (CPE)	Bottle, 4-ounce	Matheson	4335-70
Polyvinyl chloride (PVC)	Bottle, 4-ounce	Matheson	4334-10
Polystyrene	Container, 30-dram	Dynalab	2635
Polymethylpentene (TPX)	Graduated cylinder, 250-ml	Matheson	18591-30
Styrene-acrylonitrile	Graduated cylinder, 250-ml	Preisser	F28684
Methyl-methacrylate (Lucite)	Jar, 1-quart	LPF	8025

[a]From Cook et al., 1974a.

V. Construction of Plastic Suspension Culture Vessels

A. General

Polycarbonate and polypropylene were found to be most practical for this purpose. Polycarbonate is transparent; polypropylene is translucent. Both types of plastic are nontoxic, autoclavable, relatively inexpensive and readily available in various sizes of flasks and bottles from a number of commercial suppliers. Polycarbonate sheets up to 4 × 8 feet, and tubing up to 4 inches in diameter are available locally (Hyaline Plastics, Indianapolis, Indiana).

Plastic culture vessels of various types were adapted to our needs by the simple expedient of heating stainless steel cork borers and melting holes in the plastic where desired (Caution: This operation produces noxious, possibly toxic fumes and should be done under a fume hood). An electric drill was used to make holes in thicker materials, such as plastic sheets and tubing. A drill bit smaller than the desired hole was used, followed by larger drill bits. This was necessary to prevent chipping; thin-walled containers could also be drilled with care taken to prevent cracking. Holes drilled in thicker materials were threaded when desired, using conventional taps. Probes, thermometers, breathers, and glass tubing for inlet, outlet, and sample lines were then inserted into the culture vessels where desired—either through rubber stoppers sized to fit the holes melted or drilled in the vessels, or through polypropylene fittings which are available commercially (Matheson Scientific, Elk Grove Village, Illinois).

Polycarbonate materials can be cemented together, using Thickened Cement for Plexiglas Acrylic Sheet (Dayton Plastics, Dayton, Ohio). One of the two pieces to be joined was covered with cement, and both pieces were then immediately joined and held together for about 1 minute. Joints were ready for use in less than 1 hour.

B. Small Vessels with Suspended Stirring Bars

Three types of small plastic suspension culture vessels are shown in Fig. 1. Each of these is equipped with a plastic-coated magnetic stirring bar suspended by a nichrome wire from a stainless steel, ball-bearing swivel (Eagle Claw, Denver, Colorado). A staple fashioned from this wire was used to secure the swivels to rubber stoppers or plastic lids (staples were heated to penetrate plastic lids, then bradded down on top). Swivels were mounted with the large end up—resembling a child's top—to shield the ball-bearing mechanism from liquids inadvertently splashed on the swivel. These swivels have proved to be reliable in many months of continuous use.

In Fig. 1 a rubber vaccine stopper can be seen inserted in the sidewall of the 125-ml Erlenmeyer flask. This was done to permit sampling of cultures with a syringe, rather than by opening the culture vessel. Plastic cement (Dekadhese, Donald Tulloch Jr., Chadds Ford, Pennsylvania) was applied to seal rubber stoppers to the plastic vessels after autoclaving. The vaccine stopper becomes hardened after several autoclave cycles and will leak after needle punctures unless replaced periodically.

The 500-ml and 1000-ml plastic bottles shown in Fig. 1 are equipped to operate as closed systems. Glass sampling tubes run to the bottom of both

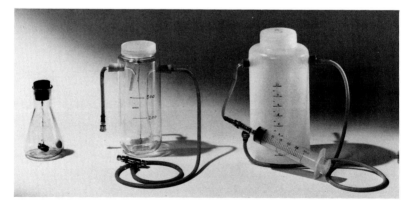

FIG. 1. Plastic suspension culture vessels with suspended stirring bars. Left to right: 125-ml polycarbonate Erlenmeyer flask, 500-ml polycarbonate centrifuge bottle, and 1000-ml polypropylene centrifuge bottle.

vessels. These tubes must be carefully positioned inside the vessels, so that the suspended stirbars do not hang up on the tubes and become inoperative. A short tube mounted on the sidewall is used for media addition and also to cycle the cell suspension from the sampling tube back into the culture, via an attached syringe—allowing a representative sample of the culture to be taken. A cotton-filled breather is mounted on the vessel wall in a location where it will not become wet during these operations. Surgical latex tubing connects the glass tubing with Luer-type stainless steel fittings (Becton-Dickinson, Rutherford, New Jersey). Tubing is autoclaved in 0.1 N NaOH for 5 minutes, rinsed, autoclaved in deionized water for 5 minutes and then thoroughly rinsed and dried prior to use.

C. Five-Gallon Carboy Stirred by Overhead-Drive Motor

Polypropylene carboys of the type shown in Fig. 2 were used to grow suspension cultures of LLC-MK$_2$ and PK-15 cells in volumes of up to 18

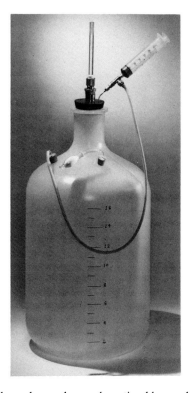

FIG. 2.   Five-gallon polypropylene carboy stirred by overhead-drive motor.

liters. This vessel was also operated as a closed-system stirrer and is shown equipped with a Teflon seal assembly, Viton-A O-ring, 24/40 ground-glass joint, glass stirring shaft, and Teflon stirring blade (Cole-Parmer, Chicago, Illinois). The tapered end of the ground-glass joint was cut off and the remainder was mounted in a hole drilled in a No. 12 stopper, which fits the neck of the carboy. This type of vessel has been useful in the preparation of batch cultures.

D. Use of Plastic Tubing and Sheets for Spin-Filter Vessels

A section of polycarbonate tubing $3\frac{1}{2}$ inches in diameter by $7\frac{1}{4}$ inches in height was cemented to a square of polycarbonate to form the suspension culture vessel shown in Fig. 3. A second, larger polycarbonate square forms the top. These squares sawed out of a large polycarbonate sheet. The top of the vessel was secured to the tubing by horizontally drilling a hole in each of 4 pieces of polycarbonate $\frac{5}{8}$ by 1 inch in diameter, $\frac{1}{4}$ inch thick, and inserting 2-inch stainless steel screws through the holes. The pieces were then cemented to the tubing. The top was drilled to accept the screws, and wingnuts

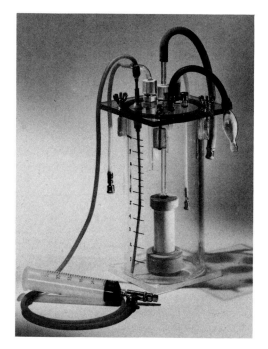

FIG 3.   One-liter spin-filter suspension culture vessel made from polycarbonate sheet and tubing.

were used to secure the top. A Neoprene rubber gasket was cut so as to slightly overlap the rim of the tubing, and provided an airtight seal. Figure 4 shows an automated cell culture system utilizing a 4-liter polycarbonate jar which has been similarly equipped with a square top fashioned from a polycarbonate sheet.

Both of the vessels shown in Figs. 3 and 4 are modified spin-filter suspension culture vessels. The spin-filter concept was originated by Arthur D. Little Co., as reported by Himmelfarb et al. (1969). The principle of the device is that boundary effects at the surface of the spinning filter allow large volumes of spent medium to be withdrawn from a suspension culture without clogging the filter with cells. Spin filter equipment is available commercially from the VirTis Co., Gardiner, New York. A disassembled filter head assembly is shown in Fig. 5. In our hands, it has been necessary to

FIG. 4. Automated 4-liter spin-filter suspension culture vessel. Section of polycarbonate sheet forms top for polycarbonate jar; modified stirbar assembly.

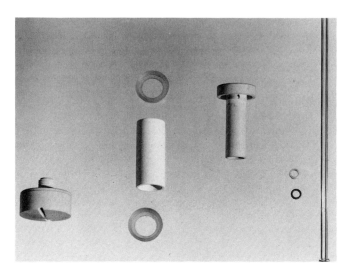

FIG 5. Disassembled filter head assembly for spin-filter device. Left to right: round magnet in plastic housing; 1-μm ceramic filter core with silicone rubber gaskets; top half of center piece—screws onto magnetic drive base; Teflon seal and Viton-A O-ring; and pilot tube.

install a rubber O-ring beneath the Teflon seal which serves as a bearing between the glass pilot tube and filter tube. This was done to eliminate air leakage—noted by the presence of air bubbles in the pilot tube during media withdrawal operations. Leakage has also occurred when medium levels were inadvertently permitted to drop, exposing the Teflon seal and O-ring and resulting in premature wear. During culture operations of several weeks' duration the effluent medium is periodically checked for the presence of cells; no problems have been encountered when new seals and O-rings were used at the time of culture initiation, and when care was taken to maintain medium levels above the seal. The round magnet in the base of the spin filter was replaced with a 3-inch stirring bar by drilling holes through the plastic housing of the base to provide greater stirring action in the 4-liter vessel shown in Fig. 4.

VI.   Conclusions

Plastic laboratory ware, tubing, and sheets can be used to construct suspension culture vessels of almost any desired configuration. In designing such vessels, consideration should be given to such variables as head space and media surface areas, since these are known to affect gas exchange rates and consequently can be limiting factors in cell growth. Minimum culture volumes are determined by the shape of the vessel and the type of stirring device utilized.

Elimination of cell adhesion to vessel walls obviates the need for enzymic detachment of cells and facilitates cell counting operations, thus giving better quantitation of cell growth. Further advantages are obtained in routine cell-handling operations. It is possible to change the medium of most cultures by allowing cells to settle out for several hours, decanting spent medium, and replenishing the culture with fresh growth medium. LLC-MK$_2$ cells handled in this manner did not adhere to the bottoms of plastic flasks after settling out at 37°C for over 18 hours (Cook et al., 1974a). Cultures grown to high population densities may suffer from oxygen deprivation during an extended settling process and are better suited to media replenishment employing the spin-filter system.

Continued exposure to culture media and to cells during settling operations apparently modifies the nonstick property of plastic surfaces in some manner, but the relatively low cost of plastic laboratory ware and materials has permitted us to replace culture vessels after extended periods of use.

LLC–MK$_2$ cells were grown in several types of plastic suspension culture vessels for over 22 months (Cook et al., 1974b). Suspension cultures of other types of cells have been grown in plastic vessels for varying periods of time; we have never encountered growth problems that were attributable to the use of plastic vessels.

Plastics thus provide a source of construction materials with which suspension culture vessels can be economically built to fit the needs of the investigator. We have found these vessels to be well suited for virus production: virus titers of parainfluenza, measles, and canine distemper viruses grown in suspension cultures of LLC-MK$_2$ cells were equal to, or greater than, those obtained in our laboratories using monolayer cultures of the same cell line (Cook et al., 1974b). It seems likely that plastic suspension culture vessels can be used to advantage in the production of other viruses, viral reagents, and cell metabolites.

ACKNOWLEDGMENTS

The authors thank Mr. John May of our Photography and Audiovisual Services Department for providing an excellent set of pictures of our equipment.

REFERENCES

Bizzini, B., Chermann, J.-C., Jasmin, C., and Raynaud, M. (1973). U.S. Patent 3,717,551.
Capstick, P. B., Telling, R. C., Chapman, W. G., and Stewart, D. L. (1962). *Nature (London)* **195**, 1163.

Cherry, W. R., and Hull, R. N. (1956). *Anat. Rec.* **124**, 483.

Cook, R. A., and Counter, F. T. (1974). U.S. Patent 3,850,748.

Cook, R. A., Counter, F. T., and McColley, J. K. (1974a). *In Vitro* **9**, 318.

Cook, R. A., Counter, F. T., and McColley, J. K. (1974b). *In Vitro* **9**, 323.

Eagle, H. (1959). *Science* **130**, 432.

Fuccillo, D. A., Catalano, L. W., Jr., Moder, F. L., Debus, D. A., and Sever, J. L. (1969). *Appl. Microbiol.* **17**, 619.

Himmelfarb, P., Thayer, P. S., and Martin, H. E. (1969). *Science* **164**, 555.

Horng, C., and McLimans, W. (1975). *Biotechnol. Bioeng.* **17**, 713.

Hull, R. N., Johnson, I. S., and Cherry, W. R. (1956). *Anat. Rec.* **124**, 490.

McLimans, W. F., Davis, E. V., Glover, F. L., and Rake, G. W. (1957). *J. Immunol.* **79**, 428.

Robb, J. A. (1970). *Science* **170**, 857.

Trager, G. W., and Rubin, H. (1966). *Virology* **30**, 275.

van Wezel, A. L. (1967). *Nature (London)* **216**, 64.

Walker, B. A., Brown, B. B., Krohmer, J. S., and Bonte, F. J. (1962). *Tex. Rep. Biol. Med.* **20**, 686.

Chapter 11

Determination of the Growth Rate in Human Fibroblasts in Cell Culture

SUSAN CURE AND ANDRÉ BOUÉ

Centres d'Etudes de Biologie Prenatale, Paris, France

I. Introduction

The methods most frequently used to measure the rate of multiplication of cells in culture involve removal of cells from the glass or plastic surface of the culture vessel using proteolytic enzymes or chelating agents, and then counting them in a hemacytometer or by means of an electronic cell counter. Determination of the protein content of a cell suspension can also give an indirect estimate of cell number, assuming that the average amount of protein per cell does not vary in different cell suspensions. For these techniques, cell suspensions are usually placed in small tubes, and the number of cells or amount of protein in 4 or 5 tubes is measured at regular intervals thereafter. While these methods are useful for many types of studies, we found them unsuitable for comparative growth studies of a large number of human fibroblast cell lines with chromosome anomalies (Boué *et al.*, 1975). Cell lines may vary in their sensitivity to enzymes or chelating agents, and the long time interval necessary in some cases to remove all the cells from the surface of the culture vessel may cause destruction of some cells, making counts of such cell suspensions invalid. Furthermore, many chromosomally abnormal cell lines are characterized by cells of variable size, and comparative growth curves based on protein determinations would be invalid because the amount of protein per cell (in triploid cells, for instance) might not be comparable.

II. Measuring Proliferation of Individual Clones or Colonies

A. Clonal Growth

The method we developed is based on a technique originally used by Puck *et al.* (1956) to measure the clonal growth of HeLa cells. After depositing a cell suspension in nutrient medium in petri dishes, they were able to follow cell multiplication by simply counting the cells in the individual clones formed from single cells.

Unlike HeLa cells, which are heteroploid and epithelial in nature, euploid and aneuploid human fibroblasts do not grow as discrete clones or colonies; daughter cells migrate away from each other immediately after cell division, and after the two- or three-cell stage individual colonies cannot be recognized at the cell concentrations needed for optimal growth.

B. Cell Proliferation within a Delimited Area

Since we could not keep track of cells in individual clones, we decided to count the cells in the same delimited area daily, thus measuring cell multiplication within a given area. We assumed that, on the average, for each cell that migrated out of our area of measurement, another cell would migrate into it.

III. Method

A. Technique

Well-dispersed cells are suspended in about 6 ml of nutrient medium and placed in small plastic flasks (Falcon Plastics) at a concentration of 15,000–25,000 cells per flask. The cells are incubated in a humidified CO_2 incubator and counted at regular intervals thereafter; in our laboratory counts are made 4–6 hours after subcultivation and daily thereafter. For cell counts, a pattern of squares (1.5 mm × 4 mm) is scratched onto the side of the flask to which the cells attach (with a diamond pencil or other sharp instrument and straightedge), and the cells in each square are counted at 100 × magnification using an inverted phase-contrast microscope. We determine the growth rate of each cell line in duplicate at least three different passage levels.

B.   Results

Data from typical growth experiments are presented in Tables I and II. Cells were randomly dispersed over the surface of the flask except at the edges, where cells became confluent sooner. Although early experiments indicated that the location or nature of the pattern of squares did not make any difference, we always used a pattern of five squares, side-by-side, half way between the middle and edge of the flask and parallel to its long axis. Duplicate growth curves obtained with inocula of 10,000–20,000 cells were parallel; larger inocula sometimes led to areas of confluency on day 3, which

TABLE I

GROWTH OF CELLS FROM A NORMAL DIPLOID EMBRYONIC
CELL LINE[a]

Day	Square					Total
	1	2	3	4	5	
0	9	10	8	7	13	47
1	17	18	14	13	21	83
2	33	45	34	38	39	189
3	52	84	72	65	80	353

[a] Cells in each square were counted after 5 hours of incubation and at 24-hour intervals thereafter.

TABLE II

GROWTH OF CELLS FROM A CELL LINE INITIATED FROM A
SKIN BIOPSY OF A CHILD WITH CYSTIC FIBROSIS

Day	Square					Total
	1	2	3	4	5	
0	12	7	14	9	5	47
1	15	13	20	14	7	69
2	18	15	17	21	10	81
3	28	27	26	36	17	134

caused difficulties in counting. The extent of cell migration is evident in Table II, which shows daily cell counts from a cell line that has a long doubling time.

Typical growth curves obtained by this method are shown in Fig. 1 and 2. Duplicate results from the same experiment were nearly always in close agreement. In some experiments, "conditioned" medium, consisting of half-normal growth medium and half- "used" medium, which had been removed from confluent monolayers of human diploid fibroblasts and filtered, was utilized. No differences in doubling times were noted when experiments using conditioned and nonconditioned medium were performed in duplicate.

It was initially feared that debris from cells that did not attach to the plastic surface might exert a toxic effect and inhibit the multiplication of attached cells, but replacing the medium in the flasks after 6 hours of incubation with either conditioned or ordinary medium had no effect on the growth rate. This experiment also indicated that none of the increase in cell numbers after the first counts were performed was due to attachment to the plastic surface of cells formerly in suspension. For comparative purposes, growth studies in small tubes were also performed. Cells were suspended in medium at concentrations of 50,000 or 75,000 cells/ml, and dispersed in 0.5-ml

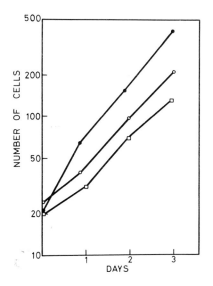

Fig. 1. Representative growth curves for a fibroblastic cell line initiated from a skin biopsy of a stillborn infant with trisomy 21. O—O, 9th passage, 20,000-cell inoculum; ●—●, 11th passage, 10,000-cell inoculum; □—□, 13th passage, 15,000-cell inoculum. (From Cure and Boué, 1973.)

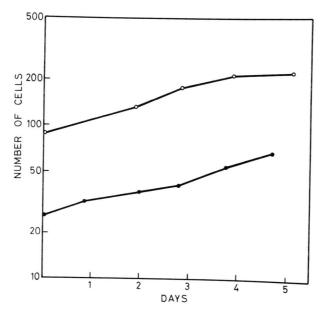

FIG. 2. Representative growth curves for a fibroblastic cell line initiated from embryonic tissue of a spontaneous abortion with trisomy 21. ●—●, 13th passage, 10,000-cell inoculum; ○—○, 14th passage, 20,000-cell inoculum. (From Cure and Boué, 1973.)

aliquots in 90 × 12 mm tubes with metal covers. These were incubated vertically in a humidified CO_2 incubator, and each day cells from 4 tubes were harvested by trypsinization, and the number of cells per tube was determined by hemacytometer counts. No significant differences in doubling times were noted when growth studies on the same cell line were performed in tubes and by counting cells in plastic flasks as described.

IV. Application

We have found this technique to be particularly useful for the determination of the doubling time of a large number of aneuploid cell lines initiated from embryos from spontaneous human abortions (Cure *et al.*, 1974), but it could also be adapted to screening various biological or chemical agents for their effects on cell multiplication or to other routine growth studies. We have also used it for measuring the growth rate of cells with epithelial morphology (BHK21 or epithelial cells from amniotic fluids) and for testing calf serum for toxicity.

V. Conclusion

This method provides an accurate means of determining the growth rate of human fibroblasts *in vitro*. The technique is rapid, simple, and inexpensive. Only 20,000–40,000 cells are required for the whole experiment, vs. several hundred thousand cells in trypsinization-resuspension methods, which involve more manipulation of cells with concomitant introduction of errors.

REFERENCES

Boué, A., Boué, J., Cure, S., Deluchat, C., and Perraudin, N. (1975). *In Vitro* **11**, 409.
Cure, S., and Boué, A. (1973). *Experientia* **29**, 907.
Cure, S., Boué, A. J., and Boué, A. (1974). *Biomedicine* **21**, 233.
Puck, T., Marcus, P., and Cieciura, J. (1956). *J. Exp. Med.* **103**, 273.

Chapter 12

Cultivation of Mammalian Cell Lines in Serum-Free Chemically Defined Medium

KIYOSHI HIGUCHI[1]

Microbiological Associates, Walkersville, Maryland

I. Introduction

Interest in chemically defined media for mammalian cell culture is based on the need for more precise tools for study of animal cell nutrition and the requirement to eliminate uncertainties arising from the use of crude serum. Serum is produced commercially from sources that cannot be adequately controlled; therefore, uniformity of the product is difficult to obtain (Boone *et al.*, 1971). Another serious problem associated with the use of serum is the hazard of microbiological contamination. Mycoplasma, animal viruses, bacteriophages, and endotoxins have been detected in commercial lots of

[1] *Present address*: 199 Meadowdale Lane, Frederick, Maryland.

serum (Fedoroff *et al.*, 1971; Kniazeff, 1968; Chu *et al.*, 1973; Molander *et al.*, 1971; Merril *et al.*, 1972; Orr *et al.*, 1975). Finally, it may be noted that availability of serum is contingent on slaughter animal supply. Any great expansion in future cell culture work may result in shortages.

Some of the difficulties mentioned above could be circumvented if it were possible to purify readily and to isolate the active substances in serum. It is clear from published results that multiple factors affecting a variety of cellular functions are present in serum, and that the growth-promoting property of serum is a summation of their activities (Sanford *et al.*, 1955; Lieberman and Ove, 1957, 1958; Fisher *et al.*, 1959; Holmes and Wolfe, 1961; Marr *et al.*, 1962; Tozer and Pirt, 1964; Healy and Parker, 1966; Paul *et al.*, 1971; Holley and Kiernan, 1971). Although many different approaches have been tried, results of work on the isolation of serum factors have been inconclusive. Another aspect of this problem is the increasing evidence that certain low-molecular-weight substances tightly bound to serum proteins may represent at least in part, the growth-promoting activity of serum. Eagle (1960) showed that some of the serum growth activity becomes dialyzable during enzymic digestion of serum proteins. De Luca *et al.* (1966) have made similar observations. Recently Pickart and Thaler (1973) described a tripeptide isolated from serum that in submicrogram levels stimulated synthesis of RNA, DNA, and protein in normal and transformed rat liver cells. However, the tripeptide, although able to induce cell division in transformed cells, failed to promote multiplication of normal cells. As seen above, the role of serum in cell nutrition is complex, and much remains to be clarified. We still look forward to the time when purified serum fractions can be used as replacement for crude serum.

Meanwhile, numerous studies on the nutrition of animal cells in chemically defined systems were undertaken with the goal of devising a serum-free culture medium (Higuchi, 1973). Probably the earliest successful defined medium was that of Healy *et al.* (1954). They propagated mouse fibroblasts, strain L, continuously in the absence of complex nutrients such as serum. Subsequently, a variety of improvements and refinements were made in the medium for L cells that have resulted in increased rates of growth and in high cell yields (Evans *et al.*, 1956a,b; Waymouth, 1959; Sanford *et al.*, 1963; Nagle *et al.*, 1963; Higuchi, 1970). During the 20 years following the successful propagation of mouse fibroblasts (strain L), many cell lines derived from a variety of other mammalian species, including cells of human origin, were cultured in defined media (Evans *et al.*, 1959, 1964; Holmes, 1959; Bakken *et al.*, 1961; Nagle *et al.*, 1963; Ham, 1965; Den Beste *et al.*, 1966; Takaoka and Katsuta, 1971; Higuchi and Robinson, 1973). Although chemically defined media developed so far have been suitable only for cultivation of continuous cell lines, the usefulness of a system in which all medium constituents can be precisely controlled cannot be overlooked. For example, Donta (1975) described cultivation of functional glial cells in a defined medium. More recently, Thompson *et al.* (1975) reported on the cultivation in defined medium of hormone-responsive rat liver tumor cells (strain HTC). Gerschenson *et al.* (1974) studied the hormonal regulation of rat liver cells cultured in chemically defined medium. Similarly, defined media were shown to be useful in studies of enzyme regulation in nonproliferating adult rat liver parenchymal cells (Bissell *et al.*, 1973; Bonney *et al.*, 1974).

TABLE I

CELL LINES GROWN IN CHEMICALLY DEFINED MEDIUM[a]

Cell lines	Yields[b] (10^6/ml)
Human origin	
HeLa (cervical carcinoma)	1.7
KB (carcinoma of nasopharynx)	1.1
HEp-2 (carcinoma of larynx)	1.4
Chang's liver	0.7
MA 160 (prostate)	1.4
Heart (Girardi)	0.9
FL (amnion)	1.5
KHOS/NP (virus-transformed nonproducer, Rhim *et al.*, 1975)	—[c]
Nonhuman primate origin	
Vero (African green monkey kidney)	0.9
BSC-1 (*Cercopithecus* monkey kidney)	1.1
LLC-MK$_2$ (rhesus monkey kidney)	0.9
Bovine origin	
AUBEK (kidney, Rossi and Kiesel, 1973)	—[c]
Rodent species origin	
L929 (mouse fibroblast)	1.3
LM (variant of L929)	1.8
AKR (mouse embryo of AKR strain)	0.8
LLC-RK$_1$ (rabbit kidney)	0.6
SIRC (rabbit cornea)	—[c]
BHK21 (baby hamster kidney)	0.8

[a] Data from Higuchi and Robinson (1973) with permission of the Tissue Culture Association, Inc.

[b] The yields represent viable cell counts made usually after 1 week's incubation after inoculation of less than 10^5 cells per milliliter of medium (Table II) per flask. Cultures were refed 3–4 times during the growth cycle. The rabbit cell (LLC-RK$_1$), however, grew very slowly and required almost 1 week for each cell doubling.

[c] No data available.

A method is described in this chapter which has permitted good growth of a variety of continuous cell lines. A list of the cell lines grown successfully in the defined medium is presented in Table I. In our experience, a majority of continuous cell lines can be cultured in the present medium. Many cells apparently require little or no period of adaptation to initiate continuous growth. The present work deals exclusively with mammalian cells grown statically in monolayer culture.

II. Materials

A.   Stock Salt Solutions, Buffers, and Saline

For convenience in preparation of culture medium, three stock salt solutions are prepared. These can be stored at room temperatures indefinitely.

1.   SALT SOLUTION I

Salt solution I is a 10-fold concentrated solution containing (in grams per liter) NaCl (74), KCl (4), $NaH_2PO_4 \cdot H_2O$ (1), and $NaHCO_3$ (3).

2.   SALT SOLUTION II

Salt solution II is a 20-fold concentrated solution containing (in grams per liter) $CaCl_2 \cdot 2 H_2O$ (5.3) and $MgCl_2 \cdot 6 H_2O$ (5.5).

3.   SALT SOLUTION III

Salt solution III is a 200-fold concentrated solution containing $0.1 M$ gluconolactone, $10^{-3} M$ $FeNH_4(SO_4)_2 \cdot 12 H_2O$, $10^{-4} M$ $ZnSO_4 \cdot 7 H_2O$, and $10^{-2} M$ HCl. Hydrochloric acid is added to inhibit fungal growth.

4.   PHOSPHATE-BUFFERED SALINE AND MODIFICATIONS

Phosphate-buffered saline (PBS) is prepared by autoclaving a solution containing $0.15 M$ NaCl, $0.01 M$ Na_2HPO_4, and phenol red (5 μg/ml). PBS modified to contain 0.1% methyl cellulose (PBSM) is prepared by addition of 17 ml of sterile 3% methyl cellulose solution (Section II,B,2) to 500 ml of PBS. PBS containing both methyl cellulose and 5×10^{-4} EDTA (PBSM–EDTA) is prepared by addition of 25 ml of $0.01 M$ EDTA (Section II,B,3) to 500 ml of PBSM.

5.   SODIUM BICARBONATE SOLUTION

The sterile stock 7.5% (w/v) $NaHCO_3$ solution can be prepared by filtration. It is important that container caps be tightly closed during storage to minimize loss of CO_2.

6. HEPES Buffer Stock

A sterile 1.0 M stock solution of HEPES buffer is prepared by autoclaving a neutralized (pH 7.0) 23.8% (w/v) solution of 4-(2-hydroxyethyl)-l-piperazineethanesulfonic acid. Use of HEPES buffer in cell cultures was described by Massie *et al.* (1972).

B. Miscellaneous Stock Solutions

1. Stock Vitamin Mixture

The vitamins for use in the medium are prepared as a single nonsterile mixture at a 1000-fold concentration and stored frozen in 5-ml aliquots in screw-capped tubes. The concentrations of the vitamins in the stock mixture can be obtained from values given in Table II. Solution of the vitamins in PBS (Section II,A,4) is facilitated by neutralization of acidic constituents to pH 7.0.

2. Methyl Cellulose Solution

A sterile 3% stock solution is prepared by autoclaving 6 gm of methyl cellulose (15 cps grade) in 200 ml of H_2O. Because methyl cellulose becomes insoluble at high temperatures, it becomes messy during autoclaving and tends to boil over. It is recommended that a large bottle (500 ml) be used to autoclave the solution. After autoclaving and cooling to room temperature, the material should be placed in a refrigerator to cool further in order to obtain a clear solution. The volume should be adjusted with sterile water to correct for any loss due to evaporation.

3. EDTA Solution

A neutral sterile 0.01 M ethylenediamine tetraacetic acid (EDTA) solution is prepared by autoclaving the neutralized (pH 7.0) solution.

4. Protamine Sulfate Solution

A 0.5% stock solution of protamine sulfate (Nutritional Biochemicals) is sterilized by membrane filtration. The sterile solution is stored in a refrigerator.

5. l-Thyroxine Solution

To prepare 100 ml of sterile 10^{-3} M stock solution of thyroxine, 80 mg of the sodium salt of l-thyroxine (Calbiochem) is dissolved in 5 ml of 1 N NaOH, diluted to 100 ml, and sterilized by filtration. The solution is stored frozen at $-20°C$.

6. Phenol Red Solution

A stock nonsterile solution is prepared by dissolving 0.5 gm of the sodium salt of phenol red in 100 ml of H_2O.

TABLE II

CHEMICALLY DEFINED MEDIUM FOR MONOLAYER CULTURES[a]

L-Amino acid	Milligrams per liter	Vitamins	Milligrams per liter
Arginine · HCl	32	Biotin	0.002
Asparagine · H_2O	150	Choline chloride	10.0
Cysteine · HCl · H_2O	22	Folic acid	1.0
Glutamine	198	Inositol	2.0
Histidine · HCl · H_2O	63	Nicotinamide	0.1
Isoleucine	33	Ca-pantothenate	0.2
Leucine	26	Pyridoxal · HCl	0.05
Lysine · HCl	28	Riboflavin	0.10
Methionine	15	Thiamine HCl	0.20
Phenylalanine	33	Vitamin B_{12}	0.002
Proline	115		
Serine	105		
Threonine	12		
Tryptophan	6		
Tyrosine	46		
Valine	35		

Inorganic salts and buffers	Milligrams per liter	Miscellaneous	Concentration
NaCl	7400	Glucose	1800 mg/ml
KCl	400	Na pyruvate	110 mg/ml
$NaH_2PO_4 \cdot H_2O$	100	Gluconolactone	89 mg/liter
$NaHCO_3$	1140[b]	Phenol red	10 mg/liter
$CaCl_2 \cdot 2H_2O$	265	Ethanol	450 mg/liter
$MgCl_2 \cdot 6H_2O$	275	Methylcellulose (15 cps)	600 mg/liter
$FeNH_4(SO_4)_2 \cdot 12H_2O$	2.4	Protamine sulfate	2.0 mg/liter
$ZnSO_4 \cdot 7H_2O$	0.14	Oleic acid	0.5 mg/liter
HEPES buffer	2975	Lecithin	1.0 mg/liter
		Cholesterol	$5 \times 10^{-6}\ M$
		L-Thyroxine	$2.5 \times 10^{-8}\ M$
		EDTA	$5 \times 10^{-6}\ M$
		Insulin (LENTE)[c]	0.05 unit/ml
		Gentamicin	25 μg/ml

[a] The medium is an adaptation of that described by Higuchi and Robinson (1973), reprinted with permission of the Tissue Culture Association Inc.

[b] The amount of $NaHCO_3$ in this formulation includes the 300 mg/liter added (as salt mixture I) during preparation of five-fold concentrated solution and the 840 mg/liter added as buffer in the final diluted medium.

[c] Commercial insulin (Eli Lilly Co.) consists of a mixture of bovine and porcine insulins.

7. Gentamicin

The 0.5% stock gentamicin sulfate solution is prepared by dilution of 10 ml of the 5% solution (Schering Co.) to 100 ml with sterile H_2O.

C. Concentrated Medium Components

1. Concentrated Basal Medium

For convenience in the formulation of the final medium, certain constituents of the defined medium are combined to form a single five-fold concentrated stock solution. Each of the amino acids listed in Table II, sufficient for 10 liters of final medium, together with 18 gm of glucose and 1.1 gm of sodium pyruvate, are placed in a 2-liter Erlenmeyer flask. To this, 1 liter of Salt Solution I (Section II,A,1) is added, and the contents are stirred until solution occurs. Next, 500 ml of Salt Solution II (Section II,A,2), 50 ml of Salt Solution III (Section II,A,3), 10 ml of stock vitamin mixture, 20 ml of stock 0.5% phenol red solution, and sufficient H_2O to make the total volume 2 liters are added and mixed. The solution is sterilized by membrane filtration and dispensed aseptically into 500-ml screw-capped bottles. The solution can be stored for several months in a refrigerator with no significant loss in quantity.

2. Serum Substitute Mixture

To 200 ml of sterile 3% methyl cellulose solution (Section II,B,2) add 185 ml of sterile H_2O, 6.25 ml of insulin (Lente, 80 units/ml, Eli Lilly Co.), 5 ml of 0.01 M EDTA (Section II,B,3), 4 ml of 0.5% protamine sulfate (Section II,B,4), and 0.25 ml of 10 ^{3} M L-thyroxine solution (Section II,B,5).

3. Lipid Mixture

The following substances are weighed into a tared, sterile 50-ml prescription bottle: oleic acid (50 mg); cholesterol (193 mg); and lecithin (100 mg) (Nutritional Biochemicals). The lipids are dissolved in 50 ml of 95% ethanol containing 0.5 (v/v) concentrated HCL. The material is stored in the freezer at -20°C.

D. Preparation of Complete Medium

The complete medium is prepared by aseptically combining and diluting the various stock solutions as follows. To a 500-ml screw-capped bottle, containing approximately 365 ml of cooled autoclaved water, are added 100 ml of Concentrated Basal Medium (Section II,C,1), 5.6 ml of 7.5% $NaHCO_3$, 6.25 ml of 1.0 M HEPES buffer, 2.5 ml of 0.5% gentamicin solution, 20 ml of the Serum Substitute (Section II,C,2), and 0.5 ml of the Lipid Mix (Section II,C,3). The complete medium can be stored for several months in the refrigerator without significant loss in ability to support cell growth.

E. Rat Tail Collagen

A convenient amount of collagen can be prepared from one adult rat tail. The amputated tail can be stored in 95% ethyl alcohol in the freezer until used. The entire procedure is conducted aseptically with sterile instruments. The tail is cut into small segments, the skin is removed, and the tendon fibers are carefully pulled out and put into a sterile dish with H_2O to avoid drying. After the tendons are collected, they are chopped into small pieces and transferred to a bottle containing 200 ml of sterile 0.1% acetic acid. The material is allowed to stand in the refrigerator for 2–3 days in order to let the fibers swell and finally form a viscous solution. The material is centrifuged in sterile tubes at 10,000 rpm for 30 minutes to remove residual solids and then dispensed aseptically into vials for storage in the freezer. The above procedure is an adaptation of a method recently described (Price, 1975).

III. Methods

A. Culture Vessels

Plastic culture flasks are recommended for cultivation of cells in defined medium. Glass flasks have been unsatisfactory. The flask with 25 cm² growth area (Falcon, type 3013) is convenient for most nutritional studies. The larger 75 cm² size flask (Falcon, type 3024) is more practical where larger cell numbers are to be produced. Approximately 1 ml of medium is used for each 5 cm² of growth surface.

Cells of most lines attach readily in the defined medium to untreated plastic growth surface; however, the rabbit corneal cell line (SIRC) and the human Kirsten virus-transformed nonproducer cell (KHOS/NP) described by Rhim *et al.* (1975), required a growth surface coated with collagen in order to obtain cell attachment. The T-75 size flasks were coated with collagen as follows. A 0.15-ml drop of sterile 1.0 M NaCl is placed adjacent to a 0.2-ml drop of collagen solution (Section II,E) on the growth surface. By means of a spreader (consisting of a rectangular block of sterile silicone rubber, 1 cm wide, attached to a handle made of doubled, twisted stainless steel wire) the two solutions are mixed on rapidly spread over the growth surface. The flask is allowed to stand at room temperature for an hour before it is lightly rinsed with 4 ml of sterile H_2O. Coated flasks can be used immediately, or they may be stored for as long as several weeks in the refrigerator before use.

B. Filtration

Animal cells in serum-free medium appear to be highly sensitive to traces of toxic substances. Certain precautions are therefore essential prior to filtration of any solution for use in cell culture. At least 200 ml of deionized H_2O should be run through a Millipore filter (0.22 μm pore size) 47 mm in diameter in a prewashing step in order to avoid possible contamination of medium with wetting agents that might be toxic.

C. Autoclaving

Autoclaving is employed in the preparation of a number of sterile solutions. The deionized H_2O used in dilution of concentrated medium stock is also autoclaved. It may be noted that an electrically operated steam autoclave (Castle, Model 999 C) is used in our work. The possibility of introducing volatile toxic substances from the steam line into the culture medium is obviated by this type of equipment.

D. Preparation of Inoculum

The stock culture to be used for preparation of inoculum is drained of spent medium and washed once with 5 ml of PBS. The cell sheet is then treated with 5 ml of PBSM–EDTA solution containing a drop of 2.5% trypsin (Microbiological Associates). The cells will disperse, usually after 2 or 3 minutes of incubation at room temperature. A drop of 7.5% $NaHCO_3$ solution may be added if necessary to provide a pH more favorable for tryptic action. The dispersed cells are transferred to a sterile 15-ml conical centrifuge tube and pelleted at approximately 1000 rpm in a tabletop clinical centrifuge. The cell pellet is washed once by resuspension in a 5 ml of PBSM solution. At this time, viable cell counts may be made by the trypan blue dye exclusion method. The cells are centrifuged again and finally resuspended in fresh growth medium.

E. Culture Procedures and Examples of Results

An inoculum of 50×10^3 cells per milliliter is adequate to yield a 20-fold cell increase per week with rapidly growing cultures such as L929, HEp-2, and HeLa. Other cultures that grow more slowly may be seeded with as many as 200×10^3 cells per milliliter. For routine propagation of most cell

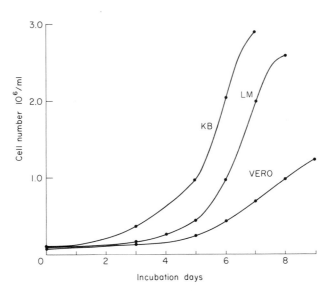

FIG. 1.   Growth curves of three cell lines in the chemically defined medium. Each culture was inoculated with approximately 10^5 cells per milliter and incubated at 36°C for the periods shown. Each culture was refed 3–5 times during the growth cycle.

lines, it is convenient to plant the cells just prior to the weekend and to refeed the cultures after 3, 5, and 6 days of incubation at 36°C. This schedule provides progressively shorter intervals between refeedings as cell numbers increase, and it allows cell transfer on the last work day of the week.

Examples of growth curves obtained with three cell lines are presented in Fig. 1. Each culture was inoculated with approximately 10^5 cells per milliliter and incubated from 7 to 9 days as shown. Both KB and LM cell lines exhibited very rapid growth rates, as indicated by cell doubling times of 24 hours or less during the steep parts of the curve. The variety of culture morphology obtained in the defined medium with several different cell lines is illustrated by the series of photomicrographs presented in Fig. 2.

IV.   Concluding Remarks on Future of Defined Medium

An important step in the future development of the chemically defined medium is the formulation of a medium permitting serum-free cultivation of normal diploid cells. In a recent discussion of this topic, Ham (1974) stressed the need for more emphasis on the study of nutrition of primary cells in culture. The growth of primary cells, however, frequently involves a complex system of specific humoral factors controlling both growth and

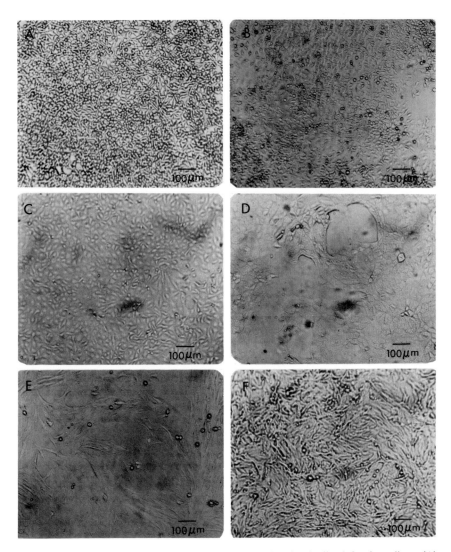

FIG. 2. Photomicrographs of cultures grown in the chemically defined medium. (A) Mouse fibroblasts, strain L929. (B) Human cervical carcinoma, strain HeLa. (C) Canine kidney, strain MDCK. (D) Bovine kidney, strain AUBEK (Rossi and Kiesel, 1973). (E) Rabbit cornea, strain SIRC. (F) Kirsten virus-transformed human nonproducer, strain KHOS/NP (Rhim *et al.*, 1975).

differentiation (Rutter *et al.*, 1973). It may be more appropriate, therefore, to initiate the proposed studies by employing in the test system a low-passage diploid cell, such as the well known cultured human diploid fibroblast, strain WI-38. According to present knowledge, serum proteins are indispensable for growth of normal diploid cells. Therefore, the variety of serum protein fractions reported to be stimulatory for cell growth (Lieberman and Ove, 1958; Tozer and Pirt, 1964; Healy and Parker, 1966; Pickart and Thaler, 1973) should be tested in a basal medium minimally dependent on serum. The defined medium described in the present chapter, capable of supporting excellent serum-free growth of many continuous cell lines, may be ideally suited for the proposed study. It is hoped that work along lines suggested above will eventually lead to the development of a completely defined medium that will support the growth of many more cell types than is now possible.

REFERENCES

Bakken, P. C., Evans, V. J., Earle, W. R., and Stevenson, R. E. (1961). *Am. J. Hyg.* **73**, 96–104.

Bissell, D. M., Hammaker, L. E., and Meyer, U. A. (1973). *J. Cell. Biol.* **59**, 722–734.

Bonney, R. J., Becker, J. E., Walker, P. R., and Potter, V. R. (1974). *In Vitro* **9**, 399–414.

Boone, C. W., Mantel, N., Caruso, T. D., Kazam, E., and Stevenson, R. E. (1971). *In Vitro* **7**, 174–189.

Chu, F. C., Johnson, J. B., Orr, H. C., Probst, P. G., and Petricciani, J. C. (1973). *In Vitro* **9**, 31–34.

De Luca, C., Habeeb, A. F. S. A., and Tritsch, G. L. (1966). *Exp. Cell Res.* **43**, 98–106.

Den Beste, H. E., Fjelde, A., Jackson, J. L., Andresen, W. F., Kerr, H. A., and Evans, V. J. (1966). *J. Natl. Cancer Inst.* **36**, 1075–1088.

Donta, S. T. (1975). *Methods Cell Biol.* **9**, 123–137.

Eagle, H. (1960). *Proc. Natl. Acad. Sci., U.S.A.* **56**, 427–432.

Evans, V. J., Bryant, J. C., Fioramonti, M. C., McQuilkin, W. T., Sanford, K. K., and Earle, W. R. (1956a). *Cancer Res.* **16**, 77–86.

Evans, V. J., Bryant, J. C., McQuilkin, W. T., Fioramonti, M. C., Sanford, K. K., Westfall, B. B., and Earle, W. R. (1956b). *Cancer Res.* **16**, 87–94.

Evans, V. J., Kerr, H. A., McQuilkin, W. T., Earle, W. R., and Hull, R. N. (1959). *Am. J. Hyg.* **70**, 297–302.

Evans, V. J., Bryant, J. C., Kerr, H. A., and Schilling, E. L. (1964). *Exp. Cell Res.* **36**, 439–474.

Fedoroff, S., Evans, V. J., Hopps, H. E., Sanford, K. K., and Boone, C. W. (1971). *In Vitro* **7**, 161–167.

Fisher, H. W., Puck, T. T., and Sato, G. (1959). *J. Exp. Med.* **109**, 649–659.

Gerschenson, L. E., Davidson, M. B., and Andersson, M. (1974). *Eur. J. Biochem.* **41**, 139–148.

Ham, R. G. (1965). *Proc. Natl. Acad. Sci. U.S.A.* **53**, 288–293.

Ham, R. G. (1974). *In Vitro* **10**, 119–129.

Healy, G. M., Fisher, D. C., and Parker, R. C. (1954). *Can. J. Biochem. Physiol.* **32**, 327–337.

Healy, G. M., and Parker, R. C. (1966). *J. Cell Biol.* **30**, 539–553.

Higuchi, K. (1970). *J. Cell. Physiol.* **75**, 65–72.

Higuchi, K. (1973). *Adv. Appl. Microbiol.* **16**, 111–136.

Higuchi, K., and Robinson, R. C. (1973). *In Vitro* **9**, 114–121.

Holley, R. W., and Kiernan, J. A. (1971). *Growth Control Cell Cult., Ciba. Found. Symp. 1970* pp. 3–15.

Holmes, R. (1959). *J. Biophys. Biochem. Cytol.* **6**, 535–536.

Holmes, R., and Wolfe, S. W. (1961). *J. Biophys. Biochem. Cytol.* **10**, 389–401.

Kniazeff, A. J. (1968). *Natl. Cancer Inst., Monogr.* **29**, 123–132.

Lieberman, I., and Ove, P. (1957). *Biochim. Biophys. Acta* **25**, 449–450.

Lieberman, I., and Ove, P. (1958). *J. Biol. Chem.* **233**, 637–642.

Marr, A. G. M., Owen, J. A., and Wilson, G. S. (1962). *Biochim. Biophys. Acta* **63**, 276–285.

Massie, H. R., Samis, H. V., and Baird, M. B. (1972). *In Vitro* **7**, 191–194.

Merril, C. R., Friedman, T. B., Attallah, A. F. M., Geier, M. R., Krell, K., and Yarkin, R. (1972). *In Vitro* **8**, 91–83.

Molander, C. W., Kniazeff, A. J.. Boone, C. W., Paley, A., and Imagawa, D. T. (1971). *In Vitro* **7**, 168–173.

Nagle, S. C., Tribble, H. R., Anderson, R. E., and Gary, N. D. (1963). *Proc. Soc. Exp. Biol. Med.* **112**, 340–344.

Orr, H. C., Sibinovic, K. H., Probst, P. G., Hochstein, H. D., and Littlejohn, D. C. (1975). *In Vitro* **11**, 230–233.

Paul, D., Lipton, A., and Klinger, I. (1971). *Proc. Natl. Acad. Sci. U.S.A.* **68**, 645–648.

Pickart, L., and Thaler, M. M. (1973). *Nature (London), New Biol.* **243**, 85–87.

Price, P. J. (1975). *TCA Manual* **1**, 43–44.

Rhim, J. S., Cho, Y., and Huebner, R. J. (1975). *Int. J. Cancer* **15**, 23–29.

Rossi, C. R. and Kiesel, G. K. (1973). *In Vitro* **9**, 147–155.

Rutter, W. J., Pictet, R. L., and Morris, P. W. (1973). *Annu. Rev. Biochem.* **42**, 601–646.

Sanford, K. K., Westfall, B. B., Fioramonti, M. C., McQuilkin, W. T., Bryant, J. C., Peppers, E. V., Evans, V. J., and Earle, W. R. (1955). *J. Natl. Cancer Inst.* **16**, 789–802.

Sanford, K. K., Dupree, L. T., and Covalesky, A. B. (1963). *Exp. Cell Res.* **31**, 345–375.

Takaoka, T., and Katsuta, H. (1971). *Exp. Cell Res.* **67**, 295–304.

Thompson, E. B., Anderson, C. U., and Lippman, M. E. (1975). *J. Cell. Physiol.* **86**, 403–412.

Tozer, B. T., and Pirt, S. J. (1964). *Nature (London)* **201**, 375–378.

Waymouth, C. (1959). *J. Natl. Cancer Inst.* **22**, 1003–1016.

Chapter 13

Improved Synthetic Media Suitable for Tissue Culture of Various Mammalian Cells

HAJIM KATSUTA AND TOSHIKO TAKAOKA

*Department of Cancer Cell Research,
Institute of Medical Science, University of Tokyo,
Shirokanedai, Minato-ku, Tokyo, Japan*

I. Introduction

After examination of nutritional requirements, published or unpublished, of mammalian cells in tissue culture (Katsuta *et al.*, 1957, 1958, 1959a,b,c, 1961; Katsuta and Takaoka, 1960; Kagawa *et al.*, 1960, 1970; Matsumura *et al.*, 1968; Takaoka *et al.*, 1968; Ohta *et al.*, 1969; Furuya *et al.*, 1971), various mixtures of synthetic media were designed, and 20 kinds of cell strains have been established in our laboratory that are capable of serially growing in protein- and lipid-free chemically defined synthetic media, e.g., the mixtures DM-120 or DM-145 (Takaoka and Katsuta, 1971; Katsuta and Takaoka, 1973). These mixtures, however, were found to be not necessarily suitable for the culture of every kind of cell, even when combined with sera.

This chapter is concerned with mixtures of multipurpose synthetic media, i.e., media in which the growth of various kinds of cells is obtainable in combination with or without serum and in a closed system as well as in an open system of culture.

TABLE I

COMPOSITION OF AMINO ACIDS AND VITAMINS IN DM SYNTHETIC MEDIA

Component	DM-120	DM-145	DM-150 DM-153	DM-160	Eagle MEM	RPMI-1640	Ham F 12
Amino acids (mg/liter)							
Ala	400	400	400	400	—	—	9
Arg	100	100	100	100	126(HCl)	200	211
Asp	25	25	25	25	—	20	13
Asn	—	—	—	25	—	50	13
Cys HCl	80	80	80	80	—	—	32
Cys Cys	—	—	—	—	24	50	—
Glu	150	150	150	150	—	20	15
Gln	100	100	300	300	292	300	146
Gly	15	15	15	15	—	10	8
His	30	30	30	30	42(HCl)	15	19(HCl)
Ile	150	150	150	150	52	50	4
Leu	400	400	400	400	52	50	13
Lys	100	100	100	100	73(HCl)	40	37
Met	80	80	80	80	15	15	4
Phe	80	80	80	80	32	20	5
Pro	12	12	12	12	—	—	35
Ser	80	80	80	80	—	30	11
Thr	100	100	100	100	48	20	12
Trp	40	40	40	40	10	5	2
Tyr	50	50	50	50	36	20	5
Val	85	85	85	85	46	20	12
Hydroxy-L-proline	—	—	—	—	—	20	—
Glutathione	—	—	—	—	—	1	—
Vitamins							
B$_1$ (thiamine)	10.0	10.0	1.0	1.0	1.0	1.0	0.3(HCl)
B$_2$ (riboflavin)	1.0	1.0	1.0	1.0	0.1	0.2	0.04
B$_6$ (pyridoxine)	1.0	1.0	1.0	1.0	1.0 (pyridoxal)	1.0	0.06(HCl)
B$_{12}$	0.005	0.005	0.005	0.005	—	0.005	1.36
Pantothenic acid	1.0	1.0	1.0	1.0	1.0	0.25	(0.26) (calcium)
Nicotinamide	5.0	50.0	1.0	1.0	1.0	1.0	0.04
Biotin	0.002	0.002	0.1	0.1	0.02	0.2	0.007
Choline HCl	250.0	250.0	5.0	5.0	1.8	3.0	14.0
C	40.0	40.0	1.0	1.0	—	—	—
Folic acid	0.01	0.01	1.0	1.0	1.0	1.0	1.32
Inositol	—	2.0	5.0	5.0	2.0	35.0	18.0
p-Aminobenzoic acid	—	—	—	—	—	1.0	—

II. Composition of Chemically Defined Synthetic Media

Briefly, synthetic media consist of three components: (1) amino acids, (2) vitamins, and (3) buffer and salts. The composition of amino acids and vitamins in some of the mixtures is shwon in Table I.

DM-145 differs from DM-120 in that it contains inositol. DM-153 (Takaoka and Katsuta, 1975) differs from both in the composition of vitamins and buffer. DM-160 is DM-153 that contains asparagine. DM-153 and DM-160 mixtures have a higher content of amido than do DM-145 and DM-120, as shown in Table II.

III. Growth of Cells in Various Media

Two kinds of cell strains were compared (Table III): (1) the strains 3T3 (mouse fibroblasts), CHO-K1 (Chinese hamster ovary fibroblasts), and RLC-10(2) (rat liver cells), which had been serially grown in media containing 10% fetal calf serum; and (2) the strains L·P3 (mouse fibroblasts L-929) and JTC-25·P3 (rat liver cells) serially grown in a protein- and lipid-free synthetic medium. In DM-150, almost similar rates of cell proliferation were found among the strains. DM-153 differs from DM-150 only in the composition of buffer and salts, as shown in Table IV.

The effect of DM-153 on the survival of human blood lymphoid cells in primary culture was compared with that of RPMI-1640, both in the presence of 10% fetal calf serum (Fig. 1). The medium was renewed once, on day 7. The number of viable cells was counted by the use of erythrosin. DM-153 accomplished a much higher preservation of the survival of lymphoid cells than did RPMI-1640, which had frequently been used by many workers for the culture of blood cells.

TABLE II

AMINO ACID CONTENTS IN GROUPS OF VARIOUS SYNTHETIC MEDIA (MG/LITER)

Groups of amino acids	DM-120	DM-153	DM-160	MEM	RPMI-1640	Ham F12
Basic (His, Lys, Arg)	230	230	230	241	255	267
Acidic (Glu, Asp)	175	175	175	0	40	28
Hydroxy (Ser, Thr)	180	180	180	48	50	23
Amido (Gln, Asn)	100	300	325	292	350	159
Others	1392	1392	1392	267	260	129
	2077	2277	2302	848	955	606

TABLE III

GROWTH OF CELLS IN VARIOUS MEDIA[a]

Cells	Serum	MEM	Ham F12	DM-120	DM-150	DM-153	LD[b]
3T3	10% FCS[c]	100[d]	85.1	—	100	—	—
CHO-K1	10% FCS[c]	82.7	100[d]	—	98.6	—	—
RLC-10(2)	10% FCS[c]	58.8	—	—	—	176.6	100[d]
L · P3	None	97.7	96.8	100[d]	97.9	—	—
JTC-25 · P3	None	100	88.9	100[d]	95.4	—	—

[a]The values indicate the rates of cell proliferation during 7 days compared to the rates (100) in control lines cultured in their usual growth media for the same period.
[b]Lactalbumin hydrolyzate, 0.4%, and buffered saline D(Takaoka, 1958).
[c]FCS, fetal calf serum.
[d]The growth medium in which the cell had been serially grown.

Previously, asparagine had not been added into our mixtures of media. The effects of asparagine and aspartic acid in the media, with or without serum, were examined. In Fig. 2, the effect of asparagine on the proliferation of JTC-25 · P3 cells grown without serum is shown. Neither asparagine nor aspartic acid were required by this strain. RLC-10(2) cells grown in a serum

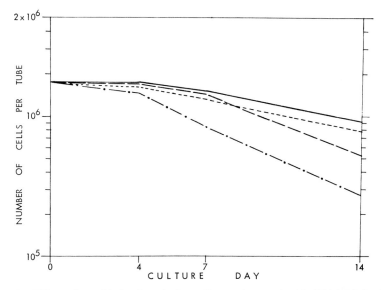

FIG. 1.   Effects of two kinds of synthetic media supplemented with 10% (v/v) fetal calf serum (FCS) on the survival of human blood lymphoid cells in primary culture. ——, total number of cells in DM-153 + 10% FCS; ----, number of viable cells in DM-153 + 10% FCS; ——, total number of cells in RPMI-1640 + 10% FCS; ·—·—, number of viable cells in RPMI-1640 + 10% FCS.

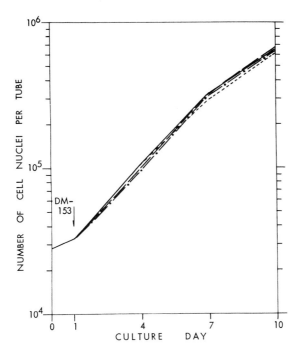

FIG. 2. Effects of aspartic acid and asparagine in synthetic media on the proliferation of JTC-25·P3 cells serially grown in the absence of serum. In this and the following figures, the arrow indicates the day when the medium was changed from the growth medium to the test medium, ———, DM-153, Asp+, Asn−; — —, DM-160, Asp+, Asn+; —··—, DM-164, Asp−, Asn+; ----, DM-165, Asp−, Asn−.

medium showed a distinct response to asparagine (Fig. 3). The cells grew fairly well in the absence of aspartic acid when supplemented with asparagine. The combined use of both, however, resulted in the highest rate of growth.

IV.   Effects of Composition of Buffer and Salts in Media

In Table IV the composition of buffer and salts in various synthetic media designed in our laboratory is shown. The compositions of DM-120, -145, and -150 are the same as in buffered saline mixture D (Table III, footnote *b*). In DM-166, -167, and -168, the concentrations of NaCl differ from each other and from DM-120, -145, and -150. The composition of DM-153 and -160 followed that of Earle's buffered saline (Earle, 1943). In DM-153, -161, -162,

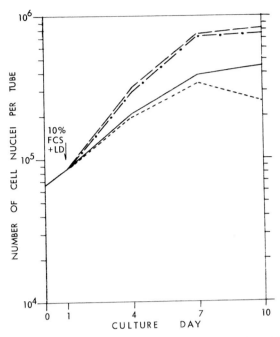

FIG. 3. Effects of aspartic acid and asparagine on the proliferation of RLC-10(2) cells serially grown in the presence of 10% fetal calf serum. ——, DM-153, Asp+, Asn−; ——, DM-160, Asp+, Asn+; ——, DM-164, Asp−, Asn+; ----, DM-165, Asp−, Asn−; FCS, fetal calf serum; LD, lactalbumin hydrolyzate and buffered saline D.

and -163, a phosphate buffer system was employed, each with a different concentration. DM-163 is characterized by containing HEPES (Eagle, 1971). The contents of $NaHCO_3$ in DM-153, DM-160, and MEM were adjusted after measurement of pH.

The effect of the concentration of NaCl in the medium is summarized in Table V. The NaCl content in each mixture in the table corresponds to that in the mixture as follows: 6000 mg/liter (RPMI-1640), 6800 mg/liter (MEM, DM-153), 7600 mg/liter (F 12), and 8000 mg/liter (DM-150). Table V indicates that cells are able to adapt to media with lower osmotic pressures but not to those with higher osmotic pressures.

The concentrations of various buffer solutions were compared with each other for their effects on the proliferation of JTC-25 · P3 cells (Fig. 4), serially grown in protein- and lipid-free synthetic media, and on that of rat liver RLC-10(2) cells (Fig. 5) serially grown in a serum medium. In the former, the cells showed the highest rate of proliferation in DM-153 and DM-159. The use of HEPES in place of $NaHCO_3$ accelerated cell proliferation when the number of cells was low in the culture, suggesting that it

TABLE IV

COMPOSITION OF BUFFER AND SALTS IN VARIOUS MEDIA (MG/LITER)

Component	DM-120 DM-145 DM-150	DM-166	DM-167	DM-168	DM-153 DM-160	DM-159	DM-161	DM-162	DM-163	MEM
NaCl	8000	6000[a]	6800	7600[b]	6800	6800	6800	6800	6800	6800
KCl	200	200	200	200	400	400	400	400	400	400
$CaCl_2$	264(2H$_2$O)	264(2H$_2$O)	264(2H$_2$O)	264(2H$_2$O)	200	200	200	200	200	200
$MgCl_2 \cdot 6H_2O$	100	100	100	100	—	—	—	—	—	—
$MgSO_4 \cdot 7H_2O$	—	—	—	—	200	200	200	200	200	200
$NaH_2PO_4 \cdot H_2O$	—	—	—	—	125	—	—	—	—	125
$Na_2HPO_4 \cdot 12H_2O$	35	35	35	35	—	35	160[c]	300[d]	160	—
KH_2PO_4	177	177	177	177	—	177	60[c]	25[d]	60	—
Glucose	1000	1000	1000	1000	1000	1000	1000	1000	1000	1000
$NaHCO_3$	1000	1000	1000	1000	1000	1000	1000	1000	1000	1000
Phenol red	—	—	—	—	6	6	6	6	6	6
HEPES	—	—	—	—	—	—	—	—	10 mM	—
Osmotic pressure of the medium	304 −307	250	270	299	279 −286	278	255	281	252	266 −282

[a] The same content of NaCl as in RPMI-1640.
[b] The same content of NaCl as in Ham F12.
[c] The same phosphate buffer mixture as in Hanks' solution (1949).
[d] The same phosphate buffer mixture as in Gey's solution (1949).

TABLE V

EFFECTS OF THE CONCENTRATION OF NaCl IN CULTURE MEDIUM ON THE PROLIFERATION OF VARIOUS CELLS[a]

		NaCl			
Cells	Serum	6000 mg/l (DM-166)	6800 mg/l (DM-167)	7600 mg/l (DM-168)	8000 mg/l (DM-150)
3T3	10% FCS[d]	100.9	100[c]	94.0	96.5
CHO-K1	10% FCS[d]	83.5	86.1	100[c]	97.1
Rat liver[b]	10% FCS[d]	75.8	92.1	107.7	100[c]
L·P3	None	108.8	114.0	104.5	100[c]
JTC-25·P3	None	101.9	107.9	118.2	100[c]

[a] The values indicate the rates of cell proliferation during 7 days compared to the rate (100) in control lines cultured in their own growth media for the same period.
[b] Secondary culture of rat liver cells.
[c] The composition in the growth medium in which the cells were serially grown.
[d] FCS, fetal calf serum.

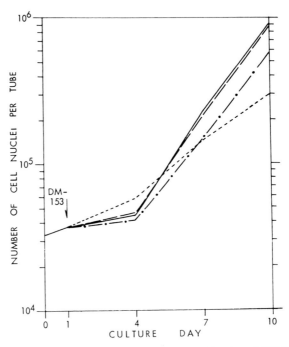

FIG. 4. Effects of the composition of buffer on the proliferation of JTC-25·P3 cells grown in the absence of serum. ——, DM-153; ——, D(DM-159); ·—·-, Hanks (DM-161); ----, HEPES (DM-163).

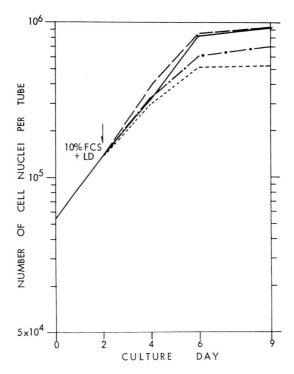

Fig. 5. Effects of the composition of buffer on the proliferation of RLC-10(2) cells serially grown in the presence of serum. ——, DM-153; ——. D(DM-159); ·—·—, Hanks (DM-161); ---, Gey(DM-162). FCS, fetal calf serum; LD, lactalbumin hydrolyzate and buffered saline D.

contributed to the maintenance of pH as confirmed by incubation in cell-free media, but cell growth was subsequently inhibited somewhat by this addition. When $NaHCO_3$ was added to the HEPES medium, cell growth was not inhibited.

In the experiments reported in Fig. 5, DM-159 and DM-153 exhibited the highest growth of cells.

The composition of buffer and salts in DM-160 is much the same as in DM-153. From these findings, we would like to recommend the use of DM-160 now as a multipurpose synthetic medium for the cultivation of various kinds of cells.

Addendum

So-called essential fatty acids have been regarded to be "essential" to the survival and respiratory functions of cells. Some of the cell strains serially

TABLE VI

PERCENTAGE COMPOSITION OF THE TOTAL FATTY ACIDS IN CELLS[a]

Fatty acids	Relative retention time	HeLa·P3 Uterus DM-145	RTH-1·P3 Thymus DM-120	RLH-1·P3 Liver DM-145	RLH-2·P3 Liver DM-145	RLH-3·P3 Liver DM-145	RLH-4·P3 Liver DM-145	RLH-5·P3 Liver DM-120	JTC-16·P3 Hepatoma DM-145	L·P3 Connective DM-120
				Cells, original tissues, and synthetic media						
14:0	0.27	1.4	2.8	6.8	5.8	5.0	2.4	3.9	1.1	1.4
16:0	0.51	17.4	25.5	17.5	21.0	19.5	20.6	16.5	16.6	23.7
16:1	0.63	16.6	12.9	20.7	16.1	19.3	17.0	6.5	4.7	9.3
18:0	1.00	7.1	18.3	5.8	8.4	8.1	6.8	17.8	15.6	10.2
18:1	1.15	50.0	39.0	41.4	41.9	41.4	46.2	51.5	52.6	55.4
x[g]	1.30	3.3	1.4	2.2	2.3	2.6	2.5	0.2	2.0	Trace
18:2 ω 6	1.45	0.0	0.0	<0.6[b]	<0.4[b]	<0.3[b]	0.0	0.0	0.0	0.0
18:3 ω 3	1.73	0.0	0.0	0.0	0.0	0.0	0.0	0.0	0.0	0.0
20:0	1.93	1.1	Trace	1.9	2.3	2.1	2.5	0.9	3.9	Trace
20:1	2.12	3.0	Trace	2.9	2.4	1.7	2.1	2.7	3.5	Trace
20:3 ω 9	2.84	Trace	Trace	0.0	0.0	Trace	Trace	0.0	0.0	Trace[c]
20:3 ω 6	3.07	0.0	0.0	0.0	0.0	0.0	0.0	0.0	0.0	Trace
20:4 ω 6	3.55	0.0	0.0	0.0	0.0	0.0	0.0	0.0	0.0	0.0
22:6 ω 3	9.31	0.0	0.0	0.0	0.0	0.0	(-)[d]	(-)[d]	(-)[d]	0.0

In serum-containing media (20% calf serum and 0.4% lactalbumin hydrolyzate in saline)

								L-929[e]
14:0	0.27	1.7	0.7	2.4	2.0	1.0	2.0	1.6	2.0
16:0	0.51	21.5	12.6	22.8	23.4	17.8	24.6	13.1	38.8
16:1	0.63	7.4	2.4	9.2	10.2	3.0	6.2	3.0	3.0
18:0	1.00	11.7	17.4	8.0	10.2	26.0	10.5	18.8	21.5
18:1	1.15	42.9	47.0	43.1	39.4	29.2	39.2	45.4	22.8
x	1.30	0.8	0.6	Trace	Trace	Trace	Trace	Trace	Trace
18:2 ω 6	1.45	8.3	11.4	10.6	11.1	9.4	12.4	11.3	7.3
18:3 ω 3	1.73	0.0	0.6	1.1	1.0	Trace	1.0	0.7	1.0
20:0	1.93	0.1	1.7	Trace	Trace	Trace	Trace	Trace	Trace
20:3 ω 9	2.84	Trace	Trace	0.0	0.0	0.0	0.0	0.0	Trace[c]
20:1	2.12	1.4	0.6	Trace	Trace	Trace	Trace	Trace	Trace
20:3 ω	3.07	Trace	0.8	Trace	Trace	Trace	Trace	Trace	0.9
20:4 ω 6	3.55	4.3	5.1	2.8	2.7	13.6	4.1	1.1	4.6
22:6 ω 3	9.31	Trace	Trace	(-)[a,f]	(-)[a,f]	(-)[a,f]	(-)[d,f]	(-)[d,f]	3.9

[a] The methylated fatty acids were analyzed by gas chromatography [F & M, Model 402, at 180°C on 2 m column containing ethylenglycol succinate as liquid phase with He as a carrier gas (Kagawa et al., 1969)]. Methyl stearate appeared 17 minutes after the sample injection. The minor peaks which appeared at the relative retention times (methyl stearate 1.00) of 0.07, 0.08, 0.09, 0.13, 0.14, 0.19, 0.21, 0.32, 0.38, 0.43, 0.72, 0.85, 3.85, 6.99, 7.53 were not taken into the percentage composition, since the amounts were too small to be accurately estimated and the methylation with 5% HCl-methanol at 80°C for 2 hours yielded dimethylacetals from plasmalogens.

[b] The peaks were on the shoulder of preceding ones.

[c] Detected by radioactivity (Kagawa et al., 1969) after culture in ^{14}C-labeled acetate containing medium.

[d] Not detected by Ag thin-layer chromatography (Kagawa et al., 1969) of methylated fatty acids.

[e] L-929 cells of the original strain maintained in a serum medium.

[f] Not detected by SE 30 gas chromatography.

[g] Chemically unidentified peak.

155

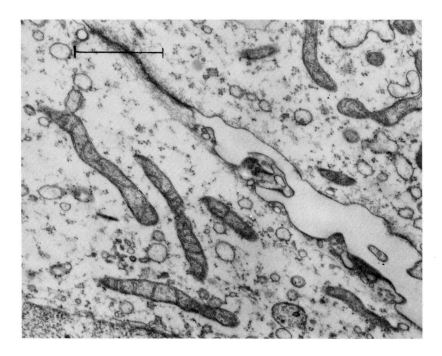

FIG. 6.　Electron micrograph of RLC-10(2), rat liver cells serially grown in a serum medium. Scale line = 1μm.

grown in protein- and lipid-free synthetic media in our laboratory were examined biochemically and morphologically. Unsaturated fatty acids were not detected biochemically in these strains, as illustrated in Table VI (Kagawa *et al.*, 1969, 1970). However, respiratory functions were unchanged in the cells. By electron microscopy, mitochondria were detected in the cytoplasm of the cells grown in a lipid-free medium (Figs. 6 and 7). This suggests that unsaturated fatty acids are required not for functions in cells, but to increase the strength of structures.

When rat ascites hepatoma cells were serially grown in protein- and lipid-free synthetic media, various rat serum proteins were confirmed to be released into the medium.

ACKNOWLEDGMENTS

The authors are indebted to Dr. Shokichi Tani of Department of Internal Medicine for the assay of osmotic pressures of synthetic media, Dr. Sumi Nagase of Sasaki Institute for the identification of rat serum proteins in the medium, and Mr. Toshitaka Akatsuka for the preparation of electron micrographs.

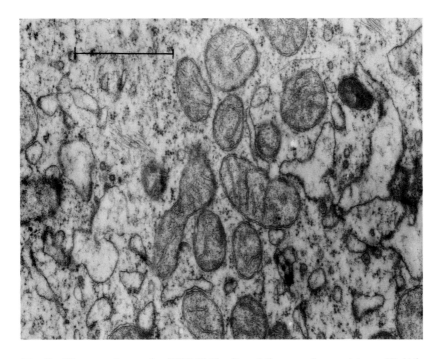

FIG. 7. Electron micrograph of JTC-25·P3 cells serially grown in a protein- and lipid-free synthetic medium. In Figs. 6 and 7 the presence of mitochondria with a healthy appearance is observed in the cytoplasm. Scale line = 1 μm.

REFERENCES

Eagle, H. (1971). *Science* **174**, 500.
Earle, W. R. (1943). *J. Natl. Cancer Inst.* **4**, 165.
Furuya, M., Takaoka, T., Nagai, Y., and Katsuta, H. (1971). *Jpn. J. Exp. Med.* **41**, 471.
Gey, G. O. (1949). Bulletin of the Tissue Culture Association, March 16.
Hanks, J. H., and Wallace, R. E. (1949). *Proc. Soc. Exp. Biol. Med.* **71**, 196.
Kagawa, Y., Kaneko, K., Takaoka, T., and Katsuta, H. (1960). *Jpn. J. Exp. Med.* **30**, 95.
Kagawa, Y., Takaoka, T., and Katsuta, H. (1969). *J. Biochem.* (*Tokyo*) **65**, 799.
Kagawa, Y., Takaoka, T., and Katsuta, H. (1970). *J. Biochem.* (*Tokyo*) **68**, 133.
Katsuta, H., and Takaoka, T. (1960). *Jpn. J. Exp. Med.* **30**, 235.
Katsuta, H., and Takaoka, T. (1973). *Methods Cell Biol.* **6**, 1–42.
Katsuta, H., Endo, H., Takaoka, T., and Oishi, Y. (1957). *Jpn. J. Exp. Med.* **27**, 343.
Katsuta, H., Takaoka, T., Hori, M., Saito, S., Suzuki, S., Someya, Y., and Ito, E. (1958). *Jpn. J. Exp. Med.* **28**, 199.
Katsuta, H., Takaoka, T., Hosaka, S., Hibino, M., Otsuki, I., Hattori, K., Suzuki, S., and Mitamura, K. (1959a). *Jpn. J. Exp. Med.* **29**, 45.
Katsuta, H., Takaoka, T., Mitamura, K., Kawada, I., Kuwabara, H., and Kuwabara, S. (1959b). *Jpn. J. Exp. Med.* **29**, 191.

Katsuta, H., Takaoka, T., Hattori, K., Kawada, I., Kuwabara, H., and Kuwabara, S. (1959c). *Jpn. J. Exp. Med.* **29**, 297.
Katsuta, H., Takaoka, T., and Kikuchi, K. (1961). *Jpn. J. Exp. Med.* **31**, 125.
Matsumura, T., Takaoka, T., and Katsuta, H. (1968). *Exp. Cell Res.* **53**, 337.
Ohta, S., Takaoka, T., and Katsuta, H. (1969). *Jpn. J. Exp. Med.* **39**, 359.
Takaoka, T. (1958). *Jpn. J. Exp. Med.* **28**, 381.
Takaoka, T., and Katsuta, H. (1971). *Exp. Cell Res.* **67**, 295.
Takaoka, T., and Katsuta, H. (1975). *Jpn. J. Exp. Med.* **45**, 11.
Takaoka, T., Katsuta, H., Ohta, S., Miyata, M., Furuya, M., and Hosokawa, A. (1968). *Jpn. J. Exp. Med.* **38**, 193.

Chapter 14

Automatic Collection of Samples from Suspension Cell Cultures for Investigation of Prolonged Metabolic Processes

GEORGE BOLCSFOLDI AND EVA ELIASSON

The Wenner–Gren Institute,
University of Stockholm,
Stockholm, Sweden

I. Introduction

Studies concerned with the kinetics of metabolic processes having a prolonged duration place certain demands on the sampling procedure, which may be difficult to handle practically. For example, investigating the pattern of synthesis of metabolites during the cell cycle necessitates the collection of samples from synchronized cell cultures at regular, frequent intervals over periods of one or more days in order to yield coherent results. During the course of our work (Bolcsfoldi and Eliasson, 1973) we assembled an apparatus that facilitates such experiments and which has been shown to give reproducible results, comparable to those obtained by manual sampling methods. Briefly, samples of a predetermined volume are automatically removed, at desired intervals of time, from a suspension cell culture and deposited in test tubes kept at $0°-2°C$ for storage until further processing can conveniently be performed.

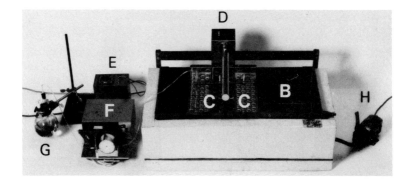

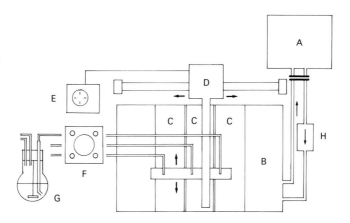

FIG. 1. Photograph and diagram of automatic cell-sampling apparatus. Manufacturers of the illustrated apparatus: Heto, Birkerød, Denmark—refrigeration unit (A) recycling heat exchanger (not shown in the photograph) with capacity at 0°C coolant temperature of 225 kcal, to be used with a circulation pump (H) delivering about 20 liters per minute at 0 m height; Kebo-Grave Labcenter, Stockholm, Sweden—water bath (B), test tube racks (C), fractionator (D) with arm modified to carry 3 Teflon tubes, timer (E). Other apparatus: peristaltic pump (F), and culture flask (G).

II.   Apparatus

The apparatus shown in the drawing and photograph of Fig. 1 consists of six components. A refrigeration unit fitted with a circulation pump maintains a 20-liter water bath containing 25% ethyl alcohol (freezing point about −10°C) at approximately 0°–2°C. Test tubes, in racks fitting into the water bath, are immersed in the cold alcohol–water solution and receive samples delivered via Teflon tubes from the cultures by a peristaltic pump and deposited by a fractionator. In the present experiments the arm of the fractionator was modified to allow collection of samples from three cell cultures simultaneously. Controlled by the timer, the fractionator moves along the row of test tubes, depositing samples at predetermined intervals of time with a volume dependent on the pump speed. In the present experiments the Teflon tubing (1 mm internal diameter) was connected to the culture by a glass capillary tube (2 mm internal diameter) mounted in the stopper of the culture flask. In order to avoid settling of the cells in the glass capillary and vertical parts of the tubing, thereby causing uneven sampling, it is essential to maintain a continuous minimum flow rate of, in our case, 25 ml/hour. In addition, in our experience, the diameter of the glass capillary tube cannot be further decreased without incurring a decreased efficiency in the withdrawal of cells.

All parts of the apparatus, excluding the refrigeration unit, can be housed in a climate chamber at 37°C.

With the present apparatus, it is possible to collect up to 80 samples from each of 3 parallel cultures with durations of a few seconds to 1 hour per sample. The maximum volume per sample, which is limited by the type of test tube used, is about 25 ml. By modifying the collecting arm, it is possible to collect up to 40 samples from 6 parallel cultures.

III.   Results and Discussion

The apparatus described has been assembled to facilitate the collection of samples at frequent, regular intervals, from cell cultures during experiments having a duration of several days. For this technique to be of use in monitoring the progress of cellular processes, it is necessary to establish that the cell samples stored in the cold, and analyzed at a later time, give the same results as samples freshly collected at these time points. Obviously such control experiments have to be carried out for each individual type of experiment to ascertain whether the metabolites to be assayed are stable under the storage conditions used. In our investigations the cell samples were stored at $0°-2°C$ in the same culture medium with which they were withdrawn from the culture flasks. In these conditions, we have seen that this collection technique gave results comparable to those obtained with freshly collected samples, with respect to cell count, radioactivity incorporated into DNA, protein content, and activity of the enzyme arginase (EC 3.5.3.1) as shown in Table I. Similarly, the results of other analyses, illustrated in Fig. 2, of the same parameters, from samples collected with this apparatus, yielded the expected temporal characteristics in the growth rate, rate of protein, and DNA accumulation normally seen in synchronized cell cultures.

TABLE I

COMPARISON OF FRESHLY COLLECTED SAMPLES (I) AND SAMPLES STORED FOR 23 HOURS (II) AT 2°C WITH RESPECT TO CELL NUMBER, INCORPORATION OF LABELED THYMIDINE, CELLULAR PROTEIN, AND ARGINASE ACTIVITY

| Exp. No. | Cell count (10^6 cells/ml) | | | | Incorporation of [Me-^3H]thymidine[b] (cpm/5 ml culture) | | Protein content[c] (mg/5 ml culture) | | Arginase activity[d] (nmoles urea per hr/ 5 ml culture) | |
| | Microscope[a] | | Celloscope[a] | | | | | | | |
	I	II	I	II	I	II	I	II	I	II
I	0.92	0.92	0.87	0.87	—	—	1.53	1.48	120	120
	0.90	0.93	0.92	0.90						
II	0.44	0.49	0.45	0.44	—	—	0.86	0.81	119	118
	0.50	0.48	0.44	0.41						
III	0.24	0.28	0.25	0.26	—	—	0.47	0.43	35	36
	0.24	0.28	0.24	0.24						
IV	—	—	—	—	516	578	—	—	—	—
	—	—	—	—	568	610	—	—	—	—
	—	—	—	—	539	590	—	—	—	—

[a]Cells were counted microscopically in a Bürcher chamber or in a Ljungberg Model 101 celloscope.
[b]Five milliliters of a cell culture (3×10^5 cells/ml) were labeled for 15 minutes with 0.02 mM [Me-^3H]thymidine (1.25 mCi/mmole), then centrifuged; the cell pellet was resuspended in Hanks' salt solution. The cells were then collected onto fiberglass filters by suction followed by three successive applications of 5% trichloroacetic acid and finally 70% ethanol. The filters were dried and placed in scintillation vials with 5 ml of Omnifluor (NEN) scintillating liquid and counted in a Beckman LS-100 liquid scintillation counter.
[c]See Weichselbaum (1946).
[d]See Schimke (1970).

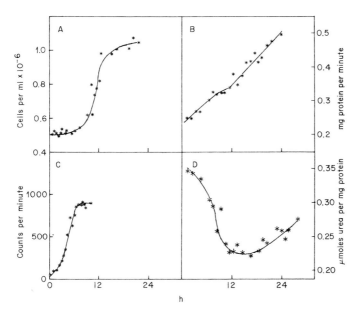

FIG. 2.   Determinations were made from automatically collected samples of a synchronized culture of Chang's liver cells. (A) Number of cells (millions/ml); (B) cellular protein (mg/ml); (C) incorporation of [Me-³H]thymidine into acid-precipitable DNA; (D) specific activity (μmoles urea/mg protein) as a function of the time elapsed since the cultures were released from a double thymidine blockade (Peterson and Anderson, 1964).

In conclusion, the virtue of the technique described above is that cell samples from several parallel cell cultures can be collected into test tubes standing in a chilled water bath and stored for later analysis of a variety of metabolites, while the whole of the apparatus, excluding the refrigeration unit, is housed in a 37° culture room. This is made possible by two features of this apparatus, which are radically different from other refrigerated fraction-collecting units. First, the fraction collector arm is mobile, instead of the test tube holders, allowing these to be immersed in a water bath; and second, the use of a liquid recycling heat-exchanger type of refrigeration unit allows the water bath to be housed in a warm environment at a distance from the refrigeration unit.

REFERENCES

Bolcsfoldi, G., and Eliasson, E. (1973). *Anal. Biochem.* **55**, 626–629.
Peterson, D. F., and Anderson, E. C. (1964). *Nature (London)* **203**, 642–643.
Schimke, R. T. (1970). *In* "Methods in Enzymology" (H. Tabor and C. W. Tabor, eds.), Vol. 17A, pp. 324–329. Academic Press, New York.
Weichselbaum, T. E. (1946). *Am. J. Clin. Pathol.* **10**, 40–49.

Chapter 15

A Simple Device and Procedure for Successful Freezing of Cells in Liquid Nitrogen Vapor

FREDERICK H. KASTEN AND DOMINIC K. YIP

Department of Anatomy,
Louisiana State University Medical Center,
New Orleans, Louisiana

I.   Introduction

Cell preservation at low temperatures has become an essential practice in cell biology, particularly in laboratories working with tissue cultures. Sufficient experience has been gained in the practical technology of cryobiology to permit highly successful preservation of cell suspensions. As an example, in 1972 there were approximately 130 animal cell lines of the American Type Culture Collection preserved in the frozen state (Shannon and Macy, 1972).

Some of the freezing equipment which has been used for cell freezing includes sophisticated controllers. For the laboratory which only occasionally is required to freeze cells, the use of elaborate and expensive equipment would likely not be justified. Alternative approaches involve the placing of ampoules for timed periods in Dry-Ice chests (Dougherty, 1962) or Dry-Ice–ethanol mixtures (Meryman, 1963). Other freezing methods involve direct cooling in vapor or liquid nitrogen (Rey, 1959). In another approach, a mechanical freezer (Ryan and Smith, 1974) or liquid nitrogen (Walker and Wilen, 1967) is used to cool an ethanol bath containing the ampoules. Kite and Doebbler (1962) placed vials in perforated cartons and suspended the latter above liquid nitrogen in cryogenic containers. A cooling plug was designed by Nagington and Greaves (1962) to be used in a Linde liquid nitrogen container; an improved commercial version, the Linde BF-5 Biological Freezer is made for use with two of the Linde containers. Finally, it should be mentioned that in many laboratories ampoules of cells are simply placed, with little control of time or temperature, in a refrigerator freezing compartment, Dry-Ice chest, or mechanical freezer at $-65°C$ for several hours or overnight, followed by storage at the coldest temperature available (Shannon and Macy, 1973).

While many of the various methods mentioned are reasonably effective, the recovery rate varies largely, according to the rate of initial freezing. The methods also differ in cost, sophistication and automation, amount of handling, and reproducibility of results. The cell freezer to be described has a number of attractive features that we feel will appeal to cell biologists and others working with isolated cell suspensions, especially where there is an infrequent need to freeze cells.

II.   Materials and Methods

A.   Cell Freezer

The cell freezer is designed to freeze cells in the vapor above liquid nitrogen in any available commercial storage tank. The freezer consists of an ampoule carrier and a base block, both made of wood (Fig. 1).[1] The base block has a central hole through which the ampoule carrier may be lowered to a controlled level in the tank. The tanks employed in the present study are the Cryenco 250 Biostat (Cryogenic Engineering Co., 4955 Bannock St., Denver, Colorado 80216) and the Minnesota Valley Engineering (New Prague, Minnesota 56071) VP-25 (Cryo-Flask. The latter is designed speci-

---

[1] It is recommended that the cell freezer be protected with several coats of paint. The freezer in our laboratory has been in use for three years after such treatment; the plywood shows no signs of fracturing and separating as a result of freezing-thawing.

fically for vapor storage of biologicals. The ampoule carrier is made from an 18-inch wooden dowel that is 1 inch in diameter. The carrier is attached to a round wooden block with a diameter equal to the neck opening of the liquid nitrogen tank. The diameter is 8 inches for the VP-25 and 4 inches for the 250 Biostat. The liquid nitrogen tank used for freezing is also used for storage of previously frozen cells supported in cannisters. Four notches are cut into the base block to accommodate the cannister handles hanging

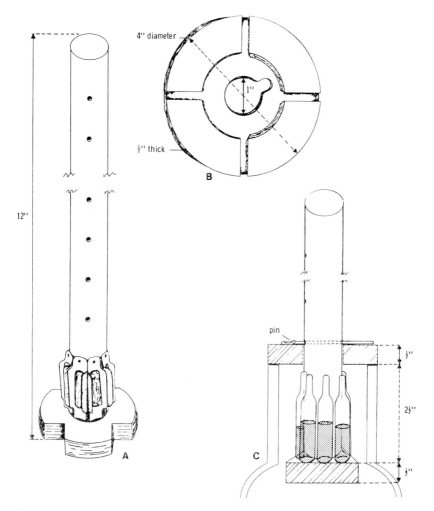

FIG. 1. Drawings of liquid nitrogen cell freezer as adapted for Cryenco 250 Biostat. Ampoule carrier is shown in (A), base block in (B), and entire unit in cryogenic tank (cutaway view) at present height (C). See text for further description.

from the top of the tank mouth so that ampoules in storage are not disturbed. Eight 2-ml ampoule holders, cut from conventional aluminum canes (Shur-Bend Mfg. Co., 5709 29th Avenue N., Minneapolis, Minnesota 55422) are fastened by nails to one end of the ampoule holder just above the round wooden block. A series of twelve 1/16-inch holes is drilled 1 inch apart in the ampoule carrier. When in use, the carrier is inserted through the central hole of the base block and lowered to a predetermined depth. This is set by the position of a metallic pin inserted through one of the holes in the carrier. Figure 1 shows drawings of the ampoule carrier, base block, and the complete cell freezer as seen in a cutaway view of a liquid nitrogen tank. The series of photographs in Fig. 2 illustrates the operational steps involved in using the cell freezer.

B. Preparation of Cells

The cells employed include two continuous tumor cell lines, mouse neuroblastoma C1300 (clone C46) and CMP, a human epithelial carcinoma (Bovis and Kasten, 1965). We also use primary cultures of contracting heart cells isolated from newborn rats (Kasten, 1973). Cells are grown in 60 mm² Falcon plastic flasks, trypsinized to single cells, centrifuged, and suspended in appropriate medium with 10% fetal calf serum and 10% dimethyl sulfoxide (DMSO) at cell densities from 1.0 to 1.5 × 10⁶ cells per milliliter. In one experiment, a comparison is made between cells frozen with 10% DMSO and 10% glycerol. Cell viability is assessed using the trypan blue-dye exclusion test (Stulberg *et al.*, 1962).

C. Temperature Calibration and Freezing Procedure

For temperature calibration, 1.2 ml of cell suspension is transferred into a 2 ml-glass ampoule (Wheaton Gold Brand Cryule, prescored, No. 651486; Wheaton Scientific, 1000 N. Tenth St., Millville, New Jersey 08332). The neck of the ampoule is broken off before the ampoule is lodged into one of the ampoule holders in the cell freezer. A glass thermometer that measures

FIG. 2. Photographs of liquid nitrogen cell freezer during calibration and freezing operations. (A) Unit with thermometer attached prior to insertion into cold nitrogen vapor of Cryenco 250 Biostat. (B) Unit immediately after lowering into tank for temperature measurements. (C) Full complement of ampoules following removal from cold room and prior to lowering to first cooling level of tank. (D) Top of unit during freezing process. Note metallic pin in ampoule carrier, which sets level. Freezing is done in the presence of cannister holders protruding through notches in baseblock. This permits ampoules already being stored in cannisters to remain undisturbed during freezing of new ampoules. See text for further description of methodology employed.

temperatures from 30° C to −200°C is inserted into a small hole drilled in the base of the cell freezer, and the bulb is put into the open ampoule. Three or more extra "dummy" ampoules, or ampoules with cells containing 1.2 ml of medium with 10% DMSO, are attached to the ampoule holders (Fig. 1A). After a precooling run in a cold room with ampoules alone, the whole unit is then lowered into the tank and set at a given level above the liquid nitrogen, according to the temperature desired. From the time the unit is placed at a specific level, temperature readings are made at 1-minute intervals until the temperature is stabilized. Cooling rates are measured at various levels with different heights of liquid nitrogen inside the tank. Also, temperature readings are made when different numbers of ampoules are frozen simultaneously. The desired cooling rate we aimed for in initial stages of freezing was of the "slow" type, i.e., approximately 1°–2°C per minute, according to the usual quoted figure in the literature for best survival (Shannon and Macy, 1973).

III.   Results

A.   Cooling Rates under Different Conditions

Initially, data are obtained on individual cooling rates of ampoules placed in a walk-in cooler at approximately 4°C and other ampoules put in liquid nitrogen vapor at two levels in each of the two storage tanks. The slowest and most desirable rate of cooling is at the top of the tank. A depth of 2.5 inches from the surface just permits the cell freezer to be accommodated. During the first 30 minutes of cooling in the cold room, the temperature of cell suspensions drops from 25° to 6° at a rate of 0.5°–3°C per minute. When cell suspensions are placed directly in the vapor phase of liquid nitrogen, 2.5 inches down from the tank mouth, the temperature drops from 25° to 11°C in a 20-minute period at rates of 0.5°–3°C per minute. At 5.5 inches down from the tank mouth, the temperature falls from 25 to −71°C in 31–34 minutes at rates of 0.5°–6°C per minute. Based on these data, it is evident that the minimum and maximum cooling rates are 0.5° and 6°C per minute. When ampoules are placed directly in cold nitrogen vapor from room temperature, the rate of temperature fall is relatively high in early stages. It therefore appeared desirable to develop a procedure in which ampoules are passed sequentially to low temperature.

In the next series of experiments, which are summarized in Table I, temperature changes and cooling rates are followed with time as ampoules of cell suspensions are moved sequentially from room temperature to the cold

TABLE I

COOLING RATES OF CELL AMPOULES IN CRYENCO 250 BIOSTAT LIQUID NITROGEN TANK[a]

Position of ampoules and thermometer	Range of time change (minutes)	Range of temperature change (°C)	Cooling rate (°C/min)
Cold room	0–10	25–13	0.5–3
2.5 Inches down from the tank mouth with liquid nitrogen level at:			
7 inches	11–20	13–6.5	0.5–1.0
10 inches	11–20	13–2	0.5–1.5
5.5 Inches down from the tank mouth with liquid nitrogen level at:			
7 inches	21–47	6.5 to − 70	0.5–5
10 inches	21–38	2 to − 70	1–7

[a]Continuous cooling run experiments in which 7 ampoules of cells (C 46 neuroblastoma), including one ampoule for thermometer, were transferred at room temperature to a cold room and then to liquid nitrogen tank at two successive heights. Temperatures were read every minute.

room and then to lower vapor levels of the Cryenco 250 tank. It is seen that 10 minutes in the cold room are required to lower the temperature in ampoules from 25 to 13°C; another 10 minutes at 2.5 inches down from the tank mouth reduces the temperature to 6.5°C or 2°C, according to whether the liquid nitrogen level is 7 or 10 inches from the bottom. When the ampoules are lowered an additional 3 inches (5.5 inches down from the tank mouth), it takes 17 to 26 minutes to reach − 70°C, according to whether the liquid nitrogen level is at 7 or 10 inches. The minimum and maximum cooling rates are 0.5° and 6°C per minute. Similar data are obtained using the Minnesota Valley VP-25 tank with ampoules placed sequentially at two different levels of cold nitrogen vapor and with the liquid nitrogen filled to three different heights. Cooling changes are similar to those observed with the Cryenco 250 tank, with cooling rates never exceeding 6.5°C per minute. However, the level of liquid nitrogen in the MV-25 tank has no effect on temperature drop. This is not surprising since this tank is designed to be a vapor phase storage tank.

B. Cooling Rate Curves

An overall view of the cooling rate curve, based on data obtained with the MV-25, is illustrated in Fig. 3. Two curves are presented, which re-

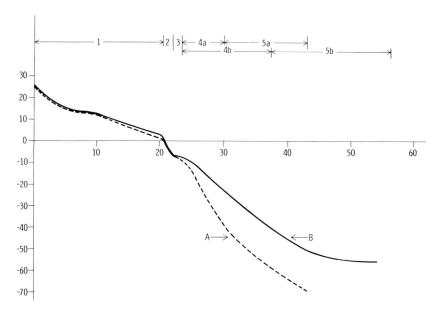

Fig. 3.   Cooling curves from 25° C to – 70° C using device and procedures described in text for nitrogen vapor freezing in the Minnesota Valley VP-25 Cryo-Flask: Abscissa: time (minutes); ordinate: temperature (° C). Curve A was derived from experiment with 4 ampoules, and curve B from 8 ampoules of C46 cells. Cooling from 0 to 10 minutes was done in cold room; from 11 to 20 minutes at 2.5 inches from top of tank mouth; and from 21 minutes until end of experiment at 5.5 inches from top of tank mouth. Five distinct slopes appear in each run, with temperature fall ranging from 0.3 to 6° C per minute. Cooling slope from room temperature to period of supercooling (slope 1) is 1° C per minute. Temperature change during eutectic period (slopes 2 and 3) is 6° C per minute or less. Rate of cooling after freezing is 2.3–5° C per minute (slopes 4 and 5 in curve A) and 1.5–2.6° C per minute (slopes 4 and 5 in curve B). See text for further discussion of results.

present the cooling sequences with 4 and with 8 ampoules. In these particular experiments, temperature readings are begun when the ampoules are left in the cold room for 10 minutes, then placed in liquid nitrogen vapor 2.5 inches down from the tank mouth, and 10 minutes later, the ampoules carrier is lowered to 5.5 inches. Temperature readings are made every minute until – 70°C is reached. There is a slow cooling of approximately 1°C per minute (slope 1) until the temperature reaches 1°–3°C, at which time there ensues a rapid fall to – 7°C (slope 2, 6°C per minute). This is followed by a plateau period (slope 3, 9.3°C per minute) for about 2 minutes, before the next temperature fall begins (slope 4, 5°C per minute, curve B). At approximately – 40°C, both curves exhibit a break in slopes, with the development of a slower temperature change (slope, 5, 2.3°C per minute, curve A; 1.5°C per minute, curve B). The major difference between the two cooling curves occurs in slopes 4 and 5; curve A (carrier with 4 ampoules) exhibits a much faster temperature fall than curve B (8 ampoules). As a result, it takes 43 minutes to reach −70°C in the first case compared with 56 minutes in the second. The fastest cooling rate (6°C per minute) occurs in both curves during a 2-minute period (slope 2) and again in slopes 4(5°C per minute) when fewer ampoules are used. The curves shown in Fig. 3 are typical of those reproduced as well in several other experiments. The basis for these temperature changes and their significance are considered in the discussion which follows.

C. Viability

As indicated in Fig. 3, our technique employed for cell freezing permits cells to be frozen to −70°C in 45–60 minutes, according to whether 4 or 8 ampoules are used. The next question concerns the viability of cells recovered from such experimental procedures. In Table II are shown viability data based on the cryogenic preservation and recovery of two tumor cell lines, CMP and C46, and primary cultures of rat heart cells. The average viability of such cells before freezing is 92%, as determined by trypan blue-dye exclusion. After freezing with 10% DMSO, as described earlier, storage at −170 to −196°C for 2–17 days, and rapid thawing, the average viability is 86%. In another set of experiments, a comparison of cell recovery is made between C46 cells frozen in 10% DMSO and 10% glycerol. In this experiment, there is only a 4% difference in recovery rates between cell populations with regard to the two cryoprotective agents.

TABLE II

RESULTS OF CELL VIABILITY TESTS AFTER FREEZING AND STORAGE

Expt. No.	Cells	Tank Model	Number of ampoules	Storage time (days)	Average viability		Cell recovery (%)
					Before (%)	After (%)	
1	Heart cells	VP-25	3	7	82	74	90
2	CMP	250	3	2	91	90	99
3	C46	250	3	7	93	89.5	96
4	C46	250	6	17	90	81	90
5	C46	VP-25	6	10	90	89	99
6	C46	VP-25	3	7	93	90	97
				Averages	90	86	95

IV. Discussion

The use of sophisticated equipment which permits programmed cooling rates with fine levels of temperature control (cf. Kalmbach and Mardiney, 1972) is essential for researchers and others interested in cryopreservation. However, the high cost of commercial instrumentation and the practical difficulties of assembling complex instrumentation seem unjustified to the average cell biologist who has only an occasional need to freeze cells. A somewhat more practical approach for slow freezing uses Dewar flasks or Dry Ice chests which contain Dry Ice or Dry Ice–ethanol mixtures, refrigerators and their freezing compartments, or commercial freezers. With these approaches it may be difficult to achieve desirable and repeatable cooling rates, although reproducible results have been reported (Leibo et al., 1970; Ryan and Smith, 1974). Also, the vials must eventually be transferred to a liquid nitrogen container for long-term preservation, since storage at $-60°$ to $-80°$ is inadequate for prolonged storage of living cells (Shannon and Macy, 1972).

The cell freezer described here is actually a significant modification of the polystyrene cooling plug designed by Nagington and Greaves (1962) and its commercial counterpart, the Linde BF-5 Biological Freezer. Both of these devices make use of cold nitrogen vapor above the liquid nitrogen of a storage tank to cool ampoules of cells down to a predetermined level prior to storage. The advantage of the modified device we describe here is its simplicity and versatility. With only a slight modification in its base, the same ampoule holder can be used on different models of liquid nitrogen tanks. Also, with this cell freezer ampoules of cells can be lowered sequentially to colder levels as freezing proceeds in order to achieve a faster and preselected rate of cooling. In practice, it seems desirable to place ampoules initially in a cold room or refrigerator for a short time and then put them in the cell freezer at two different levels of the liquid nitrogen container, before storage at $-170°$ to $-196°C$.

With the cell freezer described, the cooling rates during a complete freezing run (Fig. 3) varies with the number of ampoules employed, the depth at which they are placed, and the level of liquid nitrogen in the liquid nitrogen storage container (according to the container employed). The number of ampoules has no appreciable effect during the initial stage of cooling. At approximately $-7°C$, the cooling rate with 8 ampoules slows down to about one-half that of a 4-ampoule run. The differences are insignificant in final stages of freezing, beginning at -40 to $-45°C$. A complete freezing cycle in which cells are taken from $25°C$ to $-70°C$, takes 45–60 minutes. The freezing curve exhibits five distinct temperature slopes, which is essentially in agreement with what has been reported previously (Perry et al., 1965).

The first part of the cooling curve takes 20 minutes and involves a temperature fall from 25°C to approximately 2°C; the cooling rate is 1°C per minute. Beginning at about 20 minutes, there is a sharp rate of fall from 2° to −7°C, giving a 6°C change per minute. This drop in temperature is due to the ampoules being lowered at 20 minutes an additional 3 inches into colder vapor. The reason for dropping the temperature is to offset the latent heat of fusion which is suddenly emitted in the eutectic region. Without additional means to absorb the heat, the temperature would immediately rise a few degrees and then fall (Meryman, 1963). The plateau which begins at −7° (22 minutes) probably represents the emitted heat of fusion. Freezing is completed within several minutes and the temperature then falls rapidly at a rate of approximately 2.5–5°C per minute. At −40 to −45°C, the rate of temperature fall slows to 1.5–2.3°C per minute. It is not clear why this final change in slope occurs, but the change is not significant. It has been pointed out by Meryman (1963) that the fact that the specimen temperature does not fall at constant rate is no grounds for criticism. Most workers agree that a low and constant cooling rate is important until freezing is complete. After freezing is finished, the rate of cooling can be increased. With the technique described, maximal cooling rate encountered when cooling from 25° to −70°C is 6°C per minute and is 1°C per minute from room temperature up to the period of supercooling.

It has been popular to attempt a slow freezing rate of 1°C per minute for cryopreservation of cultured cells. This often-quoted figure is based largely on early studies with bull spermatozoa, in which it was shown that the optimal cooling rate was less than 2°C per minute (Polge and Lovelock, 1952). A study by Hauschka et al. (1959) reported that the best viability of normal and neoplastic cultured cells was achieved with slow freezing. They froze at a rate of 6°C per minute down to −25°C and then stored cells at −78°C. Later, Scherer and Hoogasian (1954) showed that strain L fibroblasts survived slow freezing better than fast freezing. According to one study (Diller et al., 1972), there is a 0% probability of intracellular freezing at cooling rates less than 6° per minute and 100% at rates greater than 17° per minute.

However, the fact that intracellular freezing occurs does not necessarily mean that cell death follows. With further studies on a variety of mammalian cells, it is now clear that maximal survival occurs not at a sharply defined cooling rate, but over a broad range of rates (Mazur et al., 1970). For example, with cultured Chinese hamster cells, peak survival occurs with cooling rates above 10°C per minute (Mazur et al., 1972). Indeed, 25% survive freezing at 300°C per minute in the absence of additive (Mazur et *al.*, 1970), and the optimal cooling velocity is approximately 100°C per minute (Mazur *et al.*, 1969). It seems clear that the optimal cooling rate varies with the cell line (Greene *et al.*, 1964). In the case of chick heart cells from primary cultures, it has been noted that cooling at rates of 2°–5°C per minute afforded nearly as good a protection as cooling at 1°C per minute (Wollenberger, 1967). With the system described here at cooling rates of 0.3–6°C per minute (from 25°C to −70°C), mammalian myocardial cells have been passed through three successive cycles of cultivation–freezing–thawing and regained spontaneous contractile function throughout (Kasten and Yip, 1974). Recovery rates for primary cultures of myocardial and endothelial cells are approximately 90%. The CMP human epithelioid tumor line and the C46 mouse neuroblastoma line have recovery rates of 90–99% (95% average) when frozen under these conditions with DMSO as a cryoprotective agent. Mouse B-16 melanoma cells, several tumor lines of nervous system origin (RN-22, N18, C6), and primary cultures of chick embryo cells have likewise been frozen and recovered successfully in our laboratory.

The uniformly high cell recovery rates obtained with this technique attest to the practical success of the cell freezer and associated freezing procedures. It is recommended that others wishing to utilize this type of cell freezer to preserve and store cells in their own liquid nitrogen container carry out the steps listed below.

V. Recommended Procedure

1. Construct the cell freezer as given here, but with dimensions of the blocks and cane notches adjusted to fit the neck opening of the particular liquid nitrogen tank available.

2. Start a trial run by placing a group of 8 ampoules (containing cells in medium with 10% DMSO) in a cold room for 10 minutes or on the highest shelf of a refrigerator (noting the thermostat setting) for this time. Read the temperature once a minute during this period. The advantage of using the cold room is that the liquid nitrogen tank can be stored here[2] and there is no rise in temperature as the ampoule holder is placed in the tank. Ampoules can be left in the ampoule holder for the entire cooling operation; space limi-

---

[2] Storing a liquid nitrogen tank in a cold room is potentially dangerous since the liquid nitrogen volatilizes and gradually displaces the oxygen in a tightly closed room. In our cold room an exhaust fan is always in operation and prevents the buildup of nitrogen vapor. Another approach is to install a recording oxygen analyzer with an alarm.

tations might make it difficult to use the ampoule holder in a refrigerator. Also, the temperature is generally better controlled in a cold room.

3. With a constant and specific level of liquid nitrogen in the tank (approximately half-full), transfer ampoules to the ampoule carrier and insert base block down 3 inches from the top of the liquid nitrogen container. One ampoule should be broken to permit a thermometer (30° to −200°C) to be inserted. Between 0° and −10°C, a brief rise in temperature occurs. Note the time at which this occurs.

4. Repeat the above step in a second experiment. One minute before anticipated heat of fusion is released, lower the base block down an additional 3 inches into cold vapor. Continue to read temperature every minute until −70°C is reached.

5. Plot temperature vs time curve to see whether the rate of temperature fall is within an acceptable range. The lower limit is relatively unimportant, except to keep the freezing time from being inconveniently long. The upper limit ideally should be less than 2°C per minute from room temperature down to 0°C. A small increase in cooling rate can be expected in the eutectic region with the procedure described until freezing is underway at close to −10°C. While a very high rate of freezing may be tolerated or even desired for certain cell lines (viz., Chinese hamster cells), in the absence of such information for other cells the slower freezing rate is recommended for nucleated mammalian cells. These rates and the ones which follow down to −70°C should be kept less than 6°C per minute.

6. If cooling rates are too high, adjust position of ampoules to higher levels in liquid nitrogen container.

7. When an acceptable cooling curve is obtained, determine viability of cells before and after cryopreservation at −170 to −196°C using the trypan blue-dye exclusion test or other method of assessing viability. For most cell lines, the recovery rate with the dye exclusion test (difference between percent viability before and after freezing divided by percent viability before freezing) can be expected to be 80–95%.

8. For routine cryopreservation, dummy ampoules with medium alone can be substituted if fewer than 8 ampoules of cells are to be frozen in a run.

9. Thawing is carried out as rapidly as possible (less than 1 minute) by immersing frozen ampoules in water previously heated to 40°C.

VI. Summary

A modified cell freezer is described for use by cell biologists and others working with cell suspensions. It is simple, essentially cost-free to put together, permits cells to be frozen in cold nitrogen vapor to $-70°C$ in 45–60 minutes before final storage in liquid nitrogen or its vapor (-170 to $-196°C$), can be done without disturbing ampoules already in storage, gives reproducible cooling rates from one run to the next, and permits an average recovery rate of 95% viable cells. The freezer consists of a base block and wooden ampoule carrier which is lowered down to a controlled level in a liquid nitrogen tank. The base block can be cut to fit any tank that is available. A procedure is described for establishing holding levels during freezing to permit optimal freezing rates for effective cryopreservation. Excellent results have been obtained using cell lines (mouse C46 and N18 neuroblastomas, rat C6 glioma and RN-22 Schwann cell tumor, human CMP tumor, mouse B-16 melanoma) and primary cultures (chick embryo, rat myocardial cells, rat endothelial cells).

ACKNOWLEDGMENTS

We thank Philip Constantin and Mary Vaughn for laboratory assistance, Linda Garner for making the drawings in Fig. 1, and William Stallworth for the photographs.

The research has been supported by US Public Health Service Grants CA-12067 from the National Cancer Institute, HL-15103 (Specialized Center of Research) from the National Heart and Lung Institute, and P.H.S. Training Grant 5-TO1-DE-0024 from the National Institute of Dental Research.

REFERENCES

Bovis, R. J., and Kasten, F. H. (1965). *J. Ultrastruct. Res.* **13**, 567.
Diller, K. R., Cravalho, E. G., and Huggins, C. E. (1972). *Cryobiology* **9**, 429.
Dougherty, R. M. (1962). *Nature (London)* **193**, 550.
Green, A. E., Silver, R. K., Krug, M., and Coriell, L. L. (1964). *Proc. Soc. Exp. Biol. Med.* **116**, 462.
Hauschka, T. S., Mitchell, J. T., and Niederpruem, D. J. (1959). *Cancer Res.* **19**, 643.
Kalmbach, K., and Mardiney, M. R., Jr. (1972). *Cryobiology* **9**, 572.
Kasten, F. H. (1973). *In* "Tissue Culture: Methods and Applications" (P. F. Kruse, Jr. and M. K. Patterson, Jr., eds.) pp. 72–81. Academic Press, New York.
Kasten, F. H., and Yip, D. K. (1974). *In Vitro* **9**, 246.
Kite, J. H., Jr., and Doebbler, G. F. (1962). *Nature (London)* **196**, 591.
Leibo, S. P., Farrant, J., Mazur, P., Hanna, M. G., Jr., and Smith, L. H. (1970). *Cryobiology* **6**, 315.
Mazur, P., Farrant, J., Leibo, S. P., and Chu, E. H. Y. (1969). *Cryobiology* **6**, 1.
Mazur, P., Leibo, S. P., and Farrant, J., Chu, E. H. Y., Hanna, M. G., Jr., and Smith, L. H. (1970). Frozen Cell, *Symp., 1969* pp. 69–85.
Mazur, P., Leibo, S. P., and Chu, E. H. Y. (1972). *Exp. Cell Res.* **71**, 345.
Meryman, H. T. (1963). *Fed. Proc. Fed. Am. Soc. Exp. Biol.* **22**, 81.
Nagington, J., and Greaves, K. I. N. (1962). *Nature (London)* **194**, 993.
Perry, V. P., Kerby, C. C., Kowalski, F. J., and Malinin, T. I. (1965). *Cryobiology* **1**, 274.
Polge, C., and Lovelock, J. E. (1952). *Vet. Rec.* **64**, 396.
Rey, L. R. (1959), *Fed. Proc. Fed. Am. Soc. Exp. Biol.* **22**, 90.
Ryan, M. M., and Smith, K. O. (1974). *Appl. Microbiol.* **27**, 616.

Scherer, W. E., and Hoogasian, A. C., (1954). *Proc. Soc. Exp. Biol. Med.* **87**, 480.

Shannon, J. E., and Macy, M. L. (1972). "Registry of Animal Cell Lines," 2nd ed., p. 20. Am. Type Cult. Collect., Rockville, Maryland.

Shannon, J. E., and Macy, M. L. (1973). *In* "Tissue Culture: Methods and Applications" (P. F. Kruse, Jr. and M. K. Patterson, Jr., eds.), pp. 712–718. Academic Press, New York.

Stulberg, C. S., Peterson, W. D., Jr., and Berman, L. J. (1962). *Natl. Cancer Inst., Monogr.* **7**, 17.

Walker, P. J., and Wilen, M. J. (1967). *Lab Pract.* **16**, 480.

Wollenberger, A. (1967). *In* "Factors Influencing Myocardial Contractility" (R. D. Tanz, R. Kavaler, and J. Roberts, eds.), pp. 317–327. Academic Press, New York.

Chapter 16

Long-Term Preservation of the Ehrlich Ascites Tumor[1]

WILLIAM A. CASSEL

Department of Microbiology, Division of Basic Health Sciences, Emory University, Atlanta, Georgia

I. Introduction

Ascites tumors, particularly of mice, have been employed in a remarkably wide variety of research endeavors, and a major symposium (Miner, 1956) attests to their usefulness. Prominent among these tumors is the Ehrlich ascites tumor of the mouse, which has become to mammalian cell research almost what *Escherichia coli* is to microbiology. Associated with its extensive use is a need for availability, in a preserved state. Soon after cell preservation techniques began to be developed, it was concluded that a cooling rate of 1°C per minute is best and that a temperature below $-130°C$ is indicated for long-term storage (Meryman, 1963). Implicit in this is the requirement for elaborate cryophilic equipment. The subject of this chapter is a simple method, within the reach of most laboratories, by which the Ehrlich ascites tumor can be preserved in a frozen state for many years.

[1] Publication No. 1317, Division of Basic Health Sciences, Emory University.

II. Initial Studies with Glycerol

The feasibility of utilizing glycerol as a preserving agent became evident from the studies of Polge *et al.* (1949) and others (Craigie, 1954; Lovelock, 1954; Lucké and Parapart, 1954; Parkes, 1954; Scherer and Hoogasian, 1954; Morgan *et al.*, 1956). Cassel (1957a) reported maintaining the Lettré, hyperdiploid, Ehrlich ascites tumor at − 78° C for 1 year, after adding glycerol to a final concentration of 20%. Additional reports confirming the value of glycerol in storing the Ehrlich tumor, and other cells, soon followed (Swim *et al.*, 1958; Hauschka *et al.*, 1959; Sugiura, 1962).

The principal findings of Cassel (1957a) are summarized in Table 1. When no glycerol was added to the tumor, the frequency of recovery improved as the storage temperature was lowered, and initial, rapid freezing in alcohol–Dry Ice (− 72° C) did not increase the survival time of cells stored at − 78° C. At a storage temperature of − 27° C, and with 20% glycerol added, no difference was found in the survival times of cells frozen directly or slowly,

TABLE I

RECOVERY OF TUMOR AFTER VARIOUS TREATMENTS[a,b]

Storage and treatment[c]	Maximal recovery time	Percentage of inoculated mice showing tumors at maximal recovery time[d]
No glycerol		
Directly to 4° C	2 Weeks (lost at 3 weeks)	100
Directly to − 27° C	6 Weeks (lost between 6 and 8 weeks)	66
Directly to − 78° C	21 Weeks (lost between 21 and 31 weeks)	33
Alcohol freeze (− 72° C), then directly to − 78° C	10 Weeks (lost between 10 and 21 weeks)	17
20% glycerol		
Directly to − 27° C	11 Weeks (lost between 11 and 25 weeks)	66
Slowly to − 27° C	11 Weeks (lost between 11 and 25 weeks)	33
20% glycerol		
Directly to − 78° C	52 Weeks (not tested further)	86 to 100
Slowly to − 78° C	52 Weeks (not tested further)	43
Alcohol freeze (− 72° C), then directly to − 78° C	10 Weeks (lost between 10 and 21 weeks)	50

[a] Modified from Cassel (1957a), reprinted by permission of The Williams & Wilkins Co.
[b] The inoculum per mouse was 0.4 ml. (4 × 10⁷ cells) of the Lettré, hyperdiploid, Ehrlich ascites tumor.
[c] Rapid, alcohol freezing involved placing the sample in an alcohol–Dry-Ice bath (− 72° C) for 10 minutes. "Slowly" indicates cooling at 1° C per minute.
[d] The animals were held under observation for 10 weeks.

although testing at shorter intervals probably would have shown that the directly frozen specimens survived longer. Similar observations were made on cells stored at $-78°C$, and rapid freezing in alcohol–Dry Ice ($-72°C$) was not beneficial. It is evident that glycerol markedly increased the cell survival at $-27°C$ and $-78°C$. Glycerol was found to be more effective at 20% than at 10% and 30%.

The hyperdiploid ($2\times$) tumor was frozen after 8–10 days of development in the mouse. A rationale for selecting this age of tumor may be seen from the data in Fig. 1. Besides a practical volume requirement, it may be noted that the tumor cells decrease after 11 days, suggesting a degenerative process at this time. In the case of the more rapidly growing hypotetraploid tumor the decline is seen after 10 days. Such data undoubtedly will vary with the tumor line, the inoculum size, and the strain, age, and sex of the mouse.

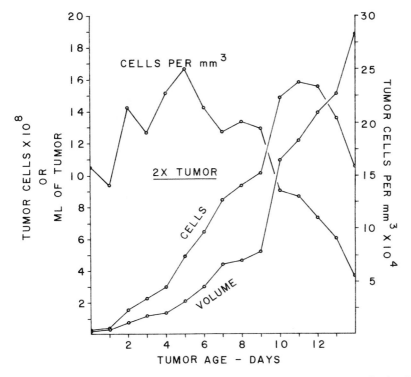

FIG. 1. Growth characteristics of the hyperdiploid, Ehrlich ascites tumor, following the inoculation of mice with 4×10^7 tumor cells. The data show: (a) the change in tumor volume, (b) the change in tumor cells per mm³, (c) the change in total tumor cells. From Cassel (1957b), reprinted by permission of The Williams & Wilkins Co.

III.   Prolonged Storage with Glycerol

A.   A Recommended Procedure

The preceding findings of Cassel (1957a) indicated that the Ehrlich ascites tumor could be held in a recoverable, frozen state for 1 year. Recently, Cassel and McCaskill (1974) reported that the hyperdiploid tumor can be recovered after 15 years. From this study, the following method can be recommended.

1. Add 1 ml of a 7-day-old tumor (2×10^8 cells) to 1 ml of 40% glycerol, in a 13×100 mm test tube. (Reagent-grade glycerol is diluted with tissue-culture distilled water and sterilized by the use of a Millipore filter.)

2. Seal the tube with a rubber stopper, mix the contents, and place at $4°C$ for 1 hour. Mix the contents again in 30 minutes.

3. Leave the tube 1 hour at $4°C$, resuspend the cells, and carry the tube directly to a freezer at $-60°C$.

4. For recovery, the specimen is thawed by rotating it in a water bath at $40°C$. This requires about 45 seconds.

5. Inoculate each recipient mouse with 0.4 ml of the specimen.

Tumor recovery is rapid, and marked tumor development becomes evident in 1–2 weeks. Some samples may show clumps of cells, which are difficult to disaggregate. This probably arises from fibrin in the ascitic plasma, but it has no obvious effect on tumor recovery. A large-gauge needle (20 gauge) may be needed for the inoculation of such material. Very likely, the addition of an anticoagulant to freshly aspirated tumor would circumvent the difficulty, without adversely affecting tumor recovery.

Unpublished investigations (W. A. Cassel) have resulted also in the recovery of the hypotetraploid Ehrlich ascites tumor (E2 clone), after preservation by the method described above. The oldest available frozen stocks, 12 and 13 years of age, were readily recovered within 2 weeks, after inoculation into mice. In this recovery, and in the case of the hyperdiploid tumor, male and female mice were inoculated with each specimen. Tumors became apparent in the male mice about 5 days earlier than in the females. This observation is in accord with the conclusion of Hartveit (1963) that females tend to be naturally more resistant to the tumor than are males.

B.   Changes during Storage

An in-depth study of the properties of recovered Ehrlich tumors was not performed by Cassel. In hyperdiploid tumor cells recovered from the third mouse passage of material frozen for 15 years, however, the size range, general cytology, and modal chromosome count were the same as prior to freezing (Cassel and McCaskill, 1974). Reports of Morgan *et al.* (1956), Hauschka *et al.* (1959), Ising (1960), and Holdridge and Hauschka (1974) state that the modal chromosome numbers of Ehrlich tumors did not change after the cells had been frozen for periods ranging from 6 months to 9 years. Chromosome morphology and frequency also were found to be stable. Thus, the indications are that the recovered tumor has essentially the same properties as before freezing.

IV. Stains As Viability Indicators

Viability staining with eosin gave little indication of the potential for the recovery of frozen tumor (Cassel, 1957a). Specimens showing 100% takes in inoculated mice, and rapid development, contained only 0.2–1.3% "viable" cells, according to the staining method. When tumor cells were heated at 55°C for 5 minutes, staining indicated that 76% of the cells were still viable. Tumor, however, could not be recovered from mice inoculated with this material. The unreliable nature of eosin and trypan blue viability staining has been reported by other investigators (Hoskins *et al.*, 1956; Hauschka *et al.*, 1959; Craven, 1960; Bouroncle, 1965). One should, therefore, be forewarned of the pitfalls in such procedures.

V. Speculations on Improving the Preservation

For reasons unknown, the generally recommended cooling at 1°C per minute is not necessary in the case of the Ehrlich tumor. This may relate to properties of the surrounding ascitic fluid and to peculiarities of the cell membrane. The surface tension of the plasma is somewhat on the low side, being 56 dynes/cm at 25°C, and a sturdy nature of the membrane is suggested by the fact that the cell is refractory to breakage by various physical methods.

Preservatives other than glycerol have been employed with the Ehrlich tumor. Based on oxygen consumption, McKee and McCarty (1973) found sucrose to be a more effective preservative than glycerol. Dimethyl sulfoxide (DMSO), first used by Lovelock and Bishop (1959), has been described as having superior cell-penetrating and preserving properties (Meryman, 1963; Klein and Trapani, 1965). Toxicity, however, has been reported for this compound (Bouroncle, 1965). Miya *et al.* (1967) found

DMSO to be a better preservative than glycerol for the Ehrlich ascites tumor. The less than $-130°C$ level, indicated for very long storage of cells (Meryman, 1963), is another variable worthy of consideration. It appears, therefore, that, while the method of Cassel (1957a; Cassel and McCaskill, 1974) is simple and adequate, the future will bring further improvements.

REFERENCES

Bouroncle, B. A. (1965). *Proc. Soc. Exp. Biol. Med.* **119**, 958.
Cassel, W. A. (1957a). *Cancer Res.* **17**, 48.
Cassel, W. A. (1957b). *Cancer Res.* **17**, 618.
Cassel, W. A., and McCaskill, K. M. (1974). *Appl. Microbiol.* **28**, 726.
Cassel, W. A., Blair, W. L., and Garrett, R. E. (1962). *Proc. Soc. Exp. Biol. Med.* **110**, 89.
Craigie, J. (1954). *Adv. Cancer Res.* **2**, 197.
Craven, C. (1960). *Exp. Cell Res.* **19**, 164.
Hartveit, F. (1963). *Acta Pathol. Microbiol. Scand.* **58**, 25.
Hauschka, T. S., Mitchell, J. T., and Niederpruem, D. J. (1959). *Cancer Res.* **19**, 643.
Holdridge, B. A., and Hauschka, T. S. (1974). *Cancer Res.* **34**, 663.
Hoskins, J. M., Meynell, G. G., and Sanders, F. K. (1956). *Exp. Cell Res.* **11**, 297.
Ising, U. (1960). *Exp. Cell Res.* **19**, 475.
Klein, I., and Trapani, R. J. (1965). *Fed. Proc. Fed. Am. Soc. Exp. Biol.* **24**, Suppl. S-108.
Lovelock, J. E. (1954). *Proc. R. Soc. Med.* **47**, 60.
Lovelock, J. E., and Bishop, M. W. H. (1959). *Nature (London)* **183**, 1394.
Lucké, B., and Parapart, A. K. (1954). *Cancer Res.* **14**, 75.
McKee, R. W., and McCarty, B. (1973). *Proc. Soc. Exp. Biol. Med.* **143**, 50.
Meryman, H. T. (1963). *Fed. Proc. Fed. Am. Soc. Exp. Biol.* **22**, 81.
Miner, R. W. (1956). *Ann. N. Y. Acad. Sci.* **63**, 637.
Miya, F., Hill, G. A., and Marcus, S. (1967). *Appl. Microbiol.* **15**, 451.
Morgan, J. F., Guerin, L. F., and Morton, H. J. (1956). *Cancer Res.* **16**, 907.
Parkes, A. S. (1954). *Lect. Sci. Basis Med.* **2**, 250.
Polge, C., Smith, A. U., and Parkes, A. S. (1949). *Nature (London)* **164**, 666.
Scherer, W. F., and Hoogasian, A. C. (1954). *Proc. Soc. Exp. Biol. Med.* **87**, 480.
Sugiura, K. (1962). *Cancer Res.* **21**, 496.
Swim, H. E., Haff, R. F., and Parker, R. F. (1958). *Cancer Res.* **18**, 711.

Chapter 17

Rapid Biochemical Screening of Large Numbers of Animal Cell Clones

MICHAEL BRENNER,[1] RANDALL L. DIMOND,[2] AND
WILLIAM F. LOOMIS[3]

I.   Introduction

Analysis of mutant strains has consistently played a key part in deciphering metabolic patterns operating in simple organisms such as bacteria and yeast. The use of mutants in studies of more complex organisms has been limited, however, primarily because of the difficulties involved in obtaining the desired mutants. Many of these difficulties can be eliminated by working with animal cells in culture rather than with whole organisms. But even so, only in certain instances has it been possible to devise selection procedures for mutant isolation (for reviews, see Basilico and Meiss, 1974; Kao and Puck, 1974; Naha, 1974; Shapiro and Varshaver, 1975); many mutant types still cannot be isolated by selective techniques because the function of the gene of interest is unknown, or because, as for most developmental functions, the activity of the gene is not important for the maintenance of the cell in culture. In these latter cases mutants can be conveniently obtained only

[1] Biological Laboratories, Harvard University, Cambridge, Massachusetts.
[2] Department of Biology, Massachusetts Institute of Technology, Cambridge, Massachusetts.
[3] Department of Biology, University of California, San Diego, La Jolla, California.

if efficient screening methods are available. In a recent article (Brenner *et al.*, 1975) we presented a procedure that should help meet this need in many cases. We described the construction and operation of a "multipipette" which can be used to clone cells into plastic multitest trays having 96 individual wells. Using this simple apparatus a team of two persons can fill more than 400 trays in an hour (40,000 wells). Once clones have grown in the wells, the multipipette can be used to add reagents for colorimetric enzyme assays. We have used this procedure successfully to screen for several developmental mutants of the cellular slime mold *Dictyostelium discoideum* (Dimond *et al.*, 1973, 1976; Dimond and Loomis, 1976; Free and Loomis, 1974). In this chapter we discuss more fully the application of this methodology to obtaining mutants of animal cells.

II. Mutagenesis and Mutation Frequency

In order for mutants to appear with a sufficient frequency that they may be isolated by screening procedures, it is necessary to mutagenize the cells. In our work with *Dictyostelium discoideum*, we found that mutagenizing to a survival rate of 0.1% yielded gene-specific mutants with a frequency of about 10^{-3} (Dimond *et al.*, 1973; Loomis *et al.*, 1977). Since this organism has a small DNA content and is haploid, one might question whether gene-specific mutations may be obtained with a similar frequency in more complex diploid animal cell lines. Fortunately, however, neither the DNA content nor the diploid state of animal cells influences the frequency of mutations that will be found for a given level of survival; the only relevant parameter is the number of genes essential for cell viability. For example, suppose that 10^4 genes are essential for cell growth. [There are no data available on the number of genes essential for viability of any animal cell line. Genetic arguments have been made to suggest that for some higher animals (e.g., man) the number of genes essential for the entire organism is on the order of 10^5 (reviewed in Lewin, 1974). The estimate here of 10^4 genes essential for a single cell type therefore seems reasonable, especially in view of the fact that many simple eukaryotes, and even nematodes and *Drosophila* survive with no more than 5×10^3 genes (Bishop, 1974; Brenner, 1974).] From the Poisson equation one can calculate that mutagenesis to a survival rate of 10^{-3} yields an average number of 6.9 lethal mutations per cell. [This assumes that killing is due to mutational events. This assumption led to reasonably accurate estimates for mutation frequencies in our studies of *D. discoideum* using *N*-methyl-*N'*-nitro-*N*-nitrosoguanidine (Dimond *et al.*, 1973; Loomis *et al.*, 1977), but may not hold for other mutagens or other cell systems

(Hsie *et al.*, 1975).] Thus the mutation frequency per gene is $6.9/10^4$. If all genes are mutated independently and at the same frequency, then among the surviving clones any given nonessential gene will be mutated also with a frequency of 6.9×10^{-4}. Hence one mutant of interest should be isolated from about every 1500 clones screened.

If, as is commonly supposed, expression of recessive mutations in diploid cells requires two independent mutational events, mutations leading to cell death would be primarily dominant mutations, or recessive mutations in genes present in only one functional copy per cell, such as on X chromosomes. Therefore the mutation frequency calculated above would refer primarily to these two classes of genes, and only these classes would yield mutations which could be readily isolated by the screening procedure. However, in the several cases where the effect of gene dosage on mutation frequencies has been determined, additional copies of a given recessive gene have decreased its mutation frequency by less than 1% of that predicted (Chasin and Urlaub, 1975). A suggested explanation for this departure from expectation is the occurrence of mitotic crossing over (Chasin and Urlaub, 1975), so it may be advisable to grow the mutagenized cells for several generations before cloning. Regardless of the actual mechanism operating, however, the practical consequence of this phenomenon is that mutations in recessive genes should occur at about the frequency given above, and so should be present in sufficiently high numbers to permit their isolation by screening procedures.

Often mutants are sought that are defective for a particular protein or RNA in order to determine the role of that molecule in some cellular process. These studies are complicated by the use of mutagenesis in that each clone isolated will carry multiple lesions. In the above example, should the cell line have 10^5 nonessential genes, mutant clones isolated would on the average carry about 69 mutations. [No direct data are available on the number of genes expressed in animal cell lines. Based on nucleic acid hybridization experiments using messenger RNA the number of genes being translated into protein has been estimated to be up to 3×10^4 for mouse L cells and HeLa cells (reviewed in Lewin, 1975). In tissues composed of mixtures of cell types, hybridization using total RNA (nuclear and cytoplasmic) has been used to obtain estimates of 4 to 8×10^4 genes being active in mouse liver, kidney, and spleen, and 3×10^5 in mouse brain (reviewed in Davidson and Britten, 1973). The value of 10^5 genes being active in a given cell line seems therefore a reasonable upper limit value.] Caution must be exercised, therefore, in assigning responsibility for any phenotype of a mutant to the gene whose product is being screened. This problem is partially alleviated by obtaining several independent mutants. The probability that one unscreened for mutation present in one strain will also be present in another is quite small; it is simply the mutation frequency itself, 6.9×10^{-4} in the above example. Hence if several independently isolated mutants all display the same phenotype, there is good assurance that the mutation causing change in the gene product being screened is responsible. However, this change need not be in the gene specifying that product, it could be in some other component affecting both the screened property and the phenotypic change, such as a processing enzyme. Thus no causal relationship need exist between the screened product and the phenotype. This problem of pleiotropy can be largely eliminated through use of conditional mutants. Correspondence of the conditionality of the screened product *in vitro* (e.g., temperature sensitivity of an enzyme), and of the associated phenotype *in vivo* strongly implies that the alteration in the screened product produces the phenotypic property noted. Conditional mutants are a necessity for obtaining strains defective in essential gene functions.

III. Screening

The screening procedure we have developed is particularly well adapted for obtaining mutants defective in enzymic activities which can be assayed colorimetrically; reagents may be added to the colonies with the multipipette, and enzyme activities be determined by visual screening. For multistep assay procedures, as many as five additions may be made using a multipipette that delivers 50-μl drops. As indicated in Table I, a surprisingly large number of enzymes may be assayed colorimetrically. For any specific enzyme and cell type it is necessary to ascertain whether the amount of enzyme produced by the cells in a multitest well is sufficient to permit its assay by the colorimetric procedure. In many cases it may be possible to make the assays extremely sensitive by employing cycling procedures analogous to those of Lowry (1962–1963).

Before conducting some of the enzyme assays it may be necessary to wash the cells free of their growth medium. This has not been possible with *D. discoideum* since the cells adhere poorly to the plastic trays; the firm adherence of most animal cell lines, however, should make washing an easy matter. The trays can probably be immersed in a large container of buffer and the wash liquid shaken or aspirated from the wells (a multipipette with narrow tubing can be used to aspirate the liquid). For some enzyme assays it is necessary to lyse the cells to obtain good activity. This may be done by adding detergent or organic solvents (e.g., Triton X-100, toluene) with the assay mixture. In these cases it will first be necessary to replicate the cells to

TABLE I

ENZYMES THAT CAN BE ASSAYED COLORIMETRICALLY[a]

Enzyme class	Color reaction
ADP producing enzymes (e.g., kinases)	Linked to NBT[b] reduction through pyruvate kinase
ATP-producing enzymes	Linked to NBT reduction through hexokinase and glucose-6-phosphate dehydrogenase
Aldolases	Linked to NBT reduction through NADH
Ammonia-producing enzymes (e.g., deaminases, amidases)	Reduction of NBT; linked to NBT reduction through glutamic dehydrogenase; indophenol formation from NH_4
Coenzyme A-producing enzymes	Formation of 2-nitro-5-mercaptobenzoate from 5,5'-dithiobis(2-nitro benzoate)
Dehydrogenases	Reduction of NBT by NADH or NADPH
Dihydroxyacetone phosphate-producing enzymes	Linked to NBT reduction through α-glycerol phosphate dehydrogenase
Esterases	Many colorimetric substrates, including napthol derivatives; many products can be assayed by the addition of color reagents
Glucose, glucose 1-phosphate, or glucose 6-phosphate-producing enzymes	Linked to NBT reduction through NADPH
Glutamate-producing enzymes	Reduction of NBT through glutamate dehydrogenase
Glyceraldehyde 3-phosphate-producing enzymes	Linked to NBT reduction through α-glycerol phosphate dehydrogenase
α-Glycerol phosphate-producing enzymes	Linked to NBT reduction through α-glycerol phosphate dehydrogenase
Glycosidases	p-Nitrophenol, naphthol derivatives, or methylumbelliferyl-linked substrates
Hydroxylases	Reduction of NBT by NADH or NADPH
Nucleotide sugar:Acceptor transferases	Linked to NBT reduction by pyruvate through nucleotide kinase and pyruvate kinase
Nucleotide sugar pyrophosphorylases	Linked to NBT reduction through NADH, NADPH, or pyruvate
Oxidases and oxygenases	Linked to NBT reduction
Peroxidases	Guaiacol; di-O-aniside
Phosphate-producing enzymes (e.g., phosphatases)	p-Nitrophenyl phosphate; formation of colored phosphomolybdate complex
Proteases	Colorimetric esterase assays; assay released tyrosine with Folin-Ciocalteau reagent
Pyruvate-producing enzymes	Reduction of NBT
Reductases	Reduction of NBT by NADH or NADPH
Transaminases	Formation of hydrazones or ninhydrin derivatives of keto acids; linked to NBT reduction through NADH or NADPH oxidation

(cont.)

TABLE I (*cont.*)

Enzyme class	Color reaction
Specific enzymes not necessarily subsumed under the above classes include the following:	
N-Acetylglucosamine-2-epimerase	Glycerol dehydratase
N-Acetylglucosamine-6-phosphate deacetylase	Glycogen synthetase
Aconitase	Guanosine-5'-phosphate pyrophosphorylase
Aldose reductase	Hyaluronidase
α-Amylase	Imidazolylacetolphosphate: glutamate amino-
Arabinose isomerase	transferase
Arylsulfatase	L-2-Keto-3-deoxyarabonate
Argininosuccinate synthetase	dehydratase
Carbonic anhydrase	Neuraminidase
Citrate-cleavage enzyme	NAD Kinase
Condensing enzyme	NAD Pyrophosphorylase
Creatine phosphokinase	Ornithine carbamoyltransferase
Crotonase	Phenol sulfokinase
γ-Cystathionase	Phosphodiesterase
Cytochrome oxidase	Phosphoribosylformimino-PRAIC
5-Dehydroquinase	ketolisomerase
Dihydrodipicolinic acid synthetase	Proline oxidase
	Rhodanese
Fructose-1,6-diphosphatase	Ribulose-5-phosphate-4-
Fumarase	epimerase
Glucosamine-6-phosphate deaminase	Serine transhydroxymethylase
	Thiol-disulfide transhydrogenase
Glucosamine phosphate isomerase	
Glutamine synthetase	Tyrosinase
γ-Glutamyltranspeptidase	UDP-N-acetylglucosamine-2'-epimerase; UDPG-4-epimerase; UDPG: fructose transglucosylase; UDP-Glucuronyltransferase

[a] Culled from various volumes of "Methods in Enzymology" (S. P. Colowick and N. O. Kaplan, eds.). Academic Press, New York; and "Methods of Enzymatic Analysis" (H. U. Bergmeyer, ed.). Academic Press, New York.

[b] NBT = nitro blue tetrazolium. It is reduced to form a mauve precipitate by NADH, NADPH, and several keto acids. It can be used to visualize either the oxidation or reduction of nicotinamide adenine dinucleotides. It may be possible to increase sensitivity by using the dinucleotide generated in the primary reaction to catalyze reduction of NBT in a second reaction in the presence of an excess of an oxidizable substrate (Lowry, 1962–1963).

a second set of trays. In our prior article (Brenner *et al.*, 1975) we described a replicator that can be used for this purpose.

In addition to colorimetric screening, assays employing radioactive substrates may be feasible in some instances; procedures have been described for flushing cells from microtest wells with trichloroacetic acid and collecting the precipitate or supernatent for scintillation counting (Harrison *et al.*,

1974). The applicability of this methodology might be expanded by using a suitable buffer to harvest the cell assay mixture followed by batch processing of the supernatant with ion-exchange resin.

Finally, screening need not be limited to enzymic activities; for example, morphological variants may also be isolated. One of us (W. F. Loomis) has used cloning of mutagenized muscle cells into multitest trays to isolate mutants defective in fusion (Loomis *et al.*, 1973).

IV. Closing Comments

The methodology we describe here allows any clonable cell line to be rapidly and inexpensively screened for mutations. The initial investment for the multipipette, replicator, and microtest plates is small, and all may be re-used repeatedly (we resterilize the microtest plates with ethylene oxide, facilities for which are available at most medical schools and hospitals). From our experience with the screening protocol, we estimate it should take under 40 man-hours to process 10,000 animal cell clones. This includes the time required for mutagenesis, cloning, replicating, washing, adding substrates, and screening. Above we calculated that only about 1500 clones should have to be screened to obtain any desired mutant. Even if this frequency is low by a factor of 10, it would still require but a few man-weeks of labor to obtain the mutants. Thus through this technique it should be possible to greatly expand the range of mutant types that can be isolated for the study of animal cells in culture.

REFERENCES

Basilico, C., and Meiss, H. K. (1974). *Methods Cell Biol.* **8**, 1–22.
Bishop, J. O. (1974). *Cell* **2**, 81–86.
Brenner, M., Tisdale, D., and Loomis, W. F., Jr. (1975). *Exp. Cell Res.* **90**, 249–252.
Brenner, S. (1974). *Genetics* **77**, 71–94.
Chasin, L. A., and Urlaub, G. (1975). *Science* **187**, 1091–1093.
Davidson, E. H., and Britten, R. J. (1973). *Q. Rev. Biol.* **48**, 565–613.
Dimond, R. L., and Loomis, W. F. (1976). *J. Biol. Chem.* **251**, 2680–2687.
Dimond, R. L., Brenner, M., and Loomis, W. F., Jr. (1973). *Proc. Natl. Acad. Sci. U.S.A.* **70**, 3356–3361.
Dimond, R. L., Farnsworth, P., and Loomis, W. F. (1976). *Dev. Biol.* **50**, 169–181.
Free, S., and Loomis, W. F. (1974). *Biochimie* **56**, 1525–1528.
Harrison, M. R., Thurman, G. B., and Thomas, G. M. (1974). *J. Immunol. Methods* **4**, 11–16.
Hsie, A. W., Brimer, P.A. , Mitchell, T. J., and Gosslee, D. G. (1975). *Somatic Cell Genet.* **1**, 247–261.
Kao, F., and Puck, T. T. (1974). *Methods Cell Biol.* **8**, 23–39.
Lewin, B. (1974). "Gene Expression," Vol. 2, pp. 148–149. Wiley, New York.

Lewin, B. (1975). *Cell* **4**, 77–93.

Loomis, W. F., Wahrman, P., and Luzzati, D. (1973). *Proc. Natl. Acad. Sci. U.S.A.* **70**, 425–429.

Loomis, W. F., Dimond, R. L., Free, S. J., and White, S. (1977). *In* "Microbes as Model Systems for Development" (D. O'Day and P. Horgan, eds.). Marcel Dekker, New York.

Lowry, O. H. (1962–1963). *Harvey Lect.* **58**, 1–19.

Naha, P. M. (1974). *Methods Cell Biol.* **8**, 41–46.

Shapiro, N. I., and Varshaver, N. B. (1975). *Methods Cell Biol.* **10**, 209–234.

Chapter 18

Use of Polyacrylamide for Cloning of Primary Tumors[1]

TOBI L. JONES AND J. STEPHEN HASKILL

*Department of Surgery, Michigan State University, East Lansing, Michigan; and
Department of Basic and Clinical Immunology and Microbiology, Medical
University of South Carolina, Charleston, South Carolina*

I. Introduction

One of the first requirements in studying the tumor biology of various neoplasias is the establishment of a line of tumor cells adapted to tissue culture. This may be more or less of a problem, depending on the type of tumor to be studied. Solid tumors have consistently presented problems since they contain stromal elements (Lesfargues, 1973; Owens and Hackett, 1972) and, often, host inflammatory elements (Evans, 1972). The viability of cells obtained from solid tumors may also present a problem, depending on the availability of nonnecrotic tumor material and the method used for preparing a cell suspension.

[1] This is publication No. 12 from the Department of Basic and Clinical Immunology and Microbiology, Medical University of South Carolina.

The outgrowth of contaminating fibroblasts once the tumor has been placed in tissue culture presents a major difficulty. Typically, both tumor cells and fibroblasts grow in a monolayer; the elimination of fibroblasts is attempted after both cell populations have proliferated. Since the rapidity with which tumor cells and fibroblasts attach to plastic may differ (Lesfargues, 1973) as well as their susceptibility to detachment by enzymic treatment (Owens and Hackett, 1972), a means is provided for separating the two populations during passage in culture.

We have attacked the problem of fibroblast contamination at the outset by eliminating their proliferative capacity. In the course of studying the use of polyacrylamide dishes for immunological tests of resistance to tumors, we used the dishes for developing lines of mouse tumor cells for *in vitro* tests. No difficulty was created by the outgrowth of fibroblasts, in contrast to our experience when plastic petri dishes were used for primary explants. We, therefore, examined the possibility that the polyacrylamide surface might be ideal for the establishment of tumor cell lines. We were successful in establishing a number of different mouse tumors in culture with this method, including mammary adenocarcinomas, a melanoma and fibrosarcomas (Jones and Haskill, 1973).

An additional problem in explanting tumor tissue into culture is the presence of host inflammatory elements. Many experimental solid tumors have been found to possess considerable numbers of macrophages and lymphocytes (Evans, 1972; Haskill *et al.*, 1975; Kerbel *et al.*, 1975). In the process of isolating cells from the tumor mass, inflammatory cells can be activated and can kill the tumor cells once they are placed in culture (Evans, 1972; Haskill *et al.*, 1975). Based on their difference in size, tumor cells and host mononuclear cells can be separated by physical means, thereby allowing a greater chance of survival and growth of the tumor cells in culture.

This chapter will describe the method of establishing tumor lines on polyacrylamide dishes, including the additional step of removing host inflammatory elements prior to placing cells in culture.

II. Methods

A. Polyacrylamide Dishes

All components of the solutions for making polyacrylamide dishes are available from Eastman Organic Chemicals, Rochester, New York. Since the final product is dialyzed extensively in water, the use of electrophoresis-grade reagents is not necessary. The solutions are made up as follows: Solution A: 36.6 gm of Trizma base in 48 ml of 1 N HCl, made up to 100 ml with H_2O, plus 0.3 ml of N,N,N',N'-tetramethylethylenediamine; Solution B: pyrogen-free distilled H_2O; Solution C: 28 gm of acrylamide in 100 ml of H_2O plus 1.47 gm of N,N'-methylenebisacrylamide; Solution D: 0.14 gm ammonium persulfate in 100 ml of distilled H_2O. Solutions A and B are mixed in equal volume. The polyacrylamide dishes are made by mixing the solutions together in a ratio of 12 parts of A + B, 9 parts C, and 15 parts D.

The polyacrylamide dishes, a variation of those used by Marbrook and Haskill (1974), are formed in custom-made plastic molds (Professor L. L. Campbell, Free Piston Development, Kingston, Ontario). After sufficient solution is added to the mold, a piece of glass is set on the top of the mold to remove excess fluid and to permit the dishes to acquire a flat bottom upon polymerization of the gel. The covered mold can be heated with a light bulb or the ratio of persulfate increased to accelerate the polymerization process. Once the process has been completed (approximately 15 minutes), the glass plate is removed from the mold, and the formed dishes either come off with the plate or can be removed from the mold with the flat end of a weighing spatula. The dishes are dialyzed in several changes of water overnight, autoclaved in isotonic saline, and stored at 4°C. The dishes are equilibrated in two changes of tissue culture medium one day before use. Each polyacrylamide dish is placed in a 35 × 100 mm plastic petri dish, and excess medium is removed with a Pasteur pipette before the addition of cells. The final product can be seen in Fig. 1.

FIG 1. Polyacrylamide dish used for the culture of tumor cells; contained in 35 × 100 mm petri dish.

B. Medium

The medium used for all our studies was CMRL 1066 (GIBCO) supplemented with 10% fetal calf serum, glutamine, pyruvate, nonessential amino acids, and antibiotics. Any tissue culture medium that supports cell growth should be adequate until the growth requirements of the particular tumor are established.

C. Preparation of Tumor Cells for Culture

We have used two methods for preparing solid tumors for culture. The first, an enzymic digestion of tumor chunks, is preferable for obtaining a cell suspension with good viability. For this method, solid tumors are removed from animals asceptically and are minced with scissors in tissue culture medium. The chunks are placed in a sterile flask containing 10–15 ml of a 0.25% trypsin solution (Sigma, St. Louis, Missouri) in Eagle's minimum essential medium, and a magnetic stirring bar. After gently stirring for 1–2 hours at room temperature, the single cells are removed, and 1–2 ml of a serum source is added to inhibit the action of the trypsin. The addition of a small quantity of DNase (100 μg) will hinder clumping that may occur. The cell suspension is then centrifuged and resuspended in tissue culture medium.

The second method for obtaining tumor cells involves the gentle teasing of clumps of tumor with small forceps into the medium, and pipetting with a Pasteur pipette to break up small clumps. The viability of cells obtained by this method tends to be low. A variation of this approach, as described by Lesfargues (1973), involves the slicing of firmly held tumor chunks very thinly, allowing tumor cells to spill into the medium. This technique can only be used on firm but noncalcified tumors. The spilled cells are then collected for culture. Cells obtained by this method have good viability (Lesfargues, 1973).

D. The Culturing of Tumor Cells

Following any of the above methods, the approximate viability of the cells is determined by dye exclusion. A 0.03–0.05 ml aliquot of the suspension (one drop from a small Pasteur pipette contains approximately 0.03 ml) containing 1 to 5×10^4 viable cells is added to each polyacrylamide dish. The cells are allowed to settle onto the surface for 30 minutes, and then 1.5 ml of tissue culture medium is added to the outer petri dish. The plates are incubated in a humidified atmosphere of 95% air, 5% CO_2. The cultures can be examined twice weekly, and fresh medium should be added every

1–1.5 weeks to the outer dish and to the polyacrylamide dish. Without disturbing the settled tumor cells, medium can be gently withdrawn from the polyacrylamide dish at its edge with a Pasteur pipette and replaced with fresh drops of medium.

When the tumor cells are established and growing, cultures are transferred to fresh polyacrylamide dishes at weekly intervals or sooner if demanded by rapid growth. The contents of one polyacrylamide dish are gently agitated and removed with a Pasteur pipette. Sufficient medium is added to the cells to allow transfer to 4 polyacrylamide dishes, thus achieving a 1:4 dilution.

The ability of fibroblasts to survive and proliferate under these culture conditions can be seen in Fig. 2. Human fibroblasts added to polyacrylamide dishes did not proliferate as did parallel samples added to the adjacent petri dish surface. However, they did retain their viability, and we have observed that fibroblasts maintained on the polyacrylamide surface for several days will proliferate when transferred to a plastic surface.

After 3–4 passages in polyacrylamide dishes, the tumor cells can be transferred to plastic petri dishes for continued maintenance. If continual proliferation of the tumor cells has ensued, their numbers should be greatly increased over that of the fibroblasts, which have been essentially diluted

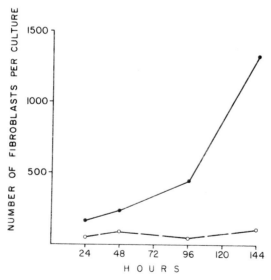

FIG. 2. Comparison of growth of human fibroblasts in polyacrylamide dishes and petri dishes. The number of viable fibroblasts in the polyacrylamide dishes (O––O) were counted on the basis of eosin dye exclusion. Well-formed, nucleated cells stained with Wright's stain were counted in the petri dishes (●—●). From Jones and Haskill (1973), by permission.

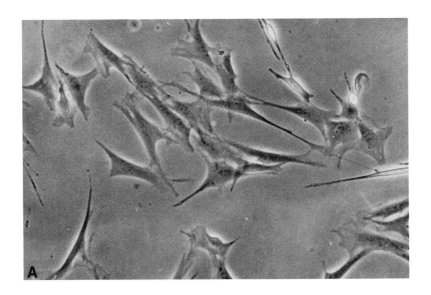

Fig. 3A.   See facing page for legend.

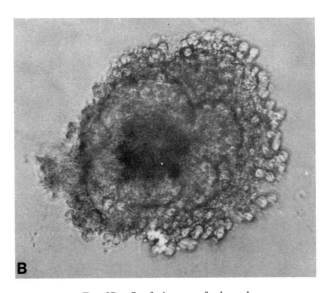

Fig. 3B.   See facing page for legend.

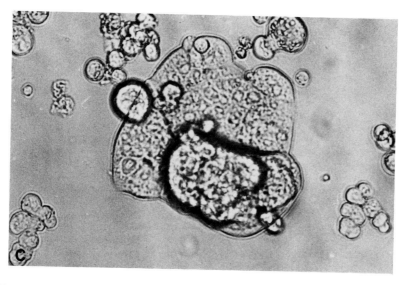

FIG. 3. Establishment of a transplantable minimal deviation rat hepatoma in culture. (A) Fibroblast overgrowth after 14 days on petri dish surface. (B) Tumor cell growth after 14 days on polyacrylamide surface. (C) Forty days after 2 passages in polyacrylamide dishes, cells were grown on petri dish surface—tumor cells with no fibroblast growth. × 400.

out. A vivid comparison of the growth of component cells of a rat hepatoma explanted on both polyacrylamide dishes and plastic petri dishes can be seen in Fig. 3. The predominant cells on the petri dishes were fibroblasts (Fig. 3A). The cells cultured on the polyacrylamide dish grew in a loose sphere (Fig. 3B), which took on the morphology seen in Fig. 3C after transfer to petri dishes. Fibroblasts were not apparent after two passages on polyacrylamide dishes.

E. Removal of Host Inflammatory Elements

An additional step to preparing tumor cells for culture is the removal of many of the lymphocytes and macrophages from the suspension prepared from the solid tumor. Since tumor cells are generally larger than the host inflammatory elements, this can be accomplished by velocity sedimentation, which separates cells primarily on the basis of size. This technique has been described in detail previously (Miller and Phillips, 1969). Briefly, a cell suspension prepared from a solid tumor by enzymic digestion is subjected to sedimentation at unit gravity in a Ficoll medium in Eagle's minimal essential medium buffered with Tris at pH 7.4. The inflammatory cells generally sediment at rates less than 10 mm/hour, while the larger tumor

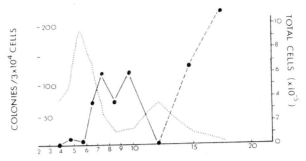

Fig. 4. Sedimentation velocity analysis of enzyme-dispersed cells in a 35-day rat sarcoma. One million cells were loaded on a unit gravity sedimentation chamber and fractionated for 2 hours. Cell fractions were counted for total cell numbers and then plated in 1% methocellulose-1066 tissue culture medium and incubated at 37°C for 8–10 days. From Haskill *et al.* (1975), with permission. · · · , Total cells; ●—●, "macrophage" colonies; ●- - -●, tumor colonies.

cells sediment at rates greater than 10 mm/hour (Haskill *et al.*, 1975). The fractions containing the larger cells can be collected and washed, and placed in polyacrylamide cultures as described above. As can be seen in Fig. 4, this technique effectively separates the tumor cells from the macrophages; lymphocytes sediment even more slowly. This technique, however, does not effectively separate fibroblasts and other contaminating connective tissue cells from tumor cells (J. S. Haskill, personal observation), so it is best combined with culture on polyacrylamide dishes.

III. Discussion

The properties of polyacrylamide that inhibit the growth of fibroblasts are not known, but may be due simply to the fact that fibroblasts do not grow unless they are able to attach and spread on a surface (Martin, 1973). The rounded appearance of tumor cells growing on the polyacrylamide surface indicates that spreading does not occur, and any attachment that does occur is weak. Thus, tumor cells that are capable of proliferation without true monolayer formation will have the advantage over fibroblasts when cultured on a polyacrylamide surface. It should be noted that this may not be suitable for all tumor cells, since some may have stringent growth requirements.

The experimental tumor lines which have been developed in polyacrylamide dishes and tested for *in vivo* growth have retained their tumor-forming capacity (Jones and Haskill, 1973). Some tumor lines were less immunogenic than their *in vivo* counterparts, but this was also observed with tumor lines developed on plastic petri dishes (Jones and Haskill, 1973). The antigenicity of the developed tumor lines remained intact, since the cells were susceptible to destruction in immunized animals and in *in vitro* tests of tumor immunity (Haskill, 1973; Jones and Haskill, 1973). Therefore, the tumor lines developed by the culture on polyacrylamide dishes retain the properties of the tumor but are free from fibroblast contamination.

REFERENCES

Evans, R. (1972). *Transplantation* **14**, 468–473.

Haskill, J. S. (1973). *J. Natl. Cancer Inst.* **51**, 1581–1588.

Haskill, J. S., Proctor, J. W., and Yamamura, Y. (1975). *J. Natl. Cancer Inst.* **54**, 387–393.

Jones, T. L., and Haskill, J. S. (1973). *J. Natl. Cancer Inst.* **51**, 1575–1580.

Kerbel, R. S., Pross, H. F., and Elliott, E. V. (1975). *Int. J. Cancer* **15**, 918–932.

Lesfargues, E. Y. (1973). *In* "Tissue Culture: Methods and Applications:: (P. F. Kruse Jr. and M. K. Patterson, Jr., eds.), pp. 45–50. Academic Press, New York.

Marbrook, J., and Haskill, J. S. (1974). *Cell. Immunol.* **13**, 12–21.

Martin, G. M. (1973). *In* "Tissue Culture: Methods and Applications" (P. F. Kruse Jr., and M. K. Patterson, Jr., eds.), pp. 39–43. Academic Press, New York.

Miller, R. G., and Phillips, R. A. (1969). *J. Cell. Physiol.* **73**, 191–201.

Owens, R. B., and Hackett, A. J. (1972). *J. Natl. Cancer Inst.* **49**, 1321–1332.

Chapter 19

Isolation of Drug-Resistant Clones of Ehrlich Ascites Tumor Cells

CHRISTOPHER A. LOMAX AND J. FRANK HENDERSON

Département de Microbiologie, Centre Hospitalier Universitaire,
Université de Sherbrooke, Sherbrooke, P.Q., Canada; and Cancer Research
Unit (McEachern Laboratory) and Department of Biochemistry,
University of Alberta, Edmonton, Alberta, Canada

I. Introduction

Drug resistance is a continuing obstacle to effective cancer chemotherapy, and several experimental models have been used in attempts to understand, and counteract or circumvent this hindrance. These model systems fall into two major classes: selection of resistant clones from cells growing in culture (Thompson and Baker, 1973), and the gradual enrichment in resistant cells of an ascites tumor population by *in vivo* selection for many weeks (e.g., Caldwell *et al.*, 1967).

The rate of spontaneous mutation to drug resistance in animal cell systems is relatively high (Brockman, 1974), and it is important to clone resistant cell lines to ensure that the population is isogeneic. Although this is easily accomplished in the case of resistant cell lines isolated *in vitro*, resistant ascitic tumors isolated *in vivo* rarely have been cloned, and this is difficult to do.

Although selection pressures used to isolate resistant ascites tumors are limited by host toxicity, relatively high drug does can usually be given because drugs administered intraperitoneally are often less toxic when an ascites tumor is present in the peritoneal cavity, and because the tumor can be transplanted to new hosts at relatively frequent intervals. Cell culture techniques exclude the host entirely and encourage the use of high selection pressures. Consequently, the resistant tumors studied in both cases have usually been highly resistant, a property that has helped enormously in the study of biochemical differences, between sensitive and resistant cells, that are responsible for the observed insensitivity to drug action.

However, there are fundamental differences between the development and study of resistance in experimental ascites tumors or cultured cell systems and in the clinical situation. In the latter case, selection pressures are low because transplantation cannot be used, intratumoral drug administration is not usually employed, drugs often appear to have lower therapeutic indexes than in experimental systems, and host toxicity is a constant limitation. Under these circumstances, then, it seems highly probable that the degree of resistance attained in clinical cases will be less than that usually studied in animal systems. Experimental studies of drug resistance relevant to the clinical situation have suffered from a lack of suitable model systems.

The procedure described here was developed for the isolation of drug-resistant clones of tumor cells *in vivo* using low selection pressures (Lomax and Henderson, 1974; Henderson *et al.*, 1975). It is based on the reports of Williams and Till (1966) and Hill and Bush (1969) that colonies of tumor cells will grow in the lungs when the cells are mixed with small plastic microspheres and injected intravenously.

II.   Procedure

To obtain clones of drug-resistant cells *in vivo*, the following procedure is used: (1) treatment of the tumor cells with a mutagen; (2) injection of a mixture of cells and plastic microspheres; (3) treatment of the tumor-bearing animal with a drug; and (4) removal of lung colonies and quantitative tests of drug sensitivity.

A.   Mutagenesis

Ehrlich ascites tumor cells are maintained by weekly intraperitoneal implantation of about 4×10^7 cells into ICR Swiss mice. Cells are usually used 5 or 6 days after implantation.

To increase the chances of obtaining drug-resistant cells, the cells are treated with the mutagen ethylmethane sulfonate (Sigma Chemical Co.) before selection. Tumor cells are removed by Pasteur pipette after laparotomy, and ascites fluid is discarded after centrifugation. Cells are washed twice by resuspension in calcium-free Krebs–Ringer phosphate medium and recentrifugation. After a final centrifugation for 7 minutes at top speed in a clinical centrifuge at room temperature, a 10% suspension in calcium-free Krebs–Ringer phosphate medium is prepared. Tumor cells finally are incubated at 37°C as a 2% suspension in 5 ml of calcium-free Krebs–Ringer phosphate medium, pH 7.4, containing 5.5 mM glucose and 200 μg of ethylmethane sulfonate per milliliter; 10-ml Erlenmeyer flasks are used. After 1 hour, the cells are collected by centrifugation, washed once, and suspended to 5 ml in calcium-free Krebs–Ringer phosphate medium. Equivalent results are obtained when cells are incubated for 3 hours with 100 μg of ethylmethane sulfonate per milliliter.

Viable cell count is estimated from the survival time of mice inoculated intracerebrally with tumor cells (Prince et al., 1957). The average survival time of a group of 6 mice is calculated, and, by comparison with a standard curve, the number of viable cells in a suspension can be determined. The average survival time of mice given 4×10^5 cells is 13 days. Viability tests are performed only to standardize the conditions of mutagenesis; they are not done routinely. Figure 1 shows the relationship of ethylmethane sulfonate concentration to cell survival.

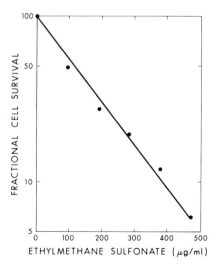

FIG. 1. Survival of Ehrlich ascites tumor cells after treatment with ethylmethane sulfonate. Cells are incubated with ethylmethane sulfonate as a 2% suspension in calcium-free Krebs–Ringer phosphate medium for 3 hours. (From Lomax and Henderson, 1974.)

.Under the conditions described above, 50–70% of the cells are killed. These cells are more sensitive to ethylmethane sulfonate than are cultured Chinese hamster ovary cells (Kao and Puck, 1968) or cultured mouse lymphoma L5178Y cells (D. Tidd and A. R. P. Paterson, personal communication).

B.   Implantation

Low-density, nonradioactive, plastic microspheres of 15 ± 5 μm diameter, are purchased as a dry powder from the 3M Company (St. Paul, Minnesota). A homogeneous suspension is prepared by sonicating for 8–10 minutes a mixture of 200-mg microspheres, 5 ml of buffered saline-glucose (5.5 mM glucose, 140 mM NaCl, 10 mM Tris-HCl, pH 7.4, and 4 mM sodium phosphate buffer, pH 7.4), and 0.02 ml of Tween 80. This gives a suspension containing about 1.2×10^7 microspheres per milliliter. The Tween 80 is required to wet the microspheres so that they will make a uniform suspension; however, the minimum amount must be used because higher concentrations are toxic to the cells.

The suspensions of mutagenized tumor cells and of microspheres are mixed as follows: 2.0 ml of cell suspension (about 2×10^8 viable cells), 1.5 ml of microsphere suspension (about 1.8×10^7 microspheres), 0.7 ml of calcium-free Krebs–Ringer phosphate.

The mixture (0.1 ml) is slowly injected into a tail vein of ICR Swiss mice. It is important that the cell–microsphere mixture be thoroughly mixed before each injection, and that only a single 0.1-ml portion be taken into the syringe each time; otherwise the microspheres will sediment and uneven inoculation will result.

Ehrlich ascites tumor cells that are not attached to microspheres will not form visible tumors in the lungs or elsewhere. However, small solid tumors may form in the tail as a result of extravasation during implantation. The incidence of these tumors decreases with practice, and they do not interfere with the subsequent selection of resistant tumor clones in the lung.

C.   Selection

In untreated mice, tumors 1–3 mm in diameter are visible on the surface of the lungs 3–4 weeks after implantation. The number of tumors is highly variable and ranges from 10 to 100 per mouse. This variability is in contrast to the results of Hill and Bush (1969) and Williams and Till (1966), who used this procedure in rats to measure cell survival quantitatively. From these previous studies with this method, single spherical tumor colonies are called "clones," although direct studies of the homogeneity of each cell population have not been carried out.

It is during the 3–4 week period of tumor growth that drug treatment is carried out to select clones that are drug resistant. The selection pressures that can be exerted are relatively low because the drug will be distributed to cell tissues, including the lungs, and the mice have to be kept alive during this period.

As an example, Fig. 2 shows the appearance of control lungs and lungs of mice treated with 6-thioguanosine. Trials involving a number of treatment schedules and drug doses had previously been carried out. In the case illustrated, 10 mg of 6-thioguanosine per kilogram was injected intraperitoneally on days 1, 2, 3, and 4 after implantation of the tumor cell–microsphere mixture, and 5 mg/kg on days 20, 22, and 24. This schedule was lethal to 20% of the mice by day 28. Lungs of drug-treated animals 28 days after tumor cell implantation had only a few large (1–3 mm in diameter) tumors and many microscopic colonies which could be detected only after fixation and staining in Bouin's fluid. Those lung tumors that were observed

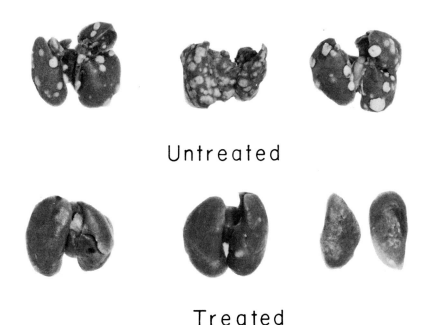

Untreated

Treated

FIG. 2.   Lungs of mice injected intravenously with suspensions of Ehrlich ascites tumor cells mixed with plastic microspheres. Top, untreated; bottom, treated with 6-thioguanosine. (From Lomax and Henderson, 1974.)

were considered as putative resistant clones. In two such experiments with 6-thioguanosine, 24 large spherical lung tumors were isolated from drug-treated animals.

D.   Isolation of Mutant Cells

After allowing 4 weeks for the development of lung colonies, the mice are killed and the lungs removed and examined for the presence of large surface tumors. These are excised and gently homogenized in 0.3 ml of buffered saline. The entire homogenate is then injected intraperitoneally into a mouse, and an ascites tumor develops in 7–14 days.

Quantitative estimates of the drug sensitivity of individual tumor lines are then made by measuring tumor packed-cell volumes after treatment with drug. In the case illustrated, groups of 20 mice were incubated intraperitoneally with 4×10^7 cells from each of 24 lung colony-derived ascites tumors. For each tumor line, 10 mice were given saline and 10 mice were treated with 10 mg/kg 6-thioguanosine on days 1, 2, and 3 after tumor implantation. On day 6 or 7 after implantation, the mice were killed, and the packed cell volume of tumor cells in the peritoneal cavity of each mouse was determined. Table I shows that two of these tumor lines (TGR-R17 and TGR-R13) are highly resistant to 6-thioguanosine at the dose and schedule used and that another three lines exhibited a low, but statistically significant degree of resistance. The other 19 cell lines tested were not resistant under the conditions of the test.

TABLE I

RESISTANCE TO 6-THIOGUANOSINE IN ASCITES TUMORS
DERIVED FROM LUNG COLONIES[a]

Cell line	Inhibition of growth (%)
EAC[b]	51.1
TGR-R1[c]	39.2
TGR-R5	21.1
TGR-R14	18.2
TGR-R17	7.7
TGR-R13	6.0

[a] Mice were treated with 10 mg of 6-thioguanosine per kilogram on days 1, 2, and 3 after tumor implantation, and packed cell volumes were measured on day 6 or 7.

[b] EAC = parent line of Ehrlich ascites tumor cells.

[c] TGR-R1 to -R17 = ascites tumors derived from lung tumors of mice treated with 6-thioguanosine.

TABLE II

ENZYMES OF GUANINE METABOLISM[a]

	Apparent enzyme activity (relative activity)		
Cell line	Guanine phosphoribosyl- transferase	Guanosine phosphorylase	Guanine deaminase
EAC	100.0	100.0	100.0
TGR-R1	55.7	86.4	84.4
TGR-R5	88.7	96.2	104.0
TGR-R13	131.0	96.1	126.0
TGR-R14	97.6	74.2	203.0
TGR-R17	106.0	83.1	149.0
RGR-R20	107.0	108.0	76.9

[a]Tumor cells were incubated with ^{14}C-labeled guanine in Fischer's medium containing 25 mM phosphate or in Krebs-Ringer phosphate medium. Apparent enzyme activities were calculated as described by Smith *et al.* (1974).

III. Biochemical Characterization

Biochemical studies can be carried out to measure changes in the purine metabolism of the tumor lines derived by the lung colony method that might contribute to their observed drug resistance.

Tumor cells were incubated *in vitro* with radioactive guanine to measure the apparent activities of several enzymes of guanine metabolism (Snyder *et al.*, 1972; Smith *et al.*, 1974). Selected results are shown in Table II. Guanine phosphoribosyltransferase activity in TGR-R1 cells was about half that in the parental line, whereas in TGR-R13 cells it was increased 30%. There were no major differences in the apparent activity of guanosine phosphorylase among the cell lines tested. Guanine deaminase activity increased in several cell lines and was elevated 2-fold in TGR-R14 cells. Some of the biochemical changes observed might be related to the observed resistance to 6-thioguanosine, but definitive conclusions on this point must await studies of 6-thioguanosine metabolism itself.

IV. Discussion

The technique described allows the isolation *in vivo* of drug-resistant clones of Ehrlich ascites tumor cells at relatively low selection pressures.

This method seems to be useful for the isolation of biochemical mutants of mammalian cells and as a model system for the study of resistance to cancer chemotherapy. It is presumably applicable to other tumor lines that will grow as lung colonies (Hill and Bush, 1969; Williams and Till, 1966) and, with suitable modifications, to tumor lines such as leukemia L1210, which forms colonies in the spleen.

We have isolated five distinct cell lines apparently resistant to 6-thioguanosine on the basis of their growth as lung colonies in treated animals. A quantitative measurement of drug sensitivity showed that most were only partially resistant. Biochemical studies indicated changes in guanine metabolism perhaps related to the observed resistance, but definitive conclusions on this point must await studies of 6-thioguanosine metabolism itself.

ACKNOWLEDGMENTS

We are grateful to Drs. D. A. Cook, R. M. Henderson, and A. Croteau for advice and to Mr. G. Zombor for expert technical assistance. This work was supported by the National Cancer Institute of Canada.

REFERENCES

Brockman, R. W. (1974). *In* Handbuch der Pharmakologie" (A. C. Sartorelli and D. G. Johns, eds.), Vol. 31, Part I, p. 352. Springer-Verlag, Berlin and New York.

Caldwell, I. C., Henderson, J. F., and Paterson, A. R. P. (1967). *Can. J. Biochem.* **45**, 753.

Henderson, J. F., Brox, L. W., Fraser, J. H., Lomax, C. A., McCoy, E. E., Snyder, F. F., and Zombor, G. (1975). "Pharmacological Basis of Cancer Chemotherapy," p. 633. Williams & Wilkins, Baltimore, Maryland.

Hill, R. P., and Bush, R. S. (1969). *J. Radiat. Biol.* **15**, 435.

Kao, F. T., and Puck, T. T. (1968). *Proc. Natl. Acad. Sci. U.S.A.* **60**, 1275.

Lomax, C. A., and Henderson, J. F. (1974). *J. Natl. Cancer Inst.* **52**, 1291.

Prince, A. M., Littell, A. S., and Gensberg, H. S. (1957). *J. Natl. Cancer Inst.* **18**, 487.

Smith, C. M., Snyder, F. F., Fontenelle, L. J., and Henderson, J. F. (1974). *Biochem. Pharmacol.* **23**, 2351.

Snyder, F. F., Henderson, J. F., and Cook, D. A. (1972). *Biochem. Pharmacol.* **21**, 2351.

Thompson, L. H., and Baker, R. H. (1973). *Methods Cell Biol.* **6**, 209.

Williams, J. F., and Till, J. E. (1966). *J. Natl. Cancer Inst.* **27**, 177.

Chapter 20

Separation of Clonogenic Cells from Stationary Phase Cultures and a Murine Fibrosarcoma by Density-Gradient Centrifugation

DAVID J. GRDINA

Section of Experimental Radiotherapy,
The University of Texas System Cancer Center,
M.D. Anderson Hospital and Tumor Institute,
Houston, Texas

I. Introduction

A variety of biophysical methods have been developed to separate selected homogeneous subpopulations of cells from complex and heterogeneous cell systems such as normal and neoplastic tissues. Based on the premise that physical differences between cells may be indicative of functional differences, these methods were designed to exploit the physical differences that exist between the various classes of cells. One such method involves the centrifugation of cells in preformed density gradients. In this system, each cell sediments to a point in the gradient where the density of the surrounding medium is equal to its own, i.e., its isopycnic point. Because the cells are in intimate contact with the surrounding medium, it is important that this material not exert a cytotoxic effect. The gradient material should also be of high density and low viscosity, easily prepared and sterilized, and stable against chemical breakdown and denaturation. It would also be advantageous if this material did not aggregate cells and was isotonic.

Renografin, an ionic derivative of triiodinated benzoic acid, is a material having most of the above requirements. It has been used successfully to separate sporulating from vegative forms of *Bacillus megaterium* (Tamir and Gilvarg, 1965) and to fractionate transformable bacteria from competent cultures of *Bacillus subtilis* (Cahn and Fox, 1968). Renografin gradients have also been used to separate viable human reticulocytes from the peripheral blood (DeSimone *et al.*, 1974) and mouse testis cells into relatively homogeneous populations of cells at well-defined stages of maturation (Meistrich and Trostle, 1975).

In this communication the use of Renografin as a supporting medium is described for the separation of homogeneous subpopulations of mammalian cells from such complex and heterogeneous populations as cultures of Chinese hamster ovary (CHO) cells grown to late stationary or plateau phase and cells derived from disaggregated murine fibrosarcoma (FSa) tumors. Stationary-phase cultures of CHO cells were used as an *in vitro* tumor model because they appear to have more properties in common with tumor populations than do cultures of exponentially growing cells (Hahn and Little, 1972; Ross and Sinclair, 1972). Both the density and clonogenic ability (i.e., *in vitro* plating efficiency, PE) of these cells were monitored as they progressed from exponential to "late" stationary phase of growth. Each of the separated tumor populations was also characterized in a similar manner. The cloning efficiency (CE) of the tumor cells, however, was determined *in vivo* using a lung colony assay.

II. Materials and Methods

A. Separation of Cells

1. Supportive Media

Renografin 60 (methylglucamine N,N'-diacetyl-3, 5-diamino-2,4,5-triiodobenzoate; E. R. Squibb and Sons, New York) is supplied as a 60% aqueous solution in sterile 50-ml ampoules. This stock solution was diluted with sterile Ringer's injection, USP (Baxter Laboratories, Division of Travenol Laboratories, Inc., Morton Grove, Illinois) to obtain solutions of

10 and 35% w/v (specific gravity of 1.061 and 1.206, respectively). Crude deoxyribonuclease (DNase; Sigma Chemical Company) was added to these solutions at a final concentration of 0.1 mg/ml. The pH of these solutions ranged from 7.1 for the 10% concentration of Renografin to 7.5 for the 60% concentration.

Bovine serum albumin (BSA; fraction V) was obtained from the Sigma Chemical Company and processed following methods described else-where (Shortman, 1968). A 20% w/v solution of BSA in distilled water containing antibiotics was dialyzed against chloroform-containing distilled water for at least 3 days. The BSA solution was lyophilized and resuspended at 26% w/w in an unbuffered balanced salt solution (Shortman, 1968). The pH was 5.1. This BSA solution was calculated to be at physiological tonicity (Williams et al., 1972).

2. Formation and Characterization of Gradients

Linear Renografin gradients were formed, under sterile conditions, at room temperature using a Glenco gradient elution apparatus (Glenco Scientific, Inc., Houston, Texas). Light material (10%) was added to the single-arm chamber, and an equal volume of heavy material (35%) was added to the double-arm chamber. The gradually diluted Renografin solution was removed from the mixing chamber through a Tygon tube by the action of a Buchler Polystaltic Pump (Model 2-6100; Buchler Instruments Division, Nuclear-Chicago Corporation, Fort Lee, New Jersey) and was allowed to flow down the side of a cellulose nitrate centrifuge tube at a rate of 1 ml per minute. Linear density gradients ranging from 5.2% w/w to 26% w/w of BSA were also generated in this manner. The glass elution tubes and Teflon valves were autoclaved. The Tygon tubing was sterilized by soaking overnight in a 70% solution of ethyl alcohol. Prior to use, the tubing was rinsed with sterile Ringer's injection solution. The cellulose nitrate centrifuge tubes were sterilized in an ethylene oxide atmosphere at 38°C for 8 hours.

Measurements of the refractive index (N) were made at 24°C with a Zeiss refractometer and related to the densities of Renografin and BSA fractions by the equations $\rho = 3.4683\ N_{24} - 3.6267$ and $\rho = 1.4129\ N_{24} - 0.8814$, respectively.

3. Centrifugation and Collection of Gradients

Either 2 to 4 × 10⁶ CHO cells or 10 to 30 × 10⁶ tumor cells were layered onto each 14-ml gradient. When larger numbers of tumor cells were to be centrifuged (50 to 80 × 10⁶ cells), 34-ml gradients were used. The gradients were centrifuged in a SW 27 rotor at 10,000 rpm (about 13,000 g) at 4°C in a Beckman model L5-50 preparative ultracentrifuge for 30 minutes. The initial rotor speed was set at 5000 rpm and then accelerated each minute by increments of 1000 rpm to the desired speed. Deceleration occurred with the brake off.

To determine the time required for the isopycnic banding of cells, human erythrocytes were spun for 30 minutes or for 2 hours. After centrifugation, the position of the erythrocyte band in the gradient was determined by measuring the refractive index of the peak fraction. Since no change in position was observed (N_{24} of 1.3776 at 30 minutes vs N_{24} of 1.3778 at 120 minutes), all spins were terminated after 30 minutes (Grdina et al., 1973).

The entire gradient was fractionated using an Auto-Densi-Flow collector (Buchler Instruments, Fort Lee, New Jersey). Fractions of about 1 ml were collected in the cold from the top of the gradient. Cells in each fraction were counted using a hemacytometer, and the refractive index of each fraction was measured.

When only a selected band of cells was to be collected, a 25-gauge needle on a 3-ml syringe was introduced through the side of the centrifuge tube at the base of the desired band. Between 2 and 4 ml of the material were removed, and the refractive index was measured. While this latter procedure of collecting cells is not as efficient as the former, it has the advantage of allowing for the rapid recovery of fractions uncontaminated by cells from either upper or lower regions of the gradient.

4. Washing Procedure

When desired, cells were washed free of Renografin by diluting each recovered cell fraction 1:5 with medium 199 (Difco Laboratories, Inc., Detroit, Michigan). The suspension was then centrifuged in a clinical bench-top centrifuge for 7 minutes at 450 g. The resulting cell pellet was resuspended in 10 ml of medium 199 and centrifuged again for 5 minutes at 90 g. The supernatant was discarded, the cell pellet was suspended in an appropriate volume of medium, and a cell count was made using a hemacytometer.

5. Determination of Cell Volume

After the cells were removed from the gradient and resuspended into Ringer's solution, cell volumes were measured with a Model B Coulter counter and automatic size distribution plotter. All calculations were based on the average of three readings.

6. Expression of the Density Profiles

Although all the density gradients used were essentially linear, the density as a function of the fraction number did vary slightly between gradients. To facilitate a comparison of results between gradients, the cell concentrations were converted to the percentage of total cells recovered found in a given density increment and plotted using density as the abscissa. In this manner the areas under the curves are normalized to equal volumes.

B. Preparation and Characterization of Chinese Hamster Ovary Cells

1. CELL LINE AND GROWTH CONDITIONS

CHO cells were grown in monolayer cultures in plastic petri dishes (60 mm ϕ) at $37°$ C in a water-saturated atmosphere of 5% CO_2 and air. The growth medium was a modified McCoy's 5A (Grand Island Biological Company, Grand Island, New York) supplemented with 20% fetal calf serum (Grand Island Biological Company, Grand Island, New York). At the initiation of each growth experiment, between 1 and 1.5×10^6 cells were plated in each plate containing 5 ml of medium. The growth medium was left unchanged throughout each experiment.

2. CELL COUNTING

To monitor cell growth, both attached and floating cells were removed from selected replicate plates and counted using a hemacytometer. Attached cells were released from the plates using a 0.025% solution of trypsin and were passed through a 25-gauge needle to disperse any clumps. The cells were centrifuged at 90 g for 5 minutes in a clinical centrifuge and pelleted, and then resuspended in Solution A (8.0 gm of NaCl, 0.4 gm of KCl, 1.0 gm of glucose, and 0.35 gm of $NaHCO_3$ in 1 liter of H_2O). The cells were counted using a phase-contrast microscope, and only refractile intact cells were scored.

3. CELL CLONOGENICITY

The clonogenicity of CHO cells was determined by scoring for their ability to form colonies. At selected, times known numbers of intact viable cells were plated into fresh medium. Cells were obtained directly from cultures, with and without storage in media at $4°$ C for 1 hour and from the peak fractions of the density gradients. The plates were incubated at $37°$ C for 7 days. The resulting colonies were fixed and stained with a 0.5% solution of crystal violet. Only colonies of 50 or more cells were counted.

4. LABELING OF CELLS AND DETERMINATION OF RADIOACTIVITY

Cells in exponential phase were labeled for 10 hours with medium containing 0.1 μCi/ml of ^{14}C-labeled thymidine (specific activity 33 mCi/m mole). The cells were removed from the plates and centrifuged in Renografin gradients. After collection, the cells were filtered onto Whatman GF/c filters and washed with 20 ml of Solution A, 20 ml of 5% cold trichloroacetic acid, and 10 ml of 95% ethanol (Meistrich, 1972). Each filter was placed on a counting vial containing 1 ml of NCS solubilizer (Amersham-Searle, Inc.) containing 9% water, and incubated for 12 hours at 37° C. Scintillation fluid was added, and radioactivity was measured by liquid scintillation counting with an efficiency of about 30%.

C. Preparation and Characterization of Murine Fibrosarcoma Cells

1. MICE AND TUMOR SYSTEM

Specific pathogen-free female C_3Hf/Bu mice from 9 to 10 weeks of age were used in each experiment. The tumor, a fibrosarcoma, was originally induced in a young female C_3H mouse by a single subcutaneous injection of 1 ml of methylcholanthrene suspended in peanut oil (Suit and Suchato, 1967). Second- to fourth-generation isotransplants of the tumor were stored in a liquid nitrogen refrigerator, and the experiments were restricted to tumors of the fifth generation.

2. TUMOR CELL SUSPENSION

Single-cell suspensions were prepared from tumors 15–20 mm in diameter. Tumor tissue containing no visible regions of necrosis or hemorrhage was finely minced and then digested with a 0.025% solution of trypsin for 20 minutes at room temperature. DNase was also added to the mixture to achieve a final concentration of 0.1 mg/ml. After removing the undigested tissue, the remaining suspension was mixed with an equal volume of medium 199 supplemented with 5% syngeneic-normal mouse serum (SNMS). The suspension was then passed through a stainless steel mesh (200 wires/inch), centrifuged 2–3 times at 225 g for 5 minutes in Solution A, and resuspended again in medium 199 with 5% SNMS. Only intact viable cells were counted as determined by phase contrast microscopy and a trypan blue dye-exclusion test. Cell viability was routinely found to be greater than 95%.

3. LUNG COLONY ASSAY

The clonogenic ability of separated FSa cells was determined using a lung colony assay (Hill and Bush, 1969). A known number of viable tumor cells, whose clonogenicity was to be measured, were mixed with 10^6 heavily irradiated (HIR) tumor cells (i.e., exposed to 10,000 rads) in 0.5 ml of medium 199 and injected into the lateral tail vein of mice that had been irradiated with 1000 rads 24 hours earlier. This procedure significantly enhanced the efficiency of the lung colony assay (Grdina *et al.*, 1975). To protect these irradiated animals from death due to bone marrow damage, 5×10^6 syngeneic bone marrow cells were injected intravenously 2 hours after irradiation. Bone marrow was obtained from the tibias and femurs of C_3Hf/Bu donor mice by a method described elsewhere (Milas and Tomljanovic, 1971).

Each experimental group contained 5 or 6 animals. After 12–14 days the mice were sacrificed, their lungs were removed, and the lobes were separated and fixed in Bouin's solution. Colonies of tumor cells were identified as white, round nodules on the surface of the yellowish lung and were easily counted with the naked eye.

4. ANALYSIS OF RESULTS

The results were statistically evaluated with the Student's *t* test.

III. Results

A. Separation of Cells

1. SEPARATION OF CHINESE HAMSTER OVARY CELLS

Presented in Fig. 1 are data illustrating the changes in the numbers of CHO cells per dish and the densities of the bands of cells formed in Renografin gradients as related to the age of the cultures. At each time interval, data from two replicate cultures are plotted. These data are representative in the description of the qualitative growth pattern of the cells, the reproducibility between the density values determined for the replicate cultures, and the appearance of distinct populations of cells in the late stationary phase cultures.

Cells from cultures grown for 10 days or less formed only a single band after centrifugation in gradients of either Renografin or BSA. Typical profiles are shown in Fig. 2. After 13 and 14 days of growth (late stationary phase), three populations of cells were routinely observed in Renografin gradients (Fig. 2c), two banding at densities of 1.102 ± 0.002 (SE), 1.148 ± 0.003 and a third pelleting at ≥ 1.197 gm/cm^3. These values represent the mean of three experiments with two replicate cultures per experiment. Intact viable cells were found in all regions of the gradient—over 90% in the light fractions and over 70% in the dense fractions and pellet. Analogous results were observed in BSA gradients. Because of the lower density

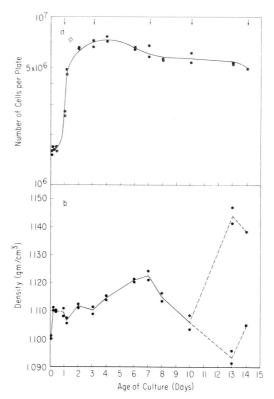

FiG. 1. (a) Growth curves of Chinese hamster ovary (CHO) cells. Cells were plated in replicate plastic petri dishes. Each point represents the number of cells per petri dish counted using a hemacytometer. $-\bigcirc-$, Transition from log to stationary phase; arrows, times at which the plating efficiency of cells were monitored. (b) The densities of CHO cells in Renografin gradients related to the age of culture. Each point represents the mean density of each band of cells formed as a result of centrifugation; these points, plotted at time intervals at which more than one band of cells appeared, are connected by a dashed line (---). Mean densities of each band were determined graphically by averaging densities at the two half-maxima points of the band. Reprinted, with permission, from Grdina et al. (1974b).

of the BSA solutions, however, the second peak is probably contained within the pellet.

A control experiment was performed to determine whether the changes in the density profiles observed for cultures after 13 days of growth were caused by dense material (lysed cells, nuclei, and other cellular debris) dragging intact cells down to greater densities. Exponentially growing cells (10^5) labeled with thymidine-^{14}C were centrifuged on two Renografin gradients in the absence or in the presence of 1.2×10^6 unlabeled cells

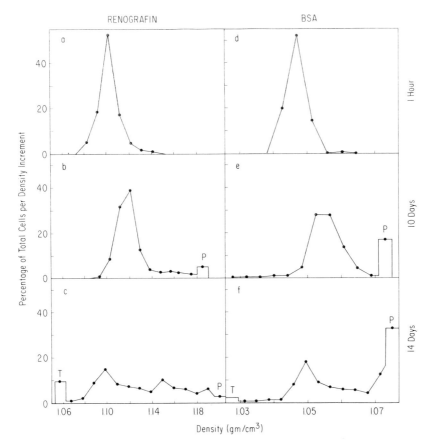

FIG. 2. Density profiles of Chinese hamster (CHO) cells centrifuged to equilibrium in gradients of Renografin (a, b, c) and bovine serum albumin (BSA) (d, e, f). Cells were obtained from replicate plates at 1 hour (a, d), 10 days (b, e), and 14 days (c, f) after plating. In each case, cells from the same plate were used for both the Renografin and BSA gradients. T, Cells that remained on top of the gradient; P, cells found in the pellet. Reprinted, with permission, from Grdina et al. (1974b).

from a late stationary phase culture. As shown in Fig. 3, the density at which these cells banded was independent of the presence of the larger numbers of cells from the stationary-phase culture, which formed two bands and a pellet. The labeled cells also appeared to band at the same density as did the cells of the lightest population present in the 14-day-old culture.

2. SEPARATION OF MURINE FIBROSARCOMA CELLS

The separation of FSa cells in 14-ml Renografin gradients is described in Fig. 4. Each of the six density profiles represents the separation of a cell

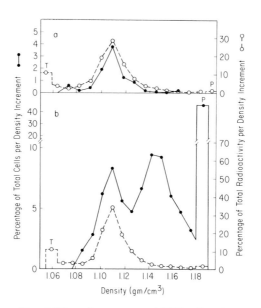

FIG. 3. Density profiles of Chinese hamster ovary (CHO) cells as determined by cell number (●—●) and radioactivity (○--○) in gradients of Renografin (a) 10^5 exponentially growing cells labeled with thymidine-^{14}C only; (b) 10^5 exponentially growing cells labeled with thymidine-^{14}C plus 1.2×10^6 unlabeled late stationary-phase cells. The same number of exponentially growing cells were layered on both gradients. Hence, to facilitate comparison of the percentage of total cells per density increment between (a) and (b), the number of cells in (b) were used as the total number for normalizing both gradients. T, Cells that remained on the top of the gradient; P, cells found in the pellet. Reprinted, with permission, from Grdina *et al.* (1974b).

suspension derived from a different FSa tumor. Three distinct regions or bands of cells were observed at densities of $\leq$ 1.064, 1.132 $\pm$0.001, and 1.170 $\pm$0.002 gm/cm³ (mean of 6 experiments $\pm$ SE). A fourth band of cells, when present, appeared at a density of 1.097 $\pm$ 0.002 gm/cm³ (mean of 4 experiments $\pm$ SE). As shown in Fig. 4, the relative number of cells banding at each of these densities varied for each tumor cell suspension. The viability of tumor cells collected at each of the four densities ranged from 95 to 100% (i.e., the cells were intact and refractile). The cells in the population collected at 1.064 gm/cm³ were homogeneous in morphology and size and had a mean volume of 733 $\pm$ 6 μm³. The second band, 1.087 gm/cm³, was also homogeneous and consisted of cells 710 $\pm$ 3 μm³ in volume. The two denser populations were heterogeneous in composition. Cells banding at 1.132 gm/cm³ were relatively small, with a mean volume of 653 $\pm$ 2 μm³. In addition to tumor cells, both lymphocytes and erythrocytes were identified in this fraction. The heaviest band contained a mixture of small and large

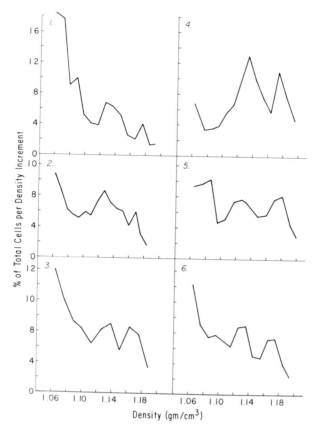

FIG. 4. The density profiles of cells separated from individual FSa tumors in continuous density gradients of Renografin. Each of the six density profiles represents a separate experiment. Reprinted, with permission, from Grdina *et al.* (1974a).

tumor cells, mean volume of 648 $\pm$ 2 μm^3, and a considerable amount of debris. The recovery of cells after separation in Renografin gradients ranged from 65 to 85%.

If a cell suspension was made from four tumors, rather than just one, and was centrifuged in 34-ml gradients of Renografin, five populations of cells were routinely formed. A representative density profile of 50 $\times$ 10^6 FSa cells separated in this manner is presented in Fig. 5. The density of each of the five populations was: band 1, 1.06 $\pm$0.01; band 2, 1.08 $\pm$0.01; band 3, 1.11 $\pm$ 0.01; band 4, 1.14 $\pm$ 0.01; and band 5, 1.17 $\pm$ 0.01 gm/cm^3 (mean of 11 experiments $\pm$ SD; cell populations were collected by the method described using a sterile needle and syringe).

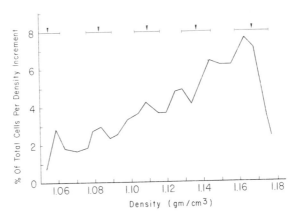

FIG. 5. A representative density profile of FSa cells from four tumors separated in a Renografin gradient. Since only selected bands were collected at times using a needle and syringe, symbols (▼) representing the mean density SD of five cell fractions collected by this method from 11 experiments are also presented at the top. Reprinted, with permission, from Grdina *et al.* (1975).

TABLE I

PLATING EFFICIENCY OF EXPONENTIAL- AND STATIONARY-PHASE CHINESE HAMSTER OVARY (CHO) CELLS[a,b]

Age of culture (days)	Plated directly (%)	Held in media at 4° C for 1 hour (%)	Recovered from peak fractions of gradient		Mean density of cells plated (gm/cm³)	
			Renografin (%)	BSA[c] (%)	Renografin	BSA
Expt. 1						
1	81	76	80	—	1.108	—
3	92	80	93	—	1.110	—
7	88	73	76	—	1.123	—
10	74	58	63	—	1.106	—
			49	—	1.105	—
14	16	16	4	—	1.139	—
			≪ 1	—	≥ 1.197	—
Expt. 2						
1	93	80	88	87	1.107	1.04(
3	88	71	92	90	1.109	1.04*
7	67	55	62	68	1.114	1.05(
10	57	51	42	42	1.113	1.05:
			60	59	1.105	1.05(
14	27	21	9	15	1.156	1.06
			≪ 1	≪ 1	≥ 1.197	≥1.07.

[a] From Grdina *et al.* (1974b).
[b] Plating efficiency of CHO cells based on 6 petri dishes per experimental point.
[c] BSA, bovine serum albumin.

B.   Clonogenicity of Separated Cells

1.   CLONOGENICITY OF CHINESE HAMSTER OVARY CELLS

The clonogenicity (i.e., plating efficiency, PE) of CHO cells was monitored at selected time intervals during their growth from exponential through late stationary phase (see Table I). Known numbers of cells were either (1) removed from culture and plated directly into fresh medium and incubated at 37° C; (2) suspended in fresh medium and stored at 4° C for 1 hour prior to plating; or (3) centrifuged in density gradients at 4° C and then collected and peak fractions plated. After the third day of growth, the plating efficiency of cells treated under each of these conditions decreased with increasing age of the culture. Comparable reductions in PE were seen for cells held in media at 4° C for 1 hour (approximate time of centrifugation), and cells recovered after centrifugation. This, coupled with the relatively high recovery of cells after centrifugation, 90% or greater, indicated that neither Renografin, BSA, nor the conditions of centrifugation were exerting any appreciable cytotoxic effect.

The three populations of cells formed after centrifugation of the late stationary-phase cultures (14 days old) differed in both density and content of clonogenic cells. There was a 3-fold enhancement of clonogenic cells in the lightest population as compared to cells of the same culture that were not separated. Based on the number of cells layered on the gradient and the PE of cells collected from all three bands, over 85% of the clonogenic cells found were present in this lightest population.

2.   CLONOGENICITY OF MURINE FIBROSARCOMA CELLS

A comparison of the clonogenicity (i.e., the ability to form pulmonary tumor nodules) of each of the separated cell populations, band 1 through band 5, to that of an unseparated control suspension is presented in Table II. The number of tumor cells injected refers only to intact viable tumor cells. To facilitate a comparison between experiments, the ratio of the cloning efficiency (CE) of the unseparated control cells to that of each of the separated populations within the same experiment is presented. The more clonogenic cell populations were collected at densities between 1.05 and 1.10 gm/cm^3 (i.e., bands 1 and 2). The difference in CE between each of these bands and that of the unseparated control population was found to be highly significant ($P < 0.001$). Of the remaining populations, however, the differences in CE were not as significant, and only the densest population, band 5, had a CE lower in magnitude than that of the unseparated control.

TABLE II

CLONOGENICITY OF TUMOR CELLS[a]

Population tested	Expt. No.	Number of tumor cells injected + 10^6 HIR[b]	Number of lung colonies[c] (10 mice/pt.: M ± SE)	Cloning efficiency (CE) (%)	CE_B/CE_{USC}[d]	P <
USC	1	1100	16.8 ± 1.8	1.53	1.0	—
	2	2980	21.7 ± 2.4	0.73	1.0	—
B_1	1	1120	44.4 ± 2.5	3.95	2.6	0.001
	2	2010	38.8 ± 1.6	1.93	2.6	0.001
B_2	1	980	52.2 ± 1.3	5.35	3.5	0.001
	2	2020	45.2 ± 2.3	2.24	3.1	0.001
B_3	1	1170	34.4 ± 4.6	2.94	1.9	0.02
	2	4060	31.5 ± 1.3	0.78	1.1	0.5
B_4	1	1280	36.8 ± 6.2	2.88	1.9	0.05
	2	4180	33.6 ± 3.6	0.80	1.1	0.5
B_5	1	1250	13.4 ± 1.9	1.07	0.7	0.01
	2	7840	50.5 ± 2.2	0.64	0.9	0.5

[a] From Grdina et al. (1975).
[b] HIR, heavily irradiated.
[c] WBI mice were used.
[d] To facilitate a comparison between experiments, the ratio of the CE of the unseparated control (USC) to that of each of the bands within the same experiment is presented.

IV. Discussion

Cell populations differing in clonogenic ability were separated in linear density gradients of Renografin from cultures of CHO cells grown to late stationary phase and suspensions of murine fibrosarcoma cells. Because of the relatively high concentration and low molecular weight of the triiodo-benzoate salts in the Renografin solutions, these gradients were hypertonic. For comparison, gradients of BSA, made to be at physiological tonicity, were also used. Although the density profiles of CHO cells separated in the two gradient systems were qualitatively similar, the buoyant density of the cells was greater in Renografin than in BSA. This is presumably due to cell shrinkage caused by the hypertonicity of the medium. The mean density determined for exponentially growing CHO cells separated in BSA (1.046 gm/cm³) is very close to the value reported using isotonic gradients of Ficoll (1.051 gm/cm³; Anderson et al., 1970). It is also apparent that neither Renografin nor BSA exerted any measurable toxic effect on the cells during the separation procedure.

Cultured Chinese hamster ovary cells, whose clonogenicity could be quantitatively assayed by a simple in vitro colony-forming unit test, were used as a model for the more complex situation that occurs in the tumor. During the first 10 days of growth only a single population of cells banded, and its density varied with the age of the culture. The shifts in density at times between 1 and 10 days after plating were greater than the variation seen in replicate cultures prepared and centrifuged at the same time. After 13–14 days of growth, three distinct populations were always observed. This was not due simply to the movement of dead cells along with trapped viable cells down the gradient (see Section III), but rather to some factor(s) that appeared to be related to the age of the culture.

The viability of the cells recovered after centrifugation, as determined by phase contrast microscopy and/or a trypan blue dye-exclusion test, was always greater than 70%. These data did not, however, accurately reflect the ability of these cells to proliferate and form colonies. The plating efficiencies decreased from 90% to 60% during the first 10 days of growth and then fell to between 27% and 16% by day 14. The possibility that the concomitant appearance at this time of three populations of cells differing in density represented cells with differing clonogenic potential was tested and confirmed. Over 85% of the total clonogenic cells recovered from the gradient were present in the lightest population.

Three to four populations of tumor cells were recovered after centrifugation when the initial cell suspension was derived from a single tumor. The densities at which the populations formed remained constant, but the relative number of cells comprising each population varied for each tumor. If cells were pooled from four or more tumors, an additional population of cells having an average density of 1.08 gm/cm^3 was resolved.

The average size or volume of the cells in each of the populations also appeared to differ. Cells banding at the heavier densities were smaller than those collected in the lighter fractions. The high osmolarity of Renografin could have accounted for this (Grdina et al., 1973). However, their size differences were still apparent after resuspension of the cells into isotonic medium.

A lung colony assay was chosen to determine the clonogenicity of the separated tumor cell populations because it is an excellent quantitative test for the transplantability of tumors. The number of lung metastases formed is in direct proportion to the number of intravenously injected tumor cells (Hill and Bush, 1969). The efficiency of this assay is enhanced by the use of whole-body irradiated animals (Withers and Milas, 1973; Brown, 1973) and by the addition of heavily irradiated cells to the tumor cell inoculum whose viability is being tested (Hill and Bush, 1969; Fidler, 1973). Presumably, each of these procedures allows for more efficient trapping of viable tumor cells in the lungs of the host animal and, therefore, an increased number of pulmonary metastases.

Using this assay, clonogenic tumor cells were found to be distributed throughout the gradient. The greatest concentrations of these cells, however, were found in populations banding between 1.05 and 1.10 gm/cm³. Only the most dense population, 1.17 gm/cm³ appeared to contain a lower concentration of clonogenic cells than did the unseparated controls. While the factor(s) leading to the distribution of cells throughout the gradient is not clear, it is probable that parameters such as cell age or state of cellular differentiation as well as the past environmental conditions in the tumor, which include changes in oxygen tension, pH, and availability of essential nutrients, are important.

Experiments were performed to determine whether heterogeneous populations such as plateau phase cultures and solid tumors contained clonogenic cells that could be separated from nonclonogenic cells by density gradient centrifugation. They do, and this was successfully demonstrated using Renografin as the separation medium.

ACKNOWLEDGMENTS

This investigation was supported by NIH Research Grant Nos. CA-06294, CA-11430 and CA-11138 from the National Cancer Institute and Institutional Grant No. RR5511-10 IN-18-151028 from the University of Texas System Cancer Center M.D. Anderson Hospital and Tumor Institute, Houston, Texas.

I gratefully acknowledge the help of Drs. I. Basic, M. L. Meistrich, and L. Milas in the performance of this research. These studies were carried out in the laboratory of Dr. H. R. Withers, and I wish to thank him for his encouragement and support. Finally, I wish to acknowledge the excellent technical assistance of Ms Kathy Mason and Ms Sharon Guzzino.

REFERENCES

Anderson, E. C., Peterson, D. F., and Tobey, R. A. (1970). *Biophys. J.* **10**, 630.
Brown, J. M. (1973). *Br. J. Radiol.* **46**, 613.
Cahn, F. G., and Fox, M. S. (1968). *J. Bacteriol.* **95**, 867.
DeSimone, J., Kleve, L., and Shaeffer, J. (1974). *J. Lab. Clin. Med.* **84**, 517.
Fidler, I. J. (1973). *Eur. J. Cancer* **9**, 223.
Grdina, D. J., Milas, L., Hewitt, R. R., and Withers, H. R. (1973). *Exp. Cell Res.* **81**, 250.
Grdina, D. J., Milas, L., Mason, K. A., and Withers, H. R. (1974a). *J. Natl. Cancer Inst.* **52**, 253.
Grdina, D. J., Meistrich, M. L., and Withers, H. R. (1974b). *Exp. Cell Res.* **85**, 15.
Grdina, D. J., Basic, I., Mason, K. A., and Withers, H. R. (1975). *Radiat. Res.* **63**, 483.
Hahn, G. M., and Little, J. B. (1972). *Radiat. Res. Q.* **8**, 39.
Hill, R. P., and Bush, R. S. (1969). *Int. J. Radiat. Biol.* **15**, 435.
Meistrich M. L. (1972). *J. Cell. Physiol.* **80**, 299.
Meistrich, M. L., and Trostle, P. K. (1975). *Exp. Cell Res.* **92**, 231.
Milas, L., and Tomljanovic, M. (1971). *Rev. Eur. Etud. Clin. Biol.* **16**, 462.
Ross, D. W., and Sinclair, W. K. (1972). *Cell Tissue Kinet.* **5**, 1.
Shortman, K. (1968). *Aust. J. Exp. Biol. Med. Sci.* **46**, 375.
Suit, H. D., and Suchato, D. (1967). *Radiology* **89**, 713.
Tamir, H., and Gilvarg, C. (1965). *J. Biol. Chem.* **241**, 1085.
Williams, N., Kraft, N., and Shortman, K. (1972). *Immunology* **22**, 885.
Withers, H. R., and Milas, L. (1973). *Cancer Res.* **33**, 1931.

Chapter 21

Rapid Screening Assay for Revertants of Murine Sarcoma Virus-Transformed Cells

S. NOMURA AND P. J. FISCHINGER

Laboratory of Viral Carcinogenesis,
National Cancer Institute,
Bethesda, Maryland

I. Introduction

Cell transformation by murine sarcoma virus (MSV) without the participation of mouse-tropic murine leukemia virus (MuLV) results in a state in which infectious virus is absent, but the MSV genome can be rescued by superinfection with MuLV. Two types of mouse cells have been obtained after MSV transformation. One type, derived from Moloney (M) MSV transformation of the 3T3FL (subline of Swiss 3T3 cells) has been designated sarcoma positive, leukemia negative Moloney-3T3FL, abbreviated S+L−(M−3T3FL) or mouse S+L− (Bassin *et al.*, 1970). It is characterized by the production of both MuLV group-specific p30 antigen and C-type particles, for which no infectivity has been described (Bassin *et al.*, 1971a). An example of the other type is the Kirsten (Ki) MSV-transformed nonproducer BALB/3T3 (Ki-BALB), which appears to be completely negative for production of MuLV particles and antigen (Aaronson and Weaver, 1971). These MSV-transformed mouse clonal cell lines give rise spontaneously and at variable rates to flat variants (revertants) with some properties

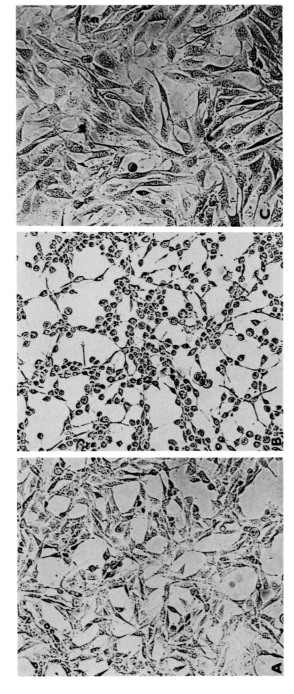

FIG. 1.   Morphology of 3T3FL, S+L−, and its variant cell lines. (A) Normal 3T3FL cells. (B) S+L− (3197-3) cells. (C) Flat variant 3D-2-14 cells (passage 20). × 100. Reproduced from Nomura et al. (1972) by courtesy of Virology.

of nontransformed cells (Fig. 1) from which MSV can no longer be rescued by superinfection with MuLV (Fischinger *et al.*, 1972; Nomura *et al.*, 1972, 1973a, 1974). During the course of investigation of reversion frequency in mouse S+L− cells, we were able to detect revertants by a rapid screening assay technique (Nomura *et al.*, 1973b) instead of using the more cumbersome cloning procedure in microtiter wells as previously described (Goldsby and Zipser, 1969; Fischinger *et al.*, 1972; Nomura *et al.*, 1972, 1973a).

The multiply cloned mouse S+L− 3197 cell line was composed of hyperrefractile, loosely attached cells and did not form a compact cell monolayer (Fig. 1). Infection of this cell line with MuLV changed the cellular morphology to a slightly more rounded cell type, and the cells were readily detached from the surface of the container. In contrast, when rescuenegative revertants derived from 3197 cells were infected with MuLV there was no observable change in cellular morphology. These characteristics were used as the basis for a rapid screening assay technique for revertants of MSV-transformed mouse cells. The rapid screening assay can also be applied to the detection of reversion frequency in Ki-BALB and S+L− revertants retransformed by M-MSV of Ki-MSV, or potentially in cat, dog, or human S+L− cells with similar properties (Fischinger *et al.*, 1974; Nomura and Fischinger, 1974).

The rapid screening assay procedure is a simple and convenient method for the detection and quantitation of revertants derived from some of MSV-transformed cells, especially in populations with very low reversion frequency, and its sensitivity is comparable to that of the cloning technique. It may also prove to be useful for the investigation of a variety of agents that increase the reversion frequency in these transformed cells.

II.   Methods

We have tested for reversion frequency in two sublines of 3197, 3-321 and 3-360 (Nomura *et al.*, 1973a), in Ki-BALB and in S+L− revertants retransformed by MSV. The method is probably applicable to any type of MSV-transformed cell that will not grow as a monolayer or that will change to a loosely attached culture after MuLV infection. All cell cultures were grown and maintained after virus infection in modified McCoy's 5a medium with 10% fetal calf serum. Cultures were maintained by routine weekly subculture by shaking.

Cells were infected in suspension with MuLV at a multiplicity of infection (m.o.i.) >5 focus-inducing units (FIU) (Bassin *et al.*, 1971b) per cell, and 1000–5000 cells in 5 ml of medium were delivered into

25 cm² flasks (Falcon), and the flasks were incubated at 37°C. After 4 days of incubation, flasks were shaken to detach the relatively loose cells, which were removed in the supernatant. Cultures were washed twice with medium and then refed with 5 ml of fresh medium. After a further 3–4 days of incubation at 37°C, flat, contact-inhibited revertant cell colonies formed and were counted. When S+L− cells were infected with MuLV at high m.o.i. (> 5 FIU), MSV was rescued from them about 12 hours after infection (Peebles *et al.*, 1971). At that time, rescue-negative revertant cells underwent MuLV infection, viral interference developed, and no transformation of revertants was possible by the rescued MSV. Revertants, therefore, remained flat and formed colonies. In an analogous manner S+L− cells of other species can be infected with leukemia viruses, which grow most readily in those cells. The same procedure then applies.

III.   Sensitivity of Assay

The sensitivity of this technique in the detection of revertants was compared with that of the cloning technique: 0.2 ml of cell suspension, containing 1 cell/ml of medium, was delivered into each microtiter well (Micro Test II tissue culture plate, Falcon); 7–9 days after plating, individual colonies were isolated and tested for rescuable MSV genome by superinfection with MuLV. Quantitatively, the results of these two assays (cloning and rapid screening, respectively) were in close agreement, with the following reversion frequencies of a rescue-negative revertant: 0.0002 and 0.0003 in 3-321 cultures, 0.005 and 0.003 in 3-360 cultures, 0.0023 and 0.0014 in Ki-BALB cultures, and 0.009 and 0.0056 in 3382 cells derived from S+L− revertent retransformed by Ki-MSV. A second cycle of reversion in M-MSV retransformed S+L− revertants was a very rare event, but did occur at a frequency of between 10^{-4} and 10^{-5} by the rapid screening technique. Results are summarized in Table I.

IV.   Reliability of Assay

The reliability of the rapid screening assay technique for revertants was assured by the following reconstruction experiment. Various numbers (400,

TABLE I

COMPARISON OF SENSITIVITY OF DETECTION OF REVERTANTS FROM
MSV-TRANSFORMED MOUSE CELLS BY CLONING AND RAPID SCREENING
ASSAY TECHNIQUES

Cell line	Assay technique	Number of cells plated	Number of revertants[a]	Average frequency
3-321[b]	Cloning			
	Microtiter wells[c]	2,300	0 ⎫	
	Plates[d]	14,100	3 ⎭	0.00018
	Rapid screening	77,200	20	0.00026
3-360[b]	Cloning			
	Microtiter wells	600	3	0.005
	Rapid screening	10,000	33	0.0033
Ki-BALB[e]	Cloning			
	Microtiter wells	800	3 ⎫	
	Plates	1,825	3 ⎭	0.0023
	Rapid screening	10,400	14	0.0014
3382[f]	Cloning			
	Microtiter wells	220	2	0.009
	Rapid screening	35,200	197	0.0056

[a] Flat and rescue-negative.

[b] Mouse S+L−subline.

[c] Plated 0.2 ml of cell suspension containing 1 cell/ml.

[d] Because of their low reversion frequency, cells were seeded into plastic plates (60 × 15 mm) at a concentration of 100 cells/plate. On day 9, flat cell colonies were isolated and recloned in microtiter wells.

[e] Kirsten MSV-transformed nonproducer BALB/3T3 cells.

[f] Kirsten MSV-retransformed S+L− revertant (3D-2-14).

200, 100, 50, and 25) of S+L− revertant 3D-2-14 cells (Nomura et al., 1972) were mixed with a constant number (10,000) of 3-321 cells and distributed into flasks in the presence or in the absence of MuLV. Four flasks were employed in each group. At 4 days, cultures were shaken, washed twice, and refed with fresh medium as described above. On day 9, cultures were fixed and stained with Giemsa, and colonies of revertant cells were counted. The average number of revertants in the infected groups was 186, 100, 48.3, 24.5, and 10, respectively. Some stained cultures are shown in Fig. 2. An average cloning efficiency of all groups was 47%, which was similar to that of 3D-2-14 cell alone (42%). These findings, nonetheless, indicated that the rapid screening assay technique allowed detection of revertant colonies in mixed population containing a large excess of transformed cells.

FIG. 2. Reconstruction experiment. Mixtures containing 200(1), 100(2), 50(3), 200(4), and 0(5) MSV-revertant cells plus 10,000 S+L− cells were seeded into flasks, and cultures (1), (2), and (3) were simultaneously infected with MuLV. On day 4, cultures (1)–(3) were shaken, washed, and refed. Photographs were taken after staining on day 9. Reproduced from Nomura *et al.* (1973b) by courtesy of *Nature* (*London*).

V.   Application of Rapid Screening Assay

The rapid screening assay procedure is not only useful for determination of the spontaneous reversion frequency in MSV-transformed cells, but also valuable for the study of the reversion effects of physical, chemical, and biological agents that enhance the frequency of reversion.

Two million 3-321 cells, whose reversion frequency was lower than 0.0003, were exposed to 20 ml of medium containing various concentrations of fluorodeoxyuridine (FUdR) for 1 hour at 37° C, or various concentrations of Colcemid for 24 hours at 37° C. At the end of the exposure period, medium containing the chemical was removed, cells were washed twice with medium and dispersed, and 100 cells were plated into 60 × 15 mm plastic plates (cloning technique), or 1000 cells were delivered, in the presence of MuLV (m.o.i. > 5 FIU), into flasks for rapid screening assay. More than 10 plates or flasks were used for each concentration of chemical. After 4 days of incubation, plates were refed and flasks were shaken, washed, and refed as described in Methods. On day 8, cultures were examined for the presence of revertant colonies. The results of dose-response of FUdR in the derivation of revertants from S+L− cells are presented in Fig. 3; the optimum dosage of FUdR was 15 μg/ml which produced a reversion frequency of 0.008, and the same efficiency of revertant detection was found by both

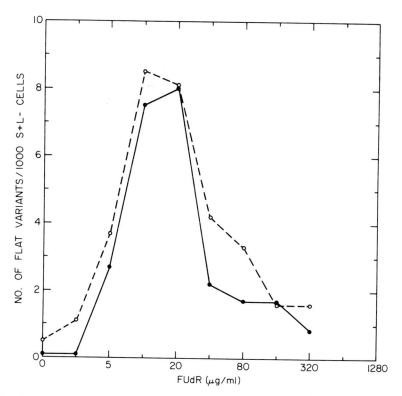

FIG. 3. Dose-response of fluorodeoxyuridine (FUdR) in the induction of revertants from S+L− cells. S+L− cells were treated with various concentrations of FUdR for 1 hour at 37° C. At the end of the exposure period, cells were washed and plated at a concentration of 100 cells per plate for cloning technique (●) or 1000 MuLV-infected cells per flask for rapid screening assay technique (○). On days 7 to 9, frequencies of revertants were determined. Reproduced from Nomura *et al.* (1973b) by courtesy of *Nature* (*London*).

techniques. FUdR, therefore, appeared to increase the reversion frequency by about 30-fold.

Treatment of 3-321 cells with Colcemid similarly demonstrated a maximum increase in appearance of revertants at the concentration of 0.04 μg/ml, which represented a 20-fold increase over spontaneous reversion frequency.

Accordingly, this test system, which is now being applied to MSV transformed cells of various species, can potentially have a wide spectrum of applicability. We recognize two specialized aspects of the system which may nonetheless be of value. The first is represented by those cells which have a rescuable MSV genome, release virus after MuLV infection, but remain attached to the surface after the usual procedure. These cells would register as phenotypic revertants. This test is then valuable in segregating from a large number of test cells this subpopulation of cells with the true revertants. Further separation of phenotypic and true revertants is possible. A second factor is that all revertants picked after this isolation procedure will produce the infecting leukemia virus. These cells may be valuable because generally they produce more virus than the original normal control cells (Nomura *et al.*, 1972).

REFERENCES

Aaronson, S. A., and Weaver, C. A. (1971). *J. Gen. Virol.* **13**, 245.

Bassin, R. H., Tuttle, N., and Fischinger, P. J. (1970). *Int. J. Cancer* **6**, 95.

Bassin, R. H., Phillips, L. A., Kramer, M. J., Haapala, D. K., Peebles, P. T., Nomura, S., and Fischinger, P. J. (1971a). *Proc. Natl. Acad. Sci. U.S.A.* **68**, 1520.

Bassin, R. H., Tuttle, N., and Fischinger, P. J. (1971b). *Nature (London)* **229**, 564.

Fischinger, P. J., Nomura, S., Peebles, P. T., Haapala, D. K., and Bassin, R. H. (1972). *Science* **176**, 1033.

Fischinger, P. J., Nomura, S., Tuttle-Fuller, N., and Dunn, K. J. (1974). *Virology* **59**, 217.

Goldsby, R. A., and Zipser, E. (1969). *Exp. Cell Res.* **54**, 271.

Nomura, S., and Fischinger, P. J. (1974). *In* "Mechanisms of Virus Disease" (W. S. Robinson and C. F. Fox, eds.), pp. 273–285. Benjamin, Menlo Park, California.

Nomura, S., Fischinger, P. J., Mattern, C. F. T., Peebles, P. T., Bassin, R. H., and Friedman, G. P. (1972). *Virology* **50**, 51.

Nomura, S., Fischinger, P. J., Mattern, C. F. T., Gerwin, B. I., and Dunn, K. J. (1973a). *Virology* **56**, 152.

Nomura, S., Dunn, K. J., and Fischinger, P. J. (1973b). *Nature (London)* **246**, 213.

Nomura, S., Dunn, K. J., Mattern, C. F. T., Hartley, J. W., and Fischinger, P. J. (1974). *J. Gen. Virol.* **25**, 207.

Peebles, P. T., Bassin, R. H., Haapala, D. K., Phillips, L. A., Nomura, S., and Fischinger, P. J. (1971). *J. Virol.* **8**, 690.

Chapter 22

Selective Techniques for the Isolation of Morphological Revertants of Sarcoma Virus-Transformed Cells

JOEL S. GREENBERGER,[1] WILLIAM I. BENSINGER,
AND STUART A. AARONSON

Laboratory of RNA Tumor Viruses,
National Cancer Institute,
Bethesda, Maryland

I. Introduction

The isolation and characterization of mutants of viruses that infect bacterial and animal cells has led to increased understanding of viral genetic functions. This approach has been increasingly used in avian and mammalian systems in attempts to determine the number and actions of genes that mediate the transforming activities of oncogenic viruses (Eckhart, 1969; Toyoshima and Vogt, 1969; Tegtmeyer and Ozer, 1971; Scolnick

---

[1] *Present address:* Joint Center for Radiation Therapy, Department of Radiation Therapy, Harvard Medical School, Boston, Massachusetts.

et al., 1972; Robb and Martin, 1972; Kimura and Dulbecco, 1972; Stephenson *et al.*, 1972a; Wong *et al.*, 1973; Wyke, 1973a). The methods used in mammalian systems to obtain virus mutants have usually involved mutagenization of virus, and screening systems have generally been designed to detect conditional lethal effects.

DNA and RNA tumor viruses that nonproductively transform and become stably associated with the cell have provided another approach with which to study both viral and cellular genes involved in the expression of viral transformation. Selection systems devised to enhance survival of revertant clones of virus transformed parental cells have yielded both temperature-dependent (Ranger and Basilico, 1973) and independent revertants (Pollock *et al.*, 1968; Rabinowitz and Sachs, 1970; Culp *et al.*, 1971; Stephenson *et al.*, 1972b; Fischinger *et al.*, 1972; Hatanaka *et al.*, 1973; Ozanne, 1973; Greenberger and Aaronson, 1974; Gazdar *et al.*, 1974) that express normal growth properties *in vitro* and *in vivo*, and yet still contain the transforming virus. This chapter describes selection techniques that have been developed for the isolation of morphological revertants of tumor virus-transformed cells and characterization of their biological properties.

II. Methods of Preparation of Transformed Cells for Selection of Revertants

A. Mutagenization of Sarcoma Virus-Transformed Nonproducer Cells

In tissue culture, murine sarcoma virus (MSV) transforms cells in the absence of type-C helper virus but requires helper leukemia virus functions for replication as an infectious virus (Aaronson and Rowe, 1970; Aaronson *et al.*, 1970; Bassin *et al.*, 1971). Cells nonproductively transformed by MSV in the absence of helper virus lack detectable leukemia virus antigens (Huebner, 1967; Aaronson and Rowe, 1970) or transplantation antigens (Stephenson and Aaronson, 1972; Strouk *et al.*, 1972; Greenberger *et al.*, 1974b) and are highly malignant *in vivo* (Stephenson and Aaronson, 1972). The sarcoma viral genome can remain stably associated with the nonproducer cell for hundreds of generations (Stephenson *et al.*, 1972b), apparently functioning like cellular genes. While certain biochemical (Salzberg and Green, 1972; Hatanaka *et al.*, 1973; Stephenson *et al.*, 1973; Fishman *et al.*, 1974) and immunological changes (Aoki *et al.*, 1973) have been detected in MSV nonproducer cells, the number of sarcoma viral functions and their gene products remain to be elucidated. Thus, the MSV

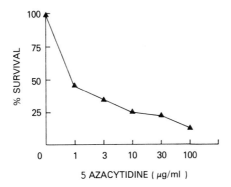

FIG. 1. Survival K-BALB cells following 24-hour treatment with increasing doses of the mutagen 5-azacytidine. Cells were allowed a 2-week recovery following drug treatment.

nonproducer appeared to be an ideal model system with which to attempt to develop selection techniques for isolation of morphological revertants of a tumor virus-transformed cell line.

Since spontaneous reversion of the MSV nonproducer cell is a rare event (Stephenson *et al.*, 1972b), we sought to increase the probability of its occurrence by treatment of the cells with a mutagen. Among the drugs used in this laboratory, three cytotoxic agents, mitomycin C, 5-azacytidine, and 8-azaguanine, which act by incorporation into and/or cross-linking DNA (Zimmerman *et al.*, 1973), have been used most extensively. To determine the optimum drug dosage, an experimental kill curve is first constructed. The effect of the mutagen is then judged by its effect on cell survival. Figure 1 shows the results of exposure to 5-azacytidine on survival of MSV-transformed nonproducer BALB/3T3 (K-BALB) cells. The optimum drug dosage was arbitrarily chosen to give a 10- to 100-fold decrease in cell survival. This amount of cell kill has led to good yields of revertant cells.

Exponentially growing cells are treated with the mutagenizing drug for 24 hours; the cells are then washed several times in drug-free medium and are allowed to recover for 1–2 weeks. For the selection of revertants, it is important that the cells be used relatively soon since continued propagation in mass culture selects for the transformed phenotype and defeats the purpose of mutagenization.

B. Suspension Culture Selection System

Methods for the isolation of revertant clones have been devised to take advantage of the differences in growth properties of the virus-transformed and normal cells. Transformed cells are capable of multiplying on contact-inhibited cell monolayers (Aaronson and Todaro, 1968), at low serum concentration (Jainchill and Todaro, 1970), or in suspension in semisolid medium (Stoker *et al.*, 1968), while the growth of normal cells is restricted under each of these conditions. The techniques for isolation of morphological revertants are based on methods used by Wyke to increase the frequency of isolation of conditional lethal mutants of avian sarcoma virus (Wyke, 1971, 1973b). The approach is to suspend single cells in semisolid medium containing agar or methyl cellulose. This allows only the transformed cells to replicate. Then an antimetabolite that kills actively dividing cells is added to the culture. After an appropriate time interval, the surviving cells are removed from the suspension, washed free of the antimetabolite, and plated into petri dishes for subsequent selection of morphologically normal-appearing cells.

Methyl cellulose in aqueous solution has a unique and useful property of being less viscous at lower temperatures than at high, making it easier to handle while cool and allowing it to become more solid at incubator temperatures. Methocel (Dow Chemical Co.), viscosity 4000 cps, comes as a fine white powder. It dissolves in water with some difficulty, but after the solution has stood for 48 hours, the methyl cellulose hydrates and develops a uniform consistency. A methyl cellulose stock solution is made up by first dissolving 5 gm of the powder in 100 ml of distilled water. This mixture is then autoclaved at 100° C and 18 psi for 20 minutes. After cooling, 100 ml of sterile 2× concentrated Dulbecco's Modification of Eagle's medium (DMEM) without serum is added. Uniformity of the suspension can be facilitated by the use of a sterile magnetic stirrer; it generally requires around 48 hours at 4° C to achieve complete hydration.

The resultant solution is quite viscous and difficult to pipette with ordinary serological pipettes. Easier handling of the viscous 2.5% stock is facilitated by the use of disposable plastic syringes for drawing up and dispensing medium. For this reason the methyl cellulose stock should be stored in wide-mouth jars. Two milliliters of the 2.5% methylcellulose stock is added to form a base in 60 mm petri dishes. The formation of a hardened base at 37° C for 12 hours helps to prevent the cells from attaching to the bottom of the plastic petri dish.

For preparation of the methyl cellulose overlay, the 2.5% stock solution is diluted to a final concentration of 0.9% using DMEM supplemented with 10% calf serum. The cells are then mixed in this solution and overlaid on the prehardened methyl cellulose base and further incubated at 37° C. Around 12 hours later the antimetabolite is added and following drug treatment, the cells are removed from the semisolid medium by dissolving it with additional DMEM at 4° C. The cells then can be easily transferred to centrifuge tubes; after two washes by centrifugation at 1000 rpm and re-suspension in complete medium, cell counts are made by hemacytometer, and the cells are plated at serial dilutions to 60-mm petri dishes.

Because cells occasionally penetrate the 2.5% methyl cellulose base and attach to the plastic substrate, we have found that agar provides an even better base. A stock suspension of 5% agar (Difco Labs special Agar-Noble) is distilled H_2O is made. The agar is heated by steaming (no pressure) to cause it to liquefy. It is then diluted to 0.7% with warm (56° C) complete DMEM and dispensed in 2-ml aliquots to 60-mm petri dishes. This suspension hardens quickly and must be dispensed while warm. Bubbles in the base are removed by light flaming. After hardening at 37° C for 12 hours, the overlay of 0.9% methyl cellulose can be added. Agar can be used at a lower concentration (0.3%), instead of methyl cellulose, as the cell suspension medium. However, cells are very difficult to remove from the agar suspension for any further manipulation.

The antimetabolite or other killer drugs used in this laboratory have included thymidine analogs, iododeoxyuridine (IdU), bromodeoxyuridine (BrdU), and ³H-labeled thymidine, which are incorporated into the DNA of replicating cells (Zimmerman et al., 1973). IdU and BrdU cause DNA-strand breaks upon excitation with high-intensity white light. In contrast, thymidine-³H kills cells by localized radiation effects (Zimmerman et al., 1973). It has been shown (Stoker et al., 1968) that transformed cells in methyl cellulose suspension culture incorporated up to 10 times more thymidine-³H than did normal cells. Another useful killer drug, 5-fluorouracil (5-Fu), both is incorporated into cellular RNA and competes with de-oxyuridylate for the enzyme thymidylate synthetase leading to impaired synthesis of both RNA and DNA (Zimmerman et al., 1973).

Figure 2 compares the effects of varying concentrations of killer drugs on colony-forming efficiency in methyl cellulose of sarcoma virus-trans-formed cells as compared to the contol BALB/3T3 line. The optimum con-centration of each drug gave a 10- to 100-fold selective kill of the transformed cells. Our technique for the selective kill of transformed cells utilizes treat-ment with either 5-fluorouracil at 4 μg/ml or thymidine-³H at 3 μCi/ml for 48 hours.

After drug treatment, the surviving cells are removed as described above, plated at 10-fold dilutions to new petri dishes, and allowed 7–10 days to recover. The petri dishes are observed daily for the appearance of morpho-logically flat colonies. Such colonies are isolated by means of cloning cylinders, sterile metal cylinders with silicone grease applied to one end. This allows the cloning cylinder to adhere to the plastic surface of the petri dish, so that localized trypsin (0.1% in phosphate-buffered saline) exposure can be used to dislodge the cells within the confines of the cylinder. The cells

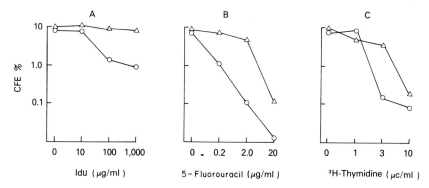

Fig. 2.   Colony-forming efficiency (CFE) of K-BALB (○) and BALB/3T3 (△) follow-ing 48-hour treatment with increasing concentrations of (A) iododeoxyuridine, (B) 5-fluorouracil, and (C) ^{3}H-labeled thymidine of cells suspended 12 hours earlier in 0.9% methyl cellulose as described in the text.

are then removed by means of a disposable pipette and transferred to new petri dishes.

The flat-appearing colonies selected in this manner tend to be contaminated by residual transformed cells. Thus, in order to isolate clonal lines of revertant cells, revertant colonies grown at high cell dilution and arising from a single cell are reisolated by the cloning cylinder method. In order to rigorously show that individual revertant clones result from separate mutagenizing events, it is necessary to obtain only a single revertant from a given stock of mutagenized transformed cells. Since a particular mutagen may have a predilection for certain sites in the DNA, we have also used different mutagens to try to increase the range of mutations, and thus increase the number of viral and/or cellular gene functions that may be affected. Figure 3 compares the morphology of the control BALB/3T3 parental, the sarcoma virus transformant, and two representative revertant clones. The appearance of the revertant cells is almost indistinguishable from that of the nontransformed line. A schematic diagram of the selection system is presented in Fig. 4.

III.   *In Vitro* Biological Properties of Morphological Revertants

The properties of morphological revertants, isolated by the methods described above, can be compared with those of the parental transformed line and with control BALB/3T3 cells. A number of *in vitro* properties have

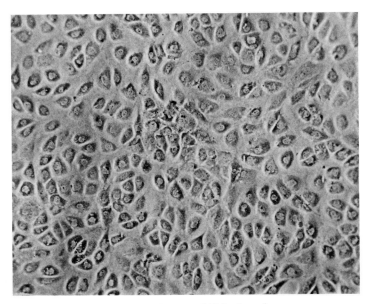

FIG. 3A.   See page 244 for legend.

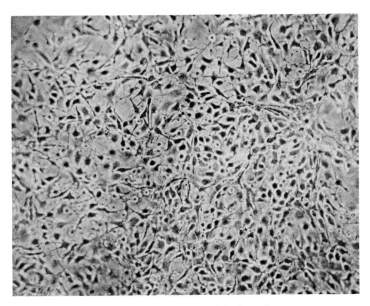

FIG. 3B.   See page 244 for legend.

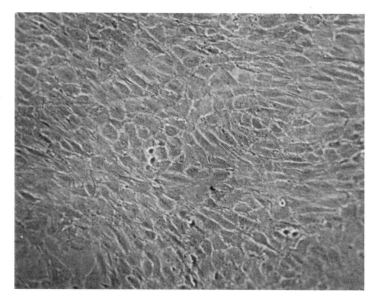

FIG. 3C.

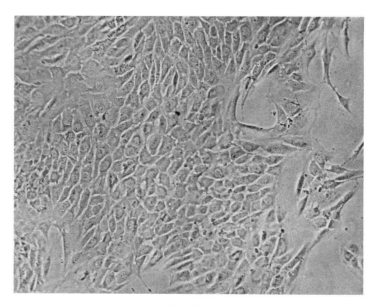

FIG. 3D.

FIG. 3.   Morphological appearance of parental and revertant clonal lines: (A) BALB/3T3;
(B) K-BALB; (C) R-26; (D) R-23.

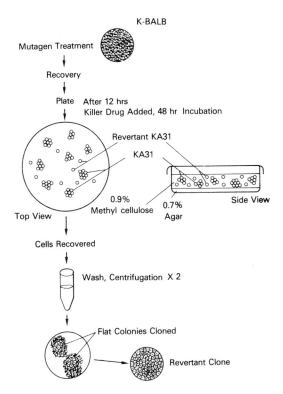

FIG. 4. Schematic diagram of the procedures used to isolate morphological revertants from a sarcoma virus-transformed nonproducer cell line (K-BALB, synonymous with KA31).

been shown to correlate with the ability of cells to form tumors in the animal. Properties associated with loss of growth control in tissue culture include the ability to grow at high density, to form colonies in semisolid medium, to form colonies on contact-inhibited monolayers, and to grow in serum-deficient medium. The saturation density of cells is determined by plating them at 1×10^4 cells/cm^2 with subsequent cell counts at daily intervals. The saturation density is taken as the cell density at which there is no further increase in three consecutive cell counts at daily intervals. As shown in Table I, the saturation densities of a series of morphological revertants range from 9.0×10^4 to 4.0×10^4 cells/cm^2, similar to that of BALB/3T3. This is at least 5- to 10-fold lower than that of the transformed nonproducer clone.

The ability of cells to form colonies on monolayers of contact-inhibited BALB/3T3 cells is another measure of loss of growth control in tissue culture. Serial dilutions of each cell to be tested are plated on confluent monolayers. The medium is changed twice weekly, and the number of dense

TABLE I

In Vitro BIOLOGICAL PROPERTIES OF K-BALB MORPHOLOGICAL
REVERTANT CLONES

Cell line	Saturation density (cells/cm^2)	Doubling time (hr)	Colony-forming efficiency[a] (% colonies/cells plated)		
			Empty petri	On BALB/3T3 monolayer	In 0.9% methyl-cellulose
BALB/3T3	5×10^4	22	10.0	< 0.001	< 0.001
K-BALB	3×10^5	23	15.0	10.0	10.0
R-20	4×10^4	21	12.6	< 0.001	< 0.001
R-21	7×10^4	24	11.6	< 0.001	< 0.001
R-23	9×10^4	22	9.2	< 0.001	< 0.001
R-24	7×10^4	23	12.4	< 0.001	< 0.001
R-26	6×10^4	21	16.0	< 0.001	< 0.001

[a] Tenfold dilutions of each cell line were transferred to empty 50-mm petri dishes and petri dishes containing a confluent monolayer of BALB/3T3 or were suspended in 0.9% methyl cellulose and layered over a prehardened base of 2.5% methyl cellulose. Growing clones were scored at 14 days.

colonies forming on the monolayer is scored at 12–14 days. This can be done either visually or with the aid of a dissecting microscope or after fixation with formalin and staining with Harris hematoxylin. As shown in Table I, most of the revertants show no detectable colony formation on monolayers of BALB/3T3. In contrast, the transformed cell line forms colonies with high efficiency (10%) under the same conditions.

Using the methyl cellulose suspension system described above, the growth of revertant cell lines can be compared with that of the parental transformed line. Cells plated at serial dilutions in this medium are incubated for 12–14 days at 37°C with the addition of 2 ml of the 0.9% methyl cellulose in complete medium every 4–5 days. While the K-BALB line readily forms colonies that can be scored visually, none of the revertants nor the BALB/3T3 parental line grow in methyl cellulose suspension.

IV.   *In Vivo* Biological Properties of Morphological Revertants

Previous studies have shown that K-BALB is highly malignant when inoculated into weanling mice (Stephenson and Aaronson, 1972), whereas BALB/3T3 is nontumorogenic under the same conditions (Aaronson and

TABLE II

In Vivo TUMORGENICITY OF K-BALB MORPHOLOGICAL
REVERTANTS, K-BALB AND BALB/3T3 CELLS

Cell dose	Fraction of tumor-bearing mice[a] following subcutaneous inoculation with				
	K-BALB	BALB/3T3	R-21	R-23	R-27
10^4	3/5	NT	NT	NT	NT
10^5	5/5	NT	NT	NT	0/5
10^6	5/5	0/5	0/5	0/5	0/5
10^7	5/5	0/5	0/5	0/5	0/5

[a] Male BALB/c mice 6–8 weeks old were inoculated with washed cells suspended in 0.2 ml of phosphate-buffered saline. The site of subcutaneous inoculation was observed weekly for 12 weeks. All mice developing K-BALB tumors were dead of tumor within 10 weeks. NT means not tested.

Todaro, 1968). The *in vivo* growth characteristics of several revertant lines are shown in Table II. When BALB/c weanlings 6–8 weeks old are inoculated subcutaneously in the intrascapular region with washed cells at 10-fold dilutions in 0.2 ml of phosphate-buffered saline, none of the revertants form tumors at doses as high as 10^7 cells. In contrast, K-BALB forms tumors in a high fraction of animals when as few as 10^4 cells are inoculated. Thus, both the *in vitro* and *in vivo* biological properties of morphological revertants isolated by the technique described in this chapter are indistinguishable from those of the control BALB/3T3 cell line and very different from those of the malignant sarcoma virus-transformed line, from which each was isolated.

V. Conclusions

A technique has been developed for the isolation of contact-inhibited morphological revertants of a clonal line of sarcoma virus-transformed nonproducer cells. The methyl cellulose technique has made it possible to obtain a large number of genetically stable clonal revertants. Other studies have shown that the majority of revertants so far isolated are altered in sarcoma viral functions necessary for expression of the transformed state (Greenberger *et al.*, 1974a). A single revertant, previously isolated by another selection method (Stephenson *et al.*, 1972b), appears to be altered in cellular functions required for the expression of transformation. The frequency of occurrence of morphological revertants is very low. Consequently the selection methods, shown here to enhance by 10- to 100-fold the survival of morphological revertants, have greatly aided efforts to isolate these variant cells. The techniques described here are also applicable to the isolation of revertants of other carcinogen-induced, virus-transformed or spontaneously transformed tumor cells. These MSV transformed nonproducer cell revertants should be of use in developing a better understanding of the virus–cell interactions involved in transformation.

REFERENCES

Aaronson, S. A., and Rowe, W. P. (1970). *Virology* **42**, 9–19.
Aaronson, S. A., and Todaro, G. J. (1968). *Science* **162**, 1024–1026.
Aaronson, S. A., Jainchill, J. L., and Todaro, G. J. (1970). *Proc. Natl. Acad. Sci. U.S.A.* **66**, 1236–1243.
Aoki, T., Stephenson, J. R., and Aaronson, S. A. (1973). *Proc. Natl. Acad. Sci. U.S.A.* **70**, 742–746.
Bassin, R. H., Phillips, L. A., Kramer, M. J., Haapala, D. K., Peebles, P. T., Nomura, S., and Fischinger, P. J. (1971). *Proc. Natl. Acad. Sci. U.S.A.* **68**, 1520–1524.
Culp, L. A., Grimes, W. J., and Black, P. H. (1971). *J. Cell Biol.* **50**, 682–690.
Eckhart, W. (1969). *Virology* **38**, 120–125.
Fischinger, P. J., Nomura, S., Peebles, P. T., Haapala, D. K., and Bassin, R. H. (1972). *Science* **176**, 1033–1035.
Fishman, P. H., Brady, R. A., Bradley, R. M., Aaronson, S. A., and Todaro, G. J. (1974). *Proc. Natl. Acad. Sci. U.S.A.* **71**, 298–301.
Gazdar, A. F., Stull, H. C., Chopra, H. C., and Ikawa, Y. (1974). *Int. J. Cancer* **13**, 219–226.
Greenberger, J. S., and Aaronson, S. A. (1974). *Virology* **57**, 339–346.
Greenberger, J. S., Anderson, G. R., and Aaronson, S. A. (1974a). *Cell* **2**, 279–286.
Greenberger, J. S., Stephenson, J. R., Aoki, T., and Aaronson, S. A. (1974b). *Int. J. Cancer* **14**, 145–152.
Hatanaka, M. Klein, R., Toni, R., Walker, J., and Gilden, R. (1973). *J. Exp. Med.* **138**, 356–363.
Huebner, R. J. (1967). *Proc. Natl. Acad. Sci. U.S.A.* **58**, 835–842.
Jainchill, J. L., and Todaro, G. J. (1970). *Exp. Cell Res.* **59**, 137–142.
Kimura, G., and Dulbecco, R. (1972). *Virology* **49**, 394–403.
Ozanne, B. (1973). *J. Virol.* **12**, 79–89.
Pollock, R. E., Green, H., and Todaro, G. J. (1968). *Proc. Natl. Acad. Sci. U.S.A.* **60**, 126–133.
Rabinowitz, Z., and Sachs, L. (1970). *Nature (London)* **225**, 136–139.
Ranger, H. C., and Basilico, C. (1973). *J. Virol.* **11**, 702–708.
Robb, J. A., and Martin, R. G. (1972). *J. Virol.* **9**, 956–968.
Salzberg, S., and Green, M. (1972). *Nature (London), New Biol.* **210**, 116–118.
Scolnick, E. M., Stephenson, J. R., and Aaronson, S. A. (1972) . *J. Virol.* **10**, 653–657.
Stephenson, J. R., and Aaronson, S. A. (1972). *J. Exp. Med.* **135**, 503–515.
Stephenson, J. R., Reynolds, R. K., and Aaronson, S. A. (1972a). *Virology* **48**, 749–756.
Stephenson, J. R., Scolnick, E. M., and Aaronson, S. A. (1972b). *Int. J. Cancer* **9**, 577–583.
Stephenson, J. R., Reynolds, R. K., and Aaronson, S. A. (1973). *J. Virol.* **11**, 218–222.

Stoker, M., O'Neill, C., Berryman, S., and Wazman, J. (1968). *Int. J. Cancer* 3, 683–690.

Strouk, V., Frunder, G., Fenyo, C. M., Lamon, E., Shuryak, H., and Klein, G. (1972). *J. Exp. Med.* 136, 344–352.

Tegtmeyer, P., and Ozer, H. (1971). *J. Virol.* 8, 516–524.

Toyoshima, K., and Vogt, P. K. (1969). *Virology* 39, 930–931.

Wong, P. K., Russ, L. J., and McCarter, J. A. (1973). *Virology* 51, 424–431.

Wyke, J. (1971). *J. Exp. Cell Res.* 66, 203–210.

Wyke, J. A. (1973a). *Virology* 52, 587–590.

Wyke, J. A. (1973b). *Virology* 54, 28–36.

Zimmerman, A. M., Cameron, I. L. and Padilla, G. M., (1973). "Drugs and the Cell Cycle." Academic Press, New York.

Chapter 23

Selection, Screening, and Isolation of Temperature-Sensitive Mutants of Avian Sarcoma Viruses

JOHN A. WYKE

Department of Tumour Virology,
Imperial Cancer Research Fund,
London, England

I. Introduction

Both conditional (temperature-sensitive, *ts*) and nonconditional mutants of RNA tumor viruses have been used increasingly in recent years to investigate viral genetics and the physiology of viral infection. Nonconditional mutants can be very stable and are highly suited for genetic experiments. The advantage of *ts* virus mutants, on the other hand, is that they can be maintained at permissive temperature, and therefore theoretically may incur mutations in any viral gene. This not only provides the possibility of mapping the whole genome, but permits different viral genes to be distinguished functionally by complementation tests. Moreover, the mode and time of action of the mutated function can be determined by studying the physiological consequences of shifts between permissive and nonpermissive temperatures.

Avian sarcoma virus (ASV) *ts* mutants have already proved to be useful tools for such studies (reviewed by Wyke, 1975), so it seems sensible to attempt to isolate more mutants on the assumption that they too will be valuable in probing the life history of the virus. In view of this, an attempt is made herein to describe, evaluate, and suggest methods for increasing the efficiency of mutant isolation. The basic features of RNA tumor virus biology are not described in detail, and for this information, and the original references on which it is based, the reader should consult Tooze (1973).

II. Theoretical Considerations in *ts* Mutant Isolation

A great deal of labor may be required to isolate confirmed *ts* virus mutants, and before embarking on such a project careful thought should be given to the virus strains to be studied, the cells to be used for growing the virus, the choice of permissive and nonpermissive temperatures, whether or not to use mutagenesis, and the type of *ts* mutant to be sought.

A. Viruses

Several laboratory strains of avian sarcoma viruses, for example, the Schmidt-Ruppin and Prague strains of Rous sarcoma virus (RSV) and B77 virus, are nondefective for both virus replication and cell transformation in embryo fibroblasts *in vitro*. In other words, a single virus particle can productively infect a cell and, since the infection is not cytocidal, the infected cell will continue to multiply, producing daughters that exhibit virus-induced transformation and shed nondefective virus progeny. These virus strains can therefore be used to isolate mutations in both replicative and transforming functions.

However, other sarcoma viruses, such as the Bryan high titer (BH) strain of RSV are nonconditionally defective in replication (*rd*) and infectious progeny can be produced only by coinfection with a helper virus. Such helpers are often transformation-defective (*td*) viruses which replicate competently in fibroblasts but cannot transform them. Naturally occurring avian leukosis viruses are included among fibroblast *td* viruses, though they clearly are capable of transforming cells of the hemopoietic series. Sarcoma viruses which cannot replicate may possess only a single nonconditional defect in replication; in the case of BH-RSV this is the inability to produce viral envelope glycoproteins. However, if their helper possesses a full set of replication functions these may complement *ts* mutations in the sarcoma virus, or the two genomes may recombine to form wild-type sarcoma viruses, so it is generally practicable to study only mutations in functions not possessed by the helper. Thus *rd* sarcoma viruses are really only suited for isolating *ts* lesions in fibroblast transformation. On the other hand, the *td* viruses, which replicate competently in solitary infection, would be quite suitable as a source of *ts* mutants defective in replication or such transforming functions as these viruses possess. Unfortunately, in the absence of fibroblast transformation, the biological assays for these viruses are generally complex and not amenable to the search for and study of *ts* mutations.

For these reasons, most *ts* mutants studied so far have been isolated from nondefective sarcoma viruses. However, care must be used in the handling even of these viruses, for after several passages they generate spontaneously *td* derivatives which, because they do not transform fibroblasts, are not detected in the focus assay for sarcoma viruses. These inapparent *td* viruses, if present in significant proportions, may contaminate clones of potential mutants and complement any *ts* defects in their replication. For this reason, it is wise to minimize the risk of *td* virus contamination by using freshly cloned sarcoma virus stocks for mutant isolation.

B. Cells

Avian RNA tumor viruses grow efficiently only in avian cells, and these host cells are another source of genes which may recombine with, or functionally complement viral *ts* mutations. All chicken cells carry at least part of the DNA genome of an RNA tumor virus, RAV-0 (Neiman, 1973). This virus is released spontaneously by some lines of chickens (Vogt and Friis, 1971), and the fibroblasts of other chicken lines may exhibit transcription and translation of this endogenous viral genome to a greater or lesser extent. Expression of some endogenous virus functions can be detected in chick embryo fibroblasts by simple tests (Section III,A). In cells that are negative in these tests, there is a good chance that virus-specific RNA is not transcribed (Bishop *et al.*, 1973; Hayward and Hanafusa, 1973), though after prolonged culture even these cells may express endogenous viral functions. Clearly, transcription would be needed if endogenous viral functions were to complement exogenously infecting mutant viruses, and there is evidence that transcription of the endogenous genome is also a prerequisite for recombination between endogenous and exogenous viral genes (Weiss *et al.*, 1973). Thus, low-passage cultures of cells negative in tests for endogenous viral gene functions should always be used when seeking or working with mutants *ts* in virus replication. Since the endogenous virus cannot transform fibroblasts, it is not necessary to be so rigorous when studying viruses *ts* for transformation maintenance (late T class mutants, Section II,E).

An alternative approach is to use cells of other birds to propagate *ts* ASV. Although all avian species studied contain DNA with some homology to RAV-0 (Tereba *et al.*, 1975), the functioning of endogenous viral genes has so far proved to be very hard or impossible to detect in cells of Japanese quail (*Coturnix coturnix japonica*) or duck (*Anas platyrhynchos*). However, the drawback of these species is that they are susceptible to only a few of the known avian RNA tumor virus envelope subgroups. Japanese quail can be reliably infected by subgroup A and E viruses whereas duck cells are susceptible only to subgroups C and D. This can be a severe limitation when mutants in more than one subgroup are required, for example, for genetic studies. (The susceptibility of chicken cells should also be tested, for some may show a genetically determined resistance to one or more envelope subgroups, though others are susceptible to all the commonly used subgroups A to E.)

It is exceptional for avian RNA tumor viruses to replicate in mammalian cells. However, mammal cells can be infected by certain ASV and viral group-specific antigens and virus-induced transformation can be detected in these hosts. Mammalian cell lines can show a homogeneity and ease of manipulation not possessed by avian cells, and the worker may wish to exploit these advantages in the study of ASV *ts* mutants. Here again the envelope subgroup is important, for mammal cells can be infected readily only by RSV of subgroup D or by ASV B77 (subgroup C), and appropriate avian hosts must therefore be chosen for propagating these viruses.

C. Temperature

Avian cells grow well at 41°C, and this is a good choice for nonpermissive temperature with 35°C to 37°C providing suitable permissive conditions. This temperature difference is sufficiently large to be readily maintained by standard incubators, enabling *ts* mutants to be detected, yet the permissive temperature is still high enough for cells and viruses to grow at reasonable rates (foci of ASV-transformed cells reach about the same size in 6 days at 41°C, 8 days at 37°C, or 9 days at 35°C). However, the choice of permissive and restrictive conditions is less clear if one wishes to study avian virus mutants in mammal cells, most of which will not grow at much above 39°C (though the NRK line of rat cells can be maintained at 40°C). One option is to use a lower temperature range for mutant selection in these circumstances, but it is possible that the host cell may itself modify the *ts* range of the virus. For instance, Graf and Friis (1973) found that an NRK clone, infected by an ASV mutant *ts* for transformation, had a transformed phenotype at 33°C (permissive temperature) and normal morphology at 37°C (nonpermissive temperature). The same ASV mutant in chick cells had permissive and restrictive temperatures of 37°C and 41°C, respectively.

These temperatures have been recommended on the assumption that it is preferable to isolate heat-sensitive mutants, since viral functions will show a natural cold sensitivity to a greater or lesser extent. No cold-sensitive ASV mutants have been reported, though Somers and Kit (1973) have isolated a mutant of this type from murine sarcoma virus.

D.   Mutagenesis

There is only one report of the isolation of *ts* mutants from nonmutagenized ASV (Temin, 1971), and though these mutants occurred at high frequency (3%) they seemed to be difficult to maintain. Other workers have used various mutagens before isolating mutants (Section III,B), but there is as yet no proof that these actually increased mutant yield. The efficacy of mutagenesis has usually been assessed by determining the lethal effect of the mutagen, but in no case was the relationship between mutant induction and degree of virus killing determined. However, it is now apparent that a number of mutants have two or more *ts* lesions, and this is probably a consequence of the mutagenesis (reviewed by Wyke, 1975). If a mutagen is used, as described in Section III,B, one should always therefore beware of the possibility that the resultant mutants may carry multiple defects.

E.   Categories of *ts* Mutants

Early studies on *ts* mutants of nondefective ASV revealed three basic classes; those with a defect in fibroblast transformation only (class T), those with defective virus replication (class R), and those coordinately incapable of both transformation and replication (class C). Each class could be subdivided by studying the result of shifting infected cultures from permissive to restrictive temperature and *vice versa* at various times after infection. Such tests showed that some *ts* functions ("early" functions) were needed only during the first day of infection, but were essential then for transformation and/or viral replication to be observed later. The expression of other functions ("late" functions) was required for the maintenance of viral replication or cell transformation, although they had no role in the initiation of these events. It seems likely that early functions are those required transiently for the DNA provirus to be formed and to become stably associated with the cell, and therefore their action, or failure of action, is probably often irreversible. Late functions are probably required after this event, need to act continuously, and thus are presumably usually reversible in effect. It is possible however, that functions may exist that are required both early and continuously or late and transiently.

This classification of viral functions is probably an oversimplification, but it defines six groups, mutations in which would show predictable behavior under different combinations of permissive and nonpermissive conditions (summarized in Table I). This in turn suggests ways in which one could seek mutants in each of these groups, but it should be remembered that in any stock of mutagenized virus there may be multiple mutants showing phenotypic traits of more than one group. On the other hand, ASV may not possess gene functions that would fit into some of these groups; there is, as yet, no evidence for early R class defects, and most, if not all, late C class mutants isolated so far have proved to be multiple mutants bearing both late R class and late T class lesions.

Using Table I as a basis, rational screening procedures for ASV *ts* mutants can be devised, and these are outlined in Section III,E. However,

TABLE 1

CLASSIFICATION OF POSSIBLE AVIAN SARCOMA VIRUS MUTANTS ON THE BASIS OF THEIR BEHAVIOR UPON TEMPERATURE SHIFT

Temperature of incubation:[a]	35°C		35°C→41°C		41°C→35°C		41°C	
Phenotypic trait:[b]	T	R	T	R	T	R	T	R
Column number:	1	2	3	4	5	6	7	8
Virus phenotype								
Not temperature sensitive								
Nondefective, wild type	+	+	+	+	+	+	+	+
Transformation defective, td	−	+	−	+	−	+	−	+
Replication defective, rd	+	−	+	−	+	−	+	−
Temperature sensitive								
T class, early, probably irreversible	+	+	+	+	−	+	−	+
R class, early, probably irreversible	+	+	+	+	+	−	+	−
C class, early, probably irreversible	+	+	+	+	−	−	−	−
T class, late, probably reversible	+	+	−	+	+	+	−	+
R class, late, probably reversible	+	+	+	−	+	+	+	−
C class, late, probably reversible	+	+	−	−	+	+	−	−

[a] Temperature shifts were performed 24–48 hours after initiating infection.
[b] T, presence of cellular transformation, R, production of infectious virus.

most workers who have isolated *ts* mutants have done so without preliminary screening, being content to isolate single foci (and thereby clones) of mutagenized ASV and to test their temperature sensitivity. Approximately 1–2% of viruses so isolated were *ts*, most of them having the late T class phenotype and showing no defect in replication. It is not clear why late T class mutants are obtained so frequently, but they are of a type that would not be complemented by endogenous or *td* viruses, their defective function is nonessential, and, by their nature, they are easier to detect than viruses defective in replication alone. In view of this it is a pity that these mutants and those of late C class, are the only ones for which a true selective method is available at present.

The late T and C class mutants will both induce reversible *ts* changes in the transformed phenotype, and hence the growth characteristics of the cell they infect (Table I, columns 3 and 5). Thus these mutants can be selected by employing, at restrictive temperature and under cultural conditions where transformed cells grow and normal cells do not, negative selection procedures against growing cells. The experimental details for a variety of negative selection techniques have been given in Volume VIII of this series (Basilico and Meiss, 1974; Kao and Puck, 1974; Naha 1974; Vogel and Pollack, 1974), and many of these may be used. The next section describes in detail a technique that has already been successful in selectively isolating ASV *ts* mutants with late transformation defects.

III. Methods Applicable to the Selection and Screening of ASV *ts* Mutants

A. Culture and Propagation of Cells and Viruses

The preparation of avian embryo fibroblast cultures, the testing of their susceptibility to various envelope subgroups of avian RNA tumor viruses, and their use in the growth and focus assay of ASV have been detailed by Vogt (1969). It is only necessary to add that a certain variation in these standard techniques is possible to suit particular laboratory conditions. Polybrene (2 μg/ml) is added to medium to enhance adsorption to the cell of all virus subgroups other than A (Toyoshima and Vogt, 1969a).

Perhaps the most widely used probe for endogenous viral functions in the cell is the chick helper factor test, fully described by Weiss *et al.* (1973), which detects viral envelope glycoproteins. The complement fixation test for endogenous viral group-specific antigens is a useful additional test if appropriate antiserum is available (Vogt and Friis, 1971).

B.   Mutagenesis

Three principal methods of chemical mutagenesis have been used.

1. Free virus (about 10^6 focus-forming units/ml) is mixed with an equal volume of N-methyl-N'-nitro-N-nitrosoguanidine (2 μg/ml in 0.2 M phosphate buffer pH 6.0, containing 0.002 M EDTA), and after 10 minutes of incubation at 30°C the mutagen is removed by extensive dialysis against Tris buffer in the presence of 10% calf serum in the cold (Martin, 1970; Kawai et al., 1972). Being unstable in solution, the mutagen is dissolved at 100 mg/ml in dimethyl sulfoxide (DMSO) immediately before use.

2. Fully transformed, virus-producing cells are treated for 24 hours with the RNA mutagens 5-azacytidine (25 μg/ml; Toyoshima and Vogt, 1969b) or 5-fluorouracil (200 μg/ml; Biquard and Vigier, 1970), collecting virus immediately afterward. Kawai and Hanafusa (1971), using 5-fluorouracil at 50 μg/ml, claimed that the mutagen does not appreciably affect virus yield at this level.

3. Sparse cell cultures are exposed to the DNA mutagen 5-bromodeoxyuridine (BUdR; 100 μg/ml) from 0.5 to 12 hours after virus infection, collecting virus at 24 hours after infection (Bader and Brown, 1971). This mutagen is presumed to act during the formation of the DNA provirus.

C.   Selective Isolation of ASV Mutants *ts* for Maintaining Cell Growth in Suspension Culture

Transformed cells are able to multiply in suspension culture, and this parameter, being independent of cell contact, is exhibited at both high and low cell densities. Negative selection directed against growth in suspension is thus very versatile, and can be applied to any number of cells (Wyke, 1971). The selective technique employed is to add to suspension cultures at nonpermissive temperature (41°) the thymidine analog BUdR, which is incorporated into the DNA of growing cells, sensitizing them to subsequent exposure to light of near-visible wavelength (Puck and Kao, 1967). Survivors of BUdR–light treatment are then grown in suspension at permissive temperature (35°C), yielding colonies of transformed cells enriched for the production of virus *ts* for transformation maintenance (Wyke, 1973).

During BUdR incorporation the cells are suspended in medium rendered semisolid by methyl cellulose (Methocel, standard grade viscosity 4000 cps, Dow Chemical Co., Midland, Michigan). Methocel is handled by methods similar to those of Stoker et al. (1968). An 8% (w/v) slurry is prepared by adding boiling water to 3.2 gm of Methocel in a 500-ml bottle with a well-fitting cap. This is autoclaved, yielding a hard, shrunken gel, and stored at room temperature. A week before use the stock is diluted to 1.6% by the addition of serum-free growth medium (Dulbecco's modified Eagle's medium, Ham's F10 medium, or M199), breaking up the gel pellet with a glass rod and shaking intermittently over a period of several days, at 4°C, until a homogeneous gel is obtained. An appropriate volume of cells in serum-containing medium is added to yield a culture containing 1.3% Methocel. At this concentration Methocel can be handled with wide-bore pipettes, particularly if cooled to 4°C, at which temperature it is less viscous. Experience soon shows how much allowance should be made for medium draining only slowly from the pipette when dispensing Methocel.

Methocel can be readily diluted, allowing easy recovery of cells from suspension culture. However, over a long period cells settle out from the medium so the medium for growth at 35°C after BUdR–light selection is solidified by agar (modified from Wyke and Linial, 1973) and plated over a preset base agar. Base agar medium (0.5%) comprises 80 ml of any of the standard minimal media mentioned above, 40 ml of double-strength minimal medium, 20 ml of tryptose phosphate broth, 12 ml of calf serum, 4 ml of inactivated chick serum, and 2 ml DMSO, all mixed with 40 ml of molten 2.5% agar. The addition to this base medium of 2 ml of folic acid (0.8 gm/ml) and 2 ml of vitamins (at 200 times their concentration in F10 medium) may marginally promote cell growth in the top agar layer. Japanese quail feeder cells (10^5/ml) are incorporated into the medium which is then dispensed in 5-ml quantities in 60-mm bacteriological grade plastic petri plates. The top layer (0.33% agar) is made by mixing 2 parts of base agar (without feeder cells) with 1 part of conditioned medium (obtained from primary cultures of Japanese quail cells) which contains the cells under study. Its initial volume can be between 1.5 and 2.5 ml, but to prevent undue drying it should be supplemented every 3–4 days by a further 1–2 ml of 0.33% agar medium.

The protocol for mutant selection is as follows. Monolayers of secondary chick embryo fibroblasts are infected with virus at a multiplicity of about 0.01 focus-forming units per cell and incubated at 35°C under agar (0.72% containing 10% tryptose phosphate broth, 5% calf serum, and 1% DMSO; Vogt, 1969) until small foci of transformation appear (about 5 days). This initial period at permissive temperature has two purposes: (1) it allows mutations in the virus to become fixed. Since the genome may be polyploid (reviewed by Wyke, 1975) and ts mutations are probably often recessive, the mutated gene may need to segregate, and, though the mechanism for this is unknown, it may require more than one cycle of infection: (2) each input virus is allowed to infect a number of cells, reducing the chance of any mutant being lost by accident during the selective manipulation. To avoid studying separate isolates of the same virus, only one confirmed mutant from each infected culture should be studied in detail, and ideally these should originate from separate stocks of mutagenized or nonmutagenized virus.

After several days the agar meniscus at the edge of the dish is loosened, the dish is inverted, allowing the agar to peel off, and the cells are trypsinized and suspended in medium such that when mixed with 1.6% Methocel medium they yield 1.3% Methocel containing 5% chick serum, 1% DMSO and up to 3×10^5 cells/ml. (To reduce thymidine levels and increase BUdR incorporation, tryptose phosphate broth is omitted.) Aliquots of this suspension (5 ml) are added to 60-mm bacteriological grade plastic petri plates containing 5 ml of a preset 0.9% agar medium base. Cultures are incubated for 40 hours at 41°C to allow nongrowing cells to reach quiescence, and 0.5 ml of $2 \times 10^{-4} M$ BUdR in unsupplemented medium is added, rocking the dishes gently to disperse the drug. Incubation continues a further 40 hours at 41°C (a time which reconstruction experiments suggest is adequate for incorporation of the analog into both strands of DNA), and the cultures are then cooled for 1 hour at 4°C. The Methocel is pipetted into four volumes of cold medium, the agar base is washed twice in cold medium, and cells are separated from the Methocel and washings by centrifugation at 300 g for 10 minutes. The cells are resuspended and counted and about 10^5 cells are cultured in plastic flasks or 60-mm petri plates for 4 hours to allow complete cell attachment. The cultures are then exposed to light from a Philips 40 W "Actinic Blue" fluorescent tube for 1 hour at a distance of about 10 cm. For the flasks this is done by inverting them and illuminating the cells from their underside; the dishes are illuminated through the lids after replacing the growth medium with 2 ml of phosphate-buffered saline. After illumination the cells are resuspended and their concentration is adjusted in conditioned medium to about 2×10^5/ml. This is mixed 1 part to 2 with 0.5% agar medium so that there are thus about 10^5 cells in each 1.5 ml of top agar, which is then plated over the 0.5% agar bases. These suspension cultures are incubated at 35°C; colonies are isolated about 2 weeks later, the cells are freeze-thawed or sonicated, and the released virus is tested for transformation at 41°C and 35°C.

Before this procedure is used, its parameters should be tested by reconstruction experiments, mixing normal cells with known proportions of cells infected with wild-type virus or a confirmed ts mutant. It may be found necessary to vary the time spent in Methocel before or after adding BUdR, or the degree and type of illumination (for example, any source of near-visible light can be used, even a glass- or plastic-shielded ultraviolet lamp). Wyke (1973) found a 20-fold selection against transformed cells in reconstruction experiments, but the increase in mutant yield was only about 5-fold under identical conditions. This discrepancy may have been due to delayed transformation of cells newly infected by wild-type virus, so that they failed to grow during the selection. An increased time in Methocel should reduce the number of such survivors.

D. Other Possible Means of Selecting for *ts* Transformation Mutants

As mentioned in Section II,E, several negative selection procedures can be applied to growing transformed cells. The BUdR–light method was chosen because it is a two-step process, neither stage of which is, on its own, lethal to the cells. The same selection has been applied on a limited scale (J. Wyke, unpublished) against the ability of transformed cells to grow in serum-depleted medium (Temin, 1966) or when in contact with a monolayer of normal chick or NRK cells (Weiss, 1970). However, these parameters cannot be manipulated so easily as growth in suspension, and their use may offer no advantages over the technique outlined above.

A different type of selection, which might also detect some replication defects, may be achieved by use of the anesthetic dibucaine, which is toxic only for certain RNA tumor virus-infected cells (Rifkin and Reich, 1971). However, the effect of dibucaine has now been known for several years and it seems likely that, if it were a useful selective agent, this would have become apparent by now.

E. Screening Procedures for Isolating ASV *ts* Mutants

Screening methods, allowing the rapid testing of a large number of potential mutants, have hardly been used in the hunt for ASV *ts* mutants. However, because of their potentially wide applications, I will outline here some approaches that may deserve further study.

The ease with which late T class mutants can be isolated, both selectively and nonselectively, prompted Wyke (1973) to search for mutants of other types by plating mutagenized virus at 35°C, then shifting cultures to 41°C 1 day after infection and isolating foci for test at this temperature. From Table I, column 3 it can be seen that all early mutants would form foci under these conditions, as would late R class mutants and wild-type virus. Reasoning that only wild-type virus would be able to reinfect at 41°C, small foci were isolated that were possibly derived solely by growth of cells infected during the first day at 35°C. Although this procedure did detect an early mutant it may not enhance detection, for later work (Wyke et al., 1975) showed that focus formation by cell division alone is an inefficient process, only about one infected cell in ten succeeding in growing to a visible focus when infection is prevented.

A more promising approach may utilize the 96-well Microtest plates and replica plating techniques described by Stephenson *et al.* (1972) for the isolation of murine leukemia virus *ts* mutants. A rational protocol is to inoculate plates, containing cells, with virus at a multiplicity of about 0.5 focus-forming units per well and to incubate these at 35° until transformation appears. The cells are then killed by treatment with mitomycin C at 10 μg per 10^6 cells for 2 hours (Macpherson and Bryden, 1971), a level that permits continued virus production. After washing, fluid from each well of transformed cells is transferred to recipient plates, (1) to (4), of cells which are treated as follows: (1) incubate at 35°; (2) incubate 41°; (3) incubate at 35°, shift to 41° 1 day post infection, kill cells with mitomycin C several hours later, wash and add fresh "indicator" cells (viable cells susceptible to infection by the viruses under study); (4) incubate at 41°C, shift to 35°C 1 day post infection, then treat as (3).

Every transformed well in the original plate which produces virus (Table I, columns 1 and 2) should transform in plate (1), whereas in plate (2) only wild-type virus and R class mutants would transform (Table I, column 7), thus identifying all other *ts* mutants. Replica (3) tests the effect of a shift from 35°C to 41°C on virus replication (Table I, column 4) and the subsequent ability of this virus to transform indicator cells at 41°C (Table I, column 7). Only wild-type virus and early R class mutants would pass this screen, thus distinguishing late R class virus from wild type. Plate (4) reveals virus that replicates on a shift from 41°C to 35°C (Table I, column 6) and can then transform indicator cells at 35°C (Table I, column 1). Wild-type virus, early T class mutants, and all late mutants would transform here, but early R and C class viruses would produce no transformation. These four replicas should thus distinguish wild-type virus from all classes of mutants and from multiple mutants, and they should give a preliminary indication of the nature of each potential mutant.

A simple screening applicable to all T class mutants would be to plate virus at 41°C under the conditions of the plaque assay of Graf (1972), so that virus that can replicate but not transform will produce a plaque of degenerate cells. This screen could be used only for viruses of envelope subgroups B and D, since they are the only subgroups that form plaques. It is obviously desirable to have a low level of non *ts td* virus in the stock, as this too would produce plaques. An even simpler scan for late T and C class mutants would involve marking foci as soon as they can be seen at 35°C, shifting cultures to 41°C, and noting which foci have disappeared 1–2 days later.

F. Treatment of Potential *ts* Mutants

Confirming the *ts* nature of a potential mutant is straightforward if a few points are remembered. First, cloning the mutant after detection often gives an apparent improvement of temperature sensitivity. It is possible that the original isolate may be contaminated, either during manipulation or by clumping of virus. As indicated in Section III,C, the probable polyploidy of the viral genome could also mean that the mutated gene is present in a heterozygote and is not fully expressed until it has segregated and is homozygous. Second, mutants may bear multiple lesions, and this can be confirmed only by studying their behavior upon reversion or recombination. Many "tight" mutants may have more than one *ts* defect, and perhaps it is now time to take notice of the "leaky" mutants that have often been ignored after isolation.

An agreed procedure has been published for the numbering and recording of avian RNA tumor virus mutants (Vogt *et al.*, 1974).

IV. Discussion

Mutants *ts* for transformation maintenance are easily obtained, both selectively and by random testing. There is now a clear incentive to obtain more mutations in other functions, but apart from the screening procedure suggested here, there seems no means to enhance the detection of such mutants. It may be possible to selectively destroy, or otherwise remove from culture, cells bearing virus-specified molecules on their surface, and agents such as dibucaine may select against virus-coded molecules within the cell. It would also be worth trying to isolate *ts* mutants of *td* viruses. The plaque assay of Graf (1972) could be used or cells could perhaps be developed that transform upon infection with *td* viruses, an analogy here being the "S^+L^-" line used by Wong *et al.* (1973) to isolate selectively murine leukemia virus *ts* mutants.

The nature and behavior of the RNA tumor virus genome are still unclear, and we have no idea of its potential response to mutagenesis, nor do we know the way in which mutated genes are propagated and expressed. It is possible, for instance, that the segregation and expression of mutants after the use of BUdR as a mutagen (which would presumably affect the DNA provirus), is different from the pattern seen after the use of an RNA mutagen. However, we do know that mutagenesis often yields multiple mutants, and for some studies single lesions would be preferable. Consideration could now be given to the use of selection to detect the (presumably) rarer mutants in nonmutagenized virus, and such selection should preferably use an agent which, unlike BUdR, has no capacity for inducing mutations during the selection procedure.

ACKNOWLEDGMENTS

I am grateful to Drs. H. Murphy and C. Norris for their help in making this manuscript intelligible.

REFERENCES

Bader, J. P., and Brown, N. R. (1971). *Nature (London), New Biol.* **234**, 11–12.

Basilico, C., and Meiss, H. K. (1974). *Methods Cell Biol.* **8**, 1–22.

Biquard, J. M., and Vigier, P. (1970). *C. R. Hebd. Seances Acad. Sci.* **271**, 2430–2433.

Bishop, J. M., Jackson, J., Quintrell, N., and Varmus, H. E. (1973). *In* "Possible Episomes in Eukaryotes" (L. G. Silvestri, ed.), pp. 61–73. North-Holland Publ., Amsterdam.

Graf, T. (1972). *Virology* **50**, 567–578.

Graf, T., and Friis, R. R. (1973). *Virology* **56**, 369–374.

Hayward, W. S., and Hanafusa, H. (1973). *J. Virol.* **11**, 157–167.

Kao, F. T., and Puck, T. T. (1974). *Methods Cell Biol.* **8**, 23–39.

Kawai, S., and Hanafusa, H. (1971). *Virology* **46**, 470–479.

Kawai, S., Metroka, C. E., and Hanafusa, H. (1972). *Virology* **49**, 302–304.

Macpherson, I. A., and Bryden, A. (1971) *Exp. Cell Res.* **69**, 240–241.

Martin, G. S. (1970). *Nature (London)* **227**, 1021–1023.

Naha, P. M. (1974). *Methods Cell Biol.* **8**, 41–46.

Neiman, P. (1973). *Virology* **53**, 196–204.

Puck, T. T., and Kao, F. T. (1967). *Proc. Natl. Acad. Sci. U.S.A.* **58**, 1227–1234.

Rifkin, D. B., and Reich, E. (1971). *Virology* **45**, 172–181.

Somers, K., and Kit, S. (1973). *Proc. Natl. Acad. Sci. U.S.A.* **70**, 2206–2210.

Stephenson, J. R., Reynolds, R. K., and Aaronson, S. A. (1972). *Virology* **48**, 749–756.

Stoker, M., O'Neill, C., Berryman, S., and Waxman, V. (1968). *Int. J. Cancer* **3**, 683–693.

Temin, H. M. (1966). *J. Natl. Cancer Inst.* **37**, 167–175.

Temin, H. M. (1971). *In* "The Biology of Oncogenic Viruses" (L. G. Silvestri, ed.), pp. 176–187. North-Holland Publ., Amsterdam.

Tereba, A., Skoog, L., and Vogt, P. K. (1975). *Virology* **65**, 524–534.

Tooze, J., ed. (1973). "The Molecular Biology of Tumour Viruses." Cold Spring Harbor Lab., Cold Spring Harbor, New York.

Toyoshima, K., and Vogt, P. K. (1969a). *Virology* **38**, 414–426.

Toyoshima, K., and Vogt, P. K. (1969b). *Virology* **39**, 930–931.

Vogel, A., and Pollack, R. (1974). *Methods Cell Biol.* **8**, 75–92.

Vogt, P. K. (1969). *In* "Fundamental Techniques in Virology" (K. Habel and N. P. Salzman, eds.), Vol. 1, pp. 198–211. Academic Press, New York.

Vogt, P. K., and Friis, R. R. (1971). *Virology* **43**, 223–234.

Vogt, P. K., Weiss, R. A., and Hanafusa, H. (1974). *J. Virol.* **13**, 551–554.

Weiss, R. A. (1970). *Exp. Cell Res.* **63**, 1–18.

Weiss, R. A., Mason, W. S., and Vogt, P. K. (1973). *Virology* **52**, 535–552.

Wong, P. K. Y., Russ, L. J., and McCarter, J. A. (1973). *Virology* **51**, 424–431.

Wyke, J. (1971). *Exp. Cell Res.* **66**, 203–208.

Wyke, J. A. (1973). *Virology* **52**, 587–590.

Wyke, J. A. (1975). *Biochim. Biophys. Acta* **417**, 91–121.

Wyke, J. A., and Linial, M. (1973). *Virology* **53**, 152–161.

Wyke, J. A., Bell, J. G., and Beamand, J. A. (1975). *Cold Spring Harbor Symp. Quant. Biol.* **39**, 897–905.

Chapter 24

Induction and Isolation of Cold-Sensitive Lines of Chinese Hamster Cells

ROSANN A. FARBER AND PAUL UNRAU

*Division of Human Genetics, Children's Hospital Medical Center,
Boston, Massachusetts; and Biology Branch, Chalk River
Nuclear Laboratories, Chalk River, Ontario, Canada*

I. Introduction

Cold-sensitive mutants of prokaryotes have been a valuable source of cells with defects in essential cellular processes (see Waldron and Roberts, 1974a). Some of these cold-sensitive mutations are in genes for which no heat-sensitive counterparts have been discovered (Ingraham and Neuhard, 1972).

Cold-sensitive strains of bacteria include an especially high frequency of ribosomal subunit assembly defective mutants (Guthrie *et al.*, 1969; Tai *et al.*, 1969). This phenomenon is probably explained by the observation that ribosome assembly is a highly temperature-dependent process (Davies and Nomura, 1972). Subunit assembly defective mutants also appear among cold-sensitive strains of fungi, but with a much lower frequency than in bacteria (Schlitt and Russell, 1974; Waldron and Roberts, 1974b).

We were interested in obtaining cold-sensitive variants of a mammalian cell line as a possible means of enriching for ribosomal mutants. These lines, whether or not they have defective ribosomes, should be useful additions to the rapidly growing list of conditional lethal mammalian cell mutants. The selection scheme which we employed (Farber and Unrau, 1975) is a modification of the bromodeoxyuridine–visible light technique as used by Scheffler and Buttin (1973) for the selection of heat-sensitive mutants of Chinese hamster cells. The mutagenesis and selection temperatures were reversed, and the times for each treatment were altered to compensate for the difference in the growth rates of wild-type cells at high and low temperatures.

II.   Cell Line

We chose the Chinese hamster ovary (CHO) (Puck *et al.*, 1958) line for these studies because of the ease with which it can be cultured and because of its near-diploid DNA content.

The generation time of wild-type CHO cells varies between 10.5 and 16 hours, depending upon the growth conditions. The plating efficiency of these cells is very high; at 37°C under our conditions it averages about 75%.

CHO cells have a near-diploid chromosome number, although there are many rearrangements of the chromosomal material (Deaven and Peterson, 1973). This property is at least theoretically an important one for a mammalian line which is to be used for obtaining mutants. Working with diploid cells is, in itself, a hardship for mutagenesis studies, but many mammalian cell lines are near-tetraploid.

It has already been possible to obtain heat-sensitive mutants of the CHO line (Thompson *et al.*, 1973), as well as many other types of mutants, such as auxotrophs (Kao and Puck, 1968; McBurney and Whitmore, 1974). We felt that these prior successes might be an indication that positive results could be obtained on the selection of cold-sensitive mutants from this line.

III.   Choice of Temperatures

In order to obtain cold-sensitive lines with phenotypes comparable to those in prokaryotes, it would be desirable to use as low a restrictive temperature as possible. We found that CHO cells would not form colonies at 32°C when plated at low density, and that the growth rate of higher density cultures was very slow at this temperature. At 33°C, however, the doubling time of these cells (14 hours) was not very different from that at 37°C (10.5 hours) (Farber and Unrau, 1975). For this reason, 33°C was chosen as the restrictive temperature for this cell line.

The permissive temperature that we used was 39°C. A higher temperature than that used for routine maintenance of wild-type cells was chosen, since the broader the span between restrictive and permissive temperatures, the greater the chance of obtaining mutants that would not grow at some temperature less than the permissive one.

IV. Mutagenesis

Cultures were routinely maintained in Eagle's basal medium supplemented with 5% fetal calf serum, nonessential amino acids, and antibiotics (penicillin, 100 units/ml; streptomycin, 100 μg/ml; and aureomycin, 50 μg/ml).

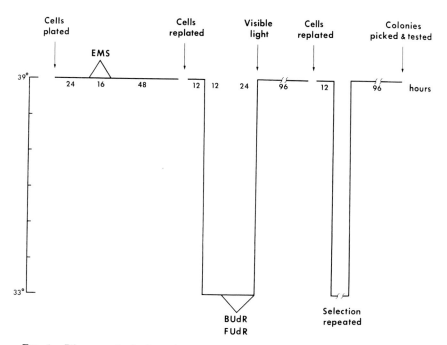

Fig. 1. Diagram of selection scheme used to obtain cold-sensitive mutants of Chinese hamster ovary cells. EMS, ethylmethane sulfonate; BUdR, bromodeoxyuridine; FUdR, fluorodeoxyuridine.

The scheme for mutagenesis and selection of cold-sensitive mutants is diagrammed in Fig. 1.

For mutagenesis, cells are seeded into large bottles at a density sufficiently low to allow several generations of growth following the treatment (approximately 10^4 cells/cm^2). During mutagenesis, the cells are kept at permissive temperature.

The cells are allowed to attach to the bottles for 24 hours, at which time ethylmethane sulfonate (EMS) is added at a concentration of 150 μg/ml for 16 hours. The mutagen is removed from the cultures at this time, and the bottles are rinsed thoroughly with medium and replenished with fresh medium. Under these conditions the cell survival rate we obtain is approximately 50%.

The cells are allowed to remain in normal medium at permissive temperature for 48 hours in order to allow for expression of any mutations that have been induced. Chu and Malling (1968) showed that the maximum frequency of 8-azaguanine-resistant mutants in Chinese hamster cells were recovered if the selective treatment was applied three generations after removal of the mutagen. This "expression time" presumably permits fixation of the mutation into both strands of the DNA and dilution of remaining wild-type gene products.

V.   Selection

The mutagenized cells are trypsinized and distributed into 50-mm petri dishes at a density of 5×10^5 cells per plate. They are allowed to attach at permissive temperature for 12 hours.

The cultures are transferred to restrictive temperature and left untreated for 12 hours, since the shutoff in DNA synthesis of mutants after the downshift might take as long as one generation time. The selective agent, bromodeoxyuridine (BUdR, 15 μg/ml), is then added for 24 hours. Fluorodeoxyuridine (FUdR, 0.2 μg/ml) is also added, in order to prevent endogenous synthesis of thymidine and, therefore, to enhance the incorporation of BUdR in growing cells. The presence of FUdR also decreases the likelihood that any mutants isolated will be simply BUdR-resistant. These drug concentrations are the same as those used by Scheffler and Buttin (1973). We found that treatment with BUdR at a concentration of 15 μg/ml resulted in 57% cell survival in the absence of visible light.

After 24 hours, the drug-containing medium is removed, and the plates are rinsed with phosphate-buffered saline (PBS). One milliliter of PBS is added to each plate, and the cells are irradiated with a long-wavelength UV lamp (365 nm) for 30 minutes at a distance of 2 inches above the plates, which have their lids left on. This light treatment, in the absence of BUdR pretreatment, has no effect on the survival of CHO cells. [If cells are irradiated in tissue-culture medium, cell killing occurs as a result of the generation of toxic photoproducts (Stoien and Wang, 1974).] This treatment presumably produces breaks in the DNA of cells which have incorporated BUdR at 33°C, so that the only survivors should be cells that failed to synthesize DNA at this temperature.

After the light treatment, normal medium is added back to the cultures, and they are maintained at permissive temperature for 4 days. We found that there were numerous colonies on the plates at this stage, so it is desirable to repeat the selection. The cells are trypsinized, pooled, and replated at the original density of 5×10^5 cells/50-mm plate. The BUdR and light treatment is repeated, and the survivors are again allowed to grow up at permissive temperature for 4 days.

VI. Isolation, Testing, and Purifying of Colonies

After the second round of selection, colonies are isolated using stainless steel cloning cylinders and transferred to Linbro well dishes. When the cells have grown up to fill the wells, each clone is split into two 50-mm petri dishes. One is put at 33°C and the other at 39°C. Lines which grow at 39°C but not at 33°C are considered cold-sensitive and are retained for further analysis.

In our first experiment nine colonies were isolated from nine different dishes. Upon retesting, six were found to be cold-sensitive. Two of the others grew normally at both temperatures, and the remaining one grew poorly at both temperatures.

Other groups working on temperature-sensitive mutants of mammalian cells have put their cell populations through four or five rounds of selection before isolating colonies and characterizing them. We were fortunate in our first experiment in finding such a high proportion of cold-sensitive lines. When we repeated the selection on a smaller scale and picked up five colonies, none were cold-sensitive. It would, therefore, probably be worthwhile to use several cycles of selection to enrich for true cold-sensitive clones.

After the colonies were tested for cold sensitivity, they were subcloned in Falcon Microtest II plates in order to eliminate any contaminating wild-type cells that might have been picked up. All subsequent analyses were done on these purified subclones.

VII.   General Characteristics of Cold-Sensitive Lines

The six cold-sensitive variants that were isolated can be divided into at least three groups on the basis of their growth properties. Two of the lines have high reversion frequencies (1×10^{-5} and 1×10^{-3}); these were excluded from further analysis. The remaining four lines have reversion frequencies of less than 2.5×10^{-6}. Of the latter four lines, two stop growing immediately after shiftdown to 33°C, whereas the other two undergo one very slow doubling before growth ceases. None of the variants exhibit density-dependent growth at restrictive temperature. They also have very high survival rates at 33°C (75–85% viability after 5 days).

On the basis of sucrose gradient sedimentation analysis of ribosomal subunits (Farber and Unrau, 1975) none of these cold-sensitive lines appears to have subunit assembly defects. One of the lines, cs-4, has recently been shown to be reversibly blocked in the G_1 stage of the cell cycle at restrictive temperature (Crane and Thomas, 1976).

VIII.   Conclusions

It is possible, using modifications of standard procedures for the selection of heat-sensitive variants, to obtain cold-sensitive mammalian cell lines. The frequency of ribosomal subunit assembly defective mutants among these lines is apparently not as high as in prokaryotes. These cold-sensitive mammalian lines should, however, provide an additional source of cells with conditional defects in essential cellular functions.

ACKNOWLEDGMENTS

The work described in this article was done in the laboratory of Dr. Robin Holliday at the National Institute for Medical Research, Mill Hill, London, England. R. A. F. was a fellow of the Jane Coffin Childs Memorial Fund for Medical Research.

REFERENCES

Chu, E. H. Y., and Malling, H. V. (1968). *Proc. Natl. Acad. Sci. U.S.A.* **61**, 1306–1312.
Crane, M. St. J., and Thomas, D. B. (1976). *Nature* **261**, 205–208.
Davies, J., and Nomura, M. (1972). *Annu. Rev. Genet.* **6**, 203–234.
Deaven, L. L., and Peterson, D. F. (1973). *Chromosoma* **41**, 129–144.
Farber, R. A., and Unrau, P. (1975). *Mol. Gen. Genet.* **138**, 233–242.
Guthrie, C., Nashimoto, H., and Nomura, M. (1969). *Proc. Natl. Acad. Sci. U.S.A.* **63**, 384–391.
Ingraham, J. L., and Neuhard, J. (1972). *J. Biol. Chem.* **247**, 6259–6265.

Kao, F. -T., and Puck, T. T. (1968). *Proc. Natl. Acad. Sci. U.S.A.* **60**, 1275–1281.
McBurney, M. W., and Whitmore, G. F. (1974). *Cell* **2**, 173–182.
Puck, T. T., Cieciura, S. J., and Robinson, A. (1958). *J. Exp. Med.* **108**, 945–956.
Scheffler, I. E., and Buttin, G. (1973). *J. Cell. Physiol.* **81**, 199–216.
Schlitt, S. C., and Russell, P. J. (1974). *J. Bacteriol.* **120**, 666–671.
Stoien, J. D., and Wang, R. J. (1974). *Proc. Natl. Acad. Sci. U.S.A.* **71**, 3961–3965.
Tai, P. C., Kessler, D. P., and Ingraham, J. (1969). *J. Bacteriol.* **97**, 1298–1304.
Thompson, L. H., Harkins, J. L., and Stanners, C. P. (1973). *Proc. Natl. Acad. Sci. U.S.A.* **70**, 3094–3098.
Waldron, C., and Roberts, C. F. (1974a). *Mol. Gen. Genet.* **134**, 99–113.
Waldron, C., and Roberts, C. F. (1974b). *Mol. Gen. Genet.* **134**, 115–132.

Chapter 25

Maintenance of Perpetual Synchrony in HeLa S3 Culture: Theoretical and Empirical Approaches

WILLIAM G. THILLY

Toxicology Group,
Department of Nutrition and Food Science,
Massachusetts Institute of Technology,
Cambridge, Massachusetts

I. Introduction

This chapter explores the possibility that the perturbing influences of the "thymidine-block" synchronization technique can be mitigated, in large part, by choosing procedures in harmony with the particular cell line in question and by taking advantage of the fact that a partially synchronized population requires a shorter period of inhibition of DNA synthesis (blocking) to achieve maximal synchrony than do asynchronous cultures. In this context, "blocking" DNA synthesis means reducing the rate of DNA synthesis to 15–25% of untreated controls and not stopping DNA synthesis entirely.

The "thymidine-block" technique is one of several similar procedures, which may be generically referred to as "temporary inhibition of DNA synthesis" techniques but are more often designated "induction" techniques. Induction techniques for synchronization differ from selection techniques in that the former involve forcing an entire cell population into a small range of cell age, while the latter depend upon isolating cells which are within a small range of cell age from an unperturbed population.

II.   A Simplistic Model of Induced Synchrony

Conceptually, the induction techniques have been approximated as shown in Fig. 1 (Bootsma *et al.*, 1964; Puck, 1964). Figure 1a shows the distribution of cell ages over the cell cycle with G_1, S, G_2, and M phases as indicated. A period of DNA inhibition equal to the duration of $G_2 + M + G_1$ phases leads to a distribution something like that shown in Fig. 1b, except that all vertical lines are meant as approximations to the "true" distribution functions. If the block is then removed, the cells traverse S phase at normal rates, and if S phase is less than one half of the total cell cycle time, a distribution, such as is approximated by Fig. 1c, in which very few cells are found in S phase, will occur. At this "magic moment," application of a second block will, in time, "collect" the cells in a distribution approximated by Fig. 1d. Mention should be made of the fact that cells are not really at the hypothetical G_1/S boundary as pictured in Fig. 1d, but have entered early S phase by virtue of slow DNA synthesis (Lambert and Studzinski, 1969). Galavazi *et al.* (1966) applied this model experimentally through three cycles of blocking and found superior division synchrony relative to culture behavior after the traditional second block. Their work was limited to three cycles by the surface area limitations of anchorage-dependent cell cultures. These data suggested to us that the improved synchrony of the third cycle came from partially synchronous cells in the distribution shown in Fig. 1d traversing another cycle and, again, being blocked in early S phase.

A little arithmetic may help to explain our reasoning. In the first moments of the first block, cells in early S phase are slowed in DNA synthesis but continue normal RNA and protein synthesis for some time (Reuckert and Mueller, 1960). We may expect that some specific RNA and proteins will accumulate (or disappear), creating an imbalance in such cells relative to cells that continue to synthesize DNA at normal rates. Thus, at the end of the blocking period for each molecular species, i, the cells will be unbalanced relative to unperturbed cells by an amount of Δx_i^1 molecules/cell. However, the cells which were just leaving S phase at the beginning of the first block will have just arrived at the G_1/S boundary when the first block is removed. Such cells will not be unbalanced, since they have not suffered the biochemical effects of blocking. Thus, the cells at the end of the first block, though distributed with regard to DNA content as shown in Fig. 1b, may be

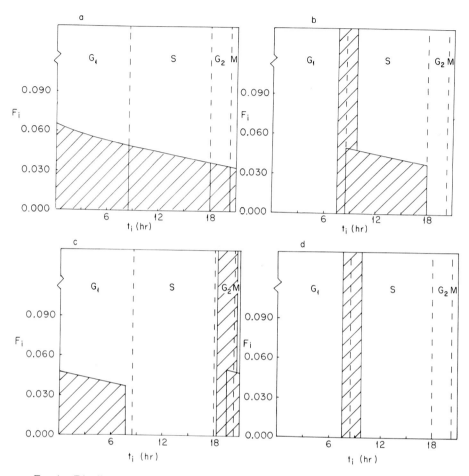

FIG. 1.   Distribution of cells over the cell cycle during synchronization. (a) Cells distributed asynchronously over a cell cycle of doubling time T. The distribution function, F_i, as a function of cell age from previous division is $F_i = \ln 2/T \exp (T - t_i) \ln 2/T$. (b) Cell age distribution relative to mitosis approximated after inhibition of DNA synthesis for a period equal to the length of $G_2 + M + G_1$ phases. (c) Cell age distribution after reversal of inhibition when "no" cells are in S phase. (d) Cell age distribution resulting from application of second period of inhibition.

thought of as effectively asynchronous with regard to many other cellular molecules. Note that this is not the same as "unbalanced growth" often discussed in the literature (Cohen and Studzinski, 1967). Unbalanced growth is measured by comparing perturbed and unperturbed cultures having the same DNA content or the same cell cycle position relative to subsequent mitosis. The point here is that unbalanced growth is predicted to be *unequally distributed* among cells of the population brought to the age distribution shown in Fig. 1b.

Subsequent cycling to the beginning of the next block will reduce imbalance in the most unbalanced cells to $\leq \frac{1}{2}\Delta X_i^1$ by simple division dilution, assuming that cells do not exacerbate imbalances after block release but, either passively or actively, move toward balance. This assumption is in agreement with Studzinski's detailed observations (Churchill and Studzinski, 1970). However, the second block creates an opportunity for further imbalance of the cells in the leading edge of Fig. 1c, experiencing a maximum imbalance, ΔX_i^2, while those of the trailing edge experience virtually no imbalance. Thus, at the end of the second block, while the cells *seem* to be highly synchronous, and *do* exhibit a high degree of division synchrony, we reason that these cells should exhibit gross variations within the population with respect to many cellular molecules: some cells will have experienced no unbalanced growth, and some will have experienced a maximum of $\leq (\frac{1}{2}\Delta X_i^1 + \Delta X_i^2)$. The upper limit of cellular differences would, thus, be $\leq (\frac{1}{2}\Delta X_i^1 + \Delta X_i^2)$.

Despite these predicted variations, it is a fact that, with regard to cell division, cells traverse the cycle with a reasonable degree of synchrony after a second thymidine block. Thus, the application of a third block should have similar effects on all cells in the population, since they will, within limits, experience the same effective blocking period. The division following the second block should result in a populations differing from each other by $\leq \frac{1}{2}(\frac{1}{2}\Delta X_i^1 + \Delta X_i^2)$ as a result of differential unbalanced growth in the first two synchronizing cycles. During the third block, all cells accumulate imbalances of approximately the same magnitude, ΔX_i^3. However, the higher degree of synchrony at the beginning of the third block permits use of a reduced blocking period to restore a satisfactory degree of synchrony. In other words, ΔX_i^3 can be made much smaller than ΔX_i^1, since the magnitude of ΔX_1^n is dependent on the length of the nth blocking period.

The range of differences among cells is, however, not increased in subsequent cycles of blocking and release, because all cells experience approximately the same amount of unbalanced growth. After the division following the third block, this range is $\frac{1}{4}(\frac{1}{2}\Delta X_i^1 + \Delta X_i^2)$, and it decreases at least geometrically with subsequent divisions. The same reasoning also leads to the prediction that the relatively large effects (imbalances) induced by the first two "long" blocking periods will be diluted by subsequent divisions, so that cells in the nth cycle will show only the effects of the n and, perhaps, the n-1 blocking periods. If the blocking period is too short to induce unbalanced growth persisting through mitosis, then only the unbalanced growth of the nth blocking period will affect the nth cycle.

This unsophisticated argument is really no more than simple arithmetic, but it leads to two important predictions about synchronizing cells:

1. The amount of unbalanced growth affecting the nth ($n > 2$) cycle can be substantially reduced relative to that observed in the first two blocking periods, which must, perforce, be long enough to synchronize cells.

2. The popular *double*-block technique may really be twice damned: once because of the known magnitude of unbalanced growth observed; and twice, because it should induce gross differences among cells in the "synchronous" population, which could lead to misinterpretation of results based on average cell behavior.

A Very Practical Note: An important consideration of planning any re-synchronization is that many cell populations have generation times differing significantly from the 24-hour cycle by which most researchers live. Thus, any steps that are not automatic (removing inhibiting agent by resuspension, for instance), would soon lead to early morning and late evening manipulations. Anticipating this problem, we adopted the expedient of choosing a batch of horse serum which produced a doubling time in our HeLa S₃ culture of just over 24 hours. Much of the success of the procedures decribed here depends on this strategem, since it is doubtful that we, as humans, would have had the energy to work continuously on a noncircadian synchronizing cycle.

III.   Choice of a Reversible Inhibitor

Up to this point, the synchronizing procedure has been considered in terms of temporal characteristics. Given a suitable reversible inhibitor of DNA synthesis, one need only empirically determine the minimal durations of initial blocking period, initial release period, second blocking period, second release period, and third blocking period to define conditions for maintaining a suspended cell population in perpetual synchrony. This could be accomplished theoretically by calculation, were cell cycle phase durations known precisely. However, since one overall effect of unbalanced growth is a shortening of G_1 phase, we have found it more expedient to define optimum periods of blocked and unblocked growth empirically.

Similarly, the identity and concentration of a suitable inhibitor must be regarded as unknown in approaching each new cell culture application.

Since inhibition of DNA synthesis is the first requirement, examining the accumulation of DNA in a culture during a period equal to the normal culture doubling time should define the lowest effective concentration for a particular compound. (The use of direct DNA analysis as opposed to the ³H-labeled thymidine incorporation is recommended, since it avoids pool prolems and the phenomenon of unstable DNA observed by Rueckert and

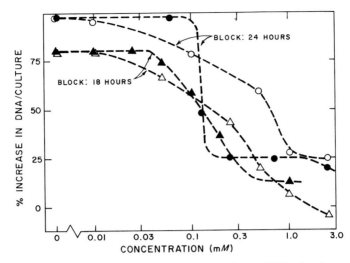

Fig. 2.   Effect of concentration on DNA synthesis. Cultures of HeLa S₃ cells were exposed
to various concentrations of test agent for the period indicated on the figure. DNA was determined in sextuplicate at the beginning and end of the exposure period. Results are expressed
as the percentage increase in DNA relative to the beginning of the exposure period. Standard
deviations of the mean are smaller than the symbols used. ●, Thymidine; ○, 5-aminouracil;
▲, hydroxyurea; △, deoxyguanosine.

Mueller, 1960.) Figure 2 summarizes our observations of the concentration
dependent effects of thymidine, 5-aminouracil, deoxyadenosine, deoxyguanosine, and hydroxyurea on the accumulation of DNA in HeLa cells.
These compounds had, at the beginning of our work, been found to be
reversible inhibitors of DNA synthesis in various cell systems (Xeros, 1962;
Duneau and Woods, 1964). Based on the responses of Fig. 2, the lowest concentration of each agent that maximally inhibited accumulation of DNA
was used to test the reversibility of its action. This was achieved by an
18-hour exposure of HeLa cells, resuspension in fresh prewarmed medium
and growth with stirring. Samples were taken and cell number was determined at 3-hour intervals. It was quickly found that deoxyadenosine had an
irreversible effect on cell growth, and that while hydroxyurea treatment
permitted cells to pass through mitosis, significant cell lysis occurred after
division. Thymidine, 5-aminouracil, and deoxyguanosine at 0.25 mM, 0.5
mM, and 1.5 mM, respectively, yielded HeLa cell growth patterns in ac-
cordance with our general expectation (Fig. 3).

    Subsequent experiments showed that deoxyguanosine had a very narrow
range of concentrations near 0.5 mM at which it was both effective and non-
toxic. 5-Aminouracil treatment was invariably accompanied by cell loss
during treatment. Thus, of the five compounds tested, thymidine appeared

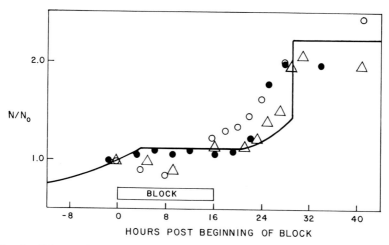

FIG. 3.   Cultures of HeLa S$_3$ cells were exposed to test agent, and growth of cultures was followed by direct counts at several points during and after the 16-hour block to DNA synthesis. The solid line represents the ideal expected behavior during and after a single block to an inhibitor of DNA synthesis. Synchronization with thymidine (●, 0.20 mM), 5-aminouracil (○, 1.5 mM), and deoxyguanosine (△, 0.50 mM).

most satisfactory for HeLa cells. Two points should be noted: (1) Our determinations were made for HeLa S$_3$ cells in Joklik-modified minimal essential medium supplemented with 10% horse serum in spinner culture and, strictly speaking, apply only to those conditions and cell population. For instance, human lymphoblasts in suspension are sensitive to minimal effective concentrations of thymidine, so other reversible inhibitors should be explored for these particular cell lines (Zielke and Littlefield, 1974). (2) The widespread use of 2 mM thymidine to synchronize HeLa cells may be criticized because this concentration is apparently excessive (Puck, 1964). We have used 0.25 mM thymidine effectively with several populations of HeLa cells and know of the success of others with this concentration.

IV.   Result of Resynchronization Procedure

A.   Division Synchrony

Once a satisfactory reversible inhibitor of DNA synthesis (thymidine), its minimal effective concentration ($\sim$0.20 mM), and a periodic schedule of growth in the presence and absence of the inhibitor (12 hours with, 12 hours without) had been defined, we were at liberty to discover whether

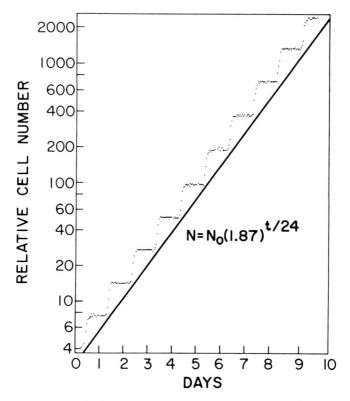

Fig. 4. Growth of HeLa S_3 cell culture in which synchrony was maintained by resynchronization with periodic exposure to 0.25 mM thymidine. A line describing exponential asynchronous growth in which 87% of the population is able to divide in a 24-hour period has been placed on the curve to emphasize the fraction of cells dividing in each cycle. Each point is the result of duplicate counts of samples on an electronic particle counter. From Thilly *et al.*, 1974a. Reproduced with the permission of John Wiley and Sons, Inc., New York.

these conditions would maintain a HeLa S_3 population in synchronous exponential growth. As Fig. 4 shows, these conditions do, indeed, result in a remarkable degree of division synchrony (Thilly *et al.*, 1974a). Figure 4 shows the first ten cycles of a fourteen-cycle experiment. We have extended these cycles for a month with no diminution in culture synchrony or vigor of growth, which serves to prove (1) that the degree of synchrony is very reproducible and (2) that we are confirmed cases of megalomania.

The points plotted in Fig. 4 are the results of hourly determinations of culture cell concentration. Hourly sampling was made practical by simply connecting a peristaltic pump and a refrigerated fraction collector to a timer in such a way that 5-ml samples were taken every hour. Cell number remained constant in the regrigerator for more than 24 hours, so a single counting of the previous day's samples was possible. We have also found that cellular RNA and DNA are sufficiently stable under refrigeration for direct determination in collected and stored samples (Thilly *et al.*, 1974b).

We have, thus, demonstrated that the cells in our culture are growing with undiminished capacity for continuous division. Variations in the inductive approach, such as the continued use of 0.25 mM thymidine in longer exposures (12–14-hour effective blocking period) leads inevitably to loss of ability to divide in the normal generation time. Even in short exposures (12 hours), 2 mM thymidine is similarly unsatisfactory.

The demonstration of undiminished capacity for cell growth is a necessary demonstration for any synchronizing procedure, if studies on the population are to provide data regarding the division cycle. Obviously, if half of the population does not divide at the end of a generation's time, the possibility exists that temporal changes observed do not reflect the sequence of events to division. Such observations could just as easily be represented as the sequence of events leading from a dividing to a nondividing state.

However necessary the demonstrations of continued capacity for cellular growth may be, it does not, per se, offer a sufficient characterization of the synchronized system. We have the responsibility of alerting the reader to the fact that the induction synchronizing procedure presented here is not without its pitfalls. There can be no argument that resynchronization by temporary inhibition of DNA synthesis does not perturb the average cell. For example, cycle processes not tightly coupled with DNA synthesis may proceed, while DNA synthesis is inhibited, for about 6 hours using the procedure described in this chapter. Our particular contribution has been to call attention to the need and to find a means to shorten the blocking period from the 12–24 hours commonly employed to some 6 hours, and thus mitigate, but not eliminate, the problem of unbalanced growth. Whether the remaining degree of perturbation is biologically significant will depend on the studies in which this system is applied.

B.   Unbalanced Growth

The same arithmetic model which suggested resynchronization as a means to reduce block length can be used to predict the behavior of parameters subject to unbalanced growth (Thilly *et al.*, 1975b). Figure 5 represents the behavior of a cellular parameter that is subject to unbalanced growth under three different limiting conditions. The first is simple accumulation of imbalance in the amount X per cycle over n cycles without any diminution of the imbalance by active cellular processes. This leads to an accumulated imbalance of $2X$. The second condition is that in which 50% of an imbalance is

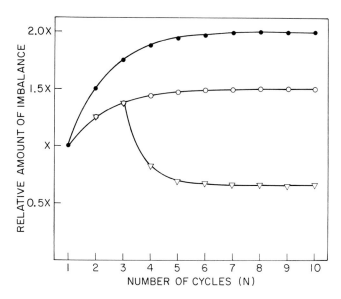

FIG. 5. Theoretical behavior of a cellular parameter which suffers unbalanced growth with certain limiting conditions. (●) Imbalance at the end of each constant period of exposure to thymidine merely remains in the cell and accumulates over successive cycles. (○) Imbalance is counteracted by a cellular relaxation process which removes 50% of excessive synthesis in each cycle having fixed periods of DNA inhibition. (▽) Imbalance balance effected by a 50% relaxation process *and* a 50% reduction in the period of DNA inhibition beginning in the fourth cycle ($n = 4$). From Thilly *et al.*, 1975b. Reproduced with the permission of John Wiley and Sons, Inc., New York.

removed by an active process between blocks. In this case, an accumulated imbalance of $1.5X$ is realized in n cycles. The third condition, meant as an approximation to our procedure, combines a 50% active removal of imbalance on a 50% decrease in the length of the effective blocking period beginning with the fourth cycle. After accumulating imbalances for three cycles, the population, in this case, is predicted to show a markedly diminished net imbalance for the hypothetical parameter in subsequent cycles.

In order to explore the behavior of diverse subcellular systems in the resynchronized HeLa system, the cellular activities of six marker enzymes, modal cell volume, and cellular RNA content were determined at the end of ten successive blocks to DNA synthesis. Figure 4 shows the increase in cell number for this particular experiment. Details of the various assays are referenced in an earlier publication, as are pertinent observations regarding the behavior of each parameter over the entire cycle and, specifically, during the period of inhibition of DNA synthesis (Thilly *et al.*, 1975a,b). For our purposes here, however, Fig. 6 summarizes the data which show that after the fifth or sixth cycle the reproducibility among successive cycles was very

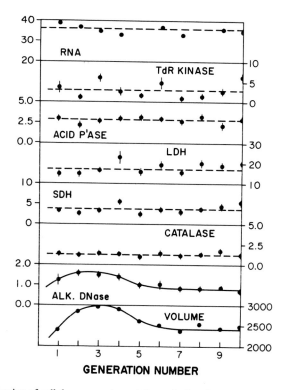

FIG. 6. Behavior of cellular parameters at the end of each period of DNA inhibition over ten successive generations of synchronized growth. Definition of units: RNA (picograms/cell); thymidine (TdR) kinase (anion exchangeable activity changing by 10^3 dpm in 15 minutes); acid phosphatase (P'ASE) (absorbance at 340 nm changing by 0.1 in 5 minutes); lactate dehydrogenase (LDH) (absorbance at 340 nm changing by 0.1 in 1 minute); succinate dehydrogenase (SDH) (absorbance at 600 nm changing by 0.1 in 12 minutes); catalase (absorbance at 240 nm changing by 0.1 in 1 minute); alkaline DNase (absorbance at 264 nm changing by 1.0 in 2 hours); modal cell volume (cubic micrometers). All assays of enzyme activity were performed immediately upon sampling. Vertical lines give standard deviations of the mean of replicates, generally triplicates. From Thilly *et al.*, 1975b. Reproduced with permission of John Wiley and Sons, Inc., New York.

satisfactory for all eight parameters. The behavior of the modal cell volume and alkaline DNase activity is remarkably similar to that predicted for parameters undergoing unbalanced growth, while the behavior of cellular RNA content and the activities of thymidine kinase, acid phosphatase, lactate dehydrogenase, succinate dehydrogenase, and catalase behave as though no accumulated imbalance affected them. Indeed, our measurement of alkaline DNase was prompted by the observations of Churchill and Studzinski (1970) that this enzyme represented the worst case of unbalanced growth uncovered in their extensive studies. That the resynchronization procedure apparently mitigated unbalanced growth to the point that alkaline DNase activity does not demonstrate cumulative imbalance was a most satisfying observation.

Behavior within individual cycles has also been characterized with regard to unbalanced growth (change defined as in a parameter during inhibition of DNA synthesis) for several parameters. Modal cellular volume, and the activities succinate dehydrogenase and catalase have been found to increase by about 15% of the normal cell cycle maximum, while cellular DNA content increases by only 5% during the effective blocking period (Thilly et al., 1975b). Therefore, those who would use this procedure to study cell cycle changes in mitochondrial or microbody physiology may be alerted to this fact and take it into account in interpreting their results. It is to be expected that other parameters will change significantly during the period of inhibition of DNA synthesis. Each investigator must be responsible for characterizing these changes. However, the great facility by which liter cultures of HeLa cells may be maintained in a state of perpetual synchronous growth greatly reduces the effort required to obtain such characterizations.

One reason for including data regarding unbalanced growth in this methodological paper is to offer an alternative procedure to the comparison of cell populations synchronized by induction and mitotic selection techniques. While the latter approach is meant to provide a comparison of unperturbed to perturbed cell systems, it also involves a comparison of anchorage-dependent to suspension-adapted cells.

Is it reasonable to assume that the demonstration of reproducible cycle-dependent behavior over successive cycles can be sufficiently rigorous for drawing certain mechanistic conclusions from observations on the system? For instance, if event A must precede event B in cell growth toward division, will the occurrence of some quantitative imbalance obscure this natural law? Noninteracting parallel series of events would *seem* to display important temporal sequences in unperturbed systems. Would such data not be liable to misinterpretation as representing causal relationships?

V.   Conclusions

The details of the development of a procedure that permits the maintenance of perpetual synchrony in HeLa S_3 suspension culture have been discussed. Some of our reasoning and observations concerning unbalanced growth in the resynchronization approach with suggestions for application to other cell suspension systems have been presented. The principal conclusion is that the inductive synchronization procedure is improved as an aid to the study of the cell cycle when synchronizing conditions are carefully designed to harmonize with the temporal and biochemical characteristics of the particular cell population. Maintenance of perpetual synchronous exponential growth is simply a demonstration that such harmony between the inducer and the induced has been achieved.

ACKNOWLEDGMENTS

The experimental work discussed here was part of my doctoral thesis accepted by the Department of Nutrition and Food Science, Massachusetts Institute of Technology, April, 1971. Dr. Gerald N. Wogan, thesis supervisor, provided the much appreciated academic and moral support in an area not traditional to toxicology, but in which he saw promise in the future study and evaluation of chemical hazards to man. The project was supported by U.S. Public Health Service Grants, Nos. T01-ES-0056-06 and P01-ES-00597 and a grant from the Massachusetts Institute of Technology Environmental Laboratory. Support for continuing work in the area of cell synchronizing technology is presently provided by the National Science Foundation (Cellular Biology Section) Grant No. BMS74-05676 A01. I gratefully acknowledge two indefatigable assistants, T. S. Nowak, Jr. and D. I. Arkin, who were supported in part by the Massachusetts Institute of Technology Undergraduate Research Opportunities Program, and Dr. Piergiorgio Righetti, now of the University of Milan, who helped in the development of the theoretical approach.

REFERENCES

Bootsma, D., Budke, L., and Vos, O. (1964). *Exp. Cell Res.* **33**, 301.
Churchill, J. R., and Studzinski, G. P. (1970). *J. Cell. Physiol.* **75**, 297.
Cohen, L. S., and Studzinski, G. P. (1967). *J. Cell. Physiol.* **69**, 331.
Duneau, R. E., and Woods, P. S. (1964). *Chromosoma* **6**, 45.
Galavazi, G., Schenk, H., and Bootsma, D. (1966). *Exp. Cell Res.* **41**, 428.
Lambert, W. D., and Studzinski, G. P. (1969). *J. Cell. Physiol.* **73**, 261.
Puck, T. T. (1964). *Science* **44**, 565.
Rueckert, R. R., and Mueller, G. C. (1960). *Cancer Res.* **20**, 1589.
Thilly, W. G., Nowak, T. S., Jr., and Wogan, G. N. (1974a). *Biotechnol. Bioeng.* **16**, 149.
Thilly, W. G., Arkin, D. I., and Wogan, G. N. (1974b). *Anal. Biochem.* **60**, 637.
Thilly, W. G., Arkin, D. I., Nowak, T. S., Jr., and Wogan, G. N. (1975a). *Biotechnol. Bioeng.* **17**, 695.
Thilly, W. G., Arkin, D. I., Nowak, T. S., Jr., and Wogan, G. N. (1975b). *Biotechnol. Bioeng.* **17**, 703.
Xeros, W. (1962). *Nature (London)* **194**, 682.
Zielke, H. R., and Littlefield, J. L. (1974). *Methods Cell Biol.* **8**, 107.

Chapter 26

A Method for Synchronization of Animal Cells in Culture[1]

Département de Biologie Générale et Appliquée,
Laboratoire Associé au C.N.R.S. No. 92,
Université Claude Bernard (Lyon-I),
Villeurbanne, France

I. Introduction

We describe in this chapter a method of preparing synchronous cell cultures. Studies of the synthesis of DNA and particularly of the enzyme systems involved, as a function of the cell cycle of eukaryotes has recently become of interest to cell biologists. However, such studies always require considerable amounts of synchronized cells.

At present several types of methods are available for the production of synchronized cell cultures. The first consists of blocking either the synthesis of DNA or mitosis by the use of chemical agents such as thymidine (Xeros,

---

[1] This work was performed at Unité de Virologie (U51), I.N.S.E.R.M., Lyon.

1962), 5-fluorodeoxyuridine (Rueckert and Mueller, 1960), amethopterin (Rueckert and Mueller, 1960), 5-aminouracil (Schindler, 1963), hydroxyurea (Plagemanne et al., 1974), Colcemid (Stubblefield et al., 1967), or vinblastine (Pfeiffer and Tolmach, 1967). Using this type of technique one can easily obtain a large quantity of synchronized cells, however, there are certain disadvantages. The use of Colcemid leads to large numbers of chromosomal abnormalities (Comings, 1971), and the inhibitors of DNA synthesis produce interference with normal nucleotide pathways of the cells since excess thymidine inhibits incorporation of uridine in RNA (Kasten et al., 1965), and only results in partial blockage of DNA synthesis (Bostock et al., 1971; Pica-Mattocia and Attardi, 1972). We have investigated the effects of excess thymidine on enzyme activities, particularly thymidine and thymidylate kinases at different points in the cell cycle (Ooka, 1976). Our results show that these enzymes are present at levels 7- to 8-fold higher in thymidine-blocked cells during the S phase than in the naturally synchronized cells. A high concentration of hydroxyurea can lead to toxic effects (lethal damage) (Sinclair, 1965; Kim et al., 1967) or chromosomal aberrations (Yu and Sinclair, 1968). These effects thus render the interpretation of results obtained by the use of such methods uncertain. A second method described by Terashima and Tolmach (1963) is applicable only to monolayer cell cultures. It is based on the observation that cells in mitosis adhere less strongly to the culture vessel and can be detached therefrom. Unfortunately this excellent method produces only a low yield of cells (1–2%) which can, however, be increased by manually shaking off the mitotic cells from mono-layers grown in low-calcium medium (Robbins and Marcus, 1964). A third method selects postmitotic cells by their reduced size. Sinclair and Bishop (1965) and Schindler et al. (1970) used low speed centrifugation in a sucrose gradient to select the lighter postmitotic cells, whereas Shall and Mc-Clelland (1971) simply allowed their cells to descend in a column of culture medium by gravity, the lighter postmitotic cells remaining near the top. This method appears to provide a good yield of synchronized cells for suspension culture. These two methods both have the unusual advantage of not sub-mitting the cells to adverse conditions (Terashima and Tolmach, 1963; Shall and McClelland, 1974) thus leading to the maximum preservation of structure and function.

Our method is based on a selection of postmitotic cells (Ooka and Daillie, 1974) by the combined method of selective detachment from a monolayer (Terashima and Tolmach, 1963) followed by a natural gravity sedimentation on a column of culture medium (Shall and McClelland, 1971). Separate use of either of the above methods achieves only moderate synchronization, whereas a combination of the two techniques produces a good yield of well-synchronized cells. The method would seem to be of general use for all cells which can be cultured as monolayers, at least temporarily.

We shall describe the technique as applied to KB cells.

II.   Cell Culture and Media

KB cells were grown in suspension at a density of 2×10^5 cells per milliliter in Eagle MEM as modified by Joklik (1968) and supplemented with 5% heat-inactivated horse serum. These cultures were maintained by replacement of one-half of old culture medium with fresh medium every 24 hours.

Monolayer cultures of the same cells were grown in Roux bottles using the same medium, and were harvested 48 hours after seeding.

III.   Preparation of Synchronous Cell Populations

Suspension cultured cells were seeded into Roux bottles at a concentration of about 8×10^6 cells per bottles. After 48 hours, old medium was removed carefully to eliminate dead cells and replaced with 30–50 ml of fresh medium. The bottles were then shaken in the horizontal plane, and the detached cells (20–30% of the total population) were recovered by centrifugation at 500 rpm for 10 minutes and resuspended in 3 ml of serum-free medium. It is essential to remove the dead cells, and to shake the cell culture gently since KB cells attach only loosely to the culture vessel. About 70×10^6 cells as a suspension in serum-free medium were then carefully deposited with a wide-mouthed pipette on the top of a sterile glass column (2.6×19 cm) containing 90 ml of complete medium at 37°C as shown in Fig. 1. The cells were allowed to settle for 40 minutes at 37°C, then the upper third (30 ml) of medium containing light cells was removed gently with a pipette, mixed with 170 ml of completed medium and incubated at 37°C. If a large quantity of synchronous cells is required, another type of glass column (4.4×28 cm) containing 300 ml of complete medium can be used. In this case, we can deposit about 500×10^6 cells on the top of column, and the light cells can again be recovered from the upper third of medium (100 ml). We also examine cells that remained in column; they did not exhibit any synchrony.

We also attempted to produce synchronized cells from a suspension culture of KB cells without previous selective detachment from glass. The

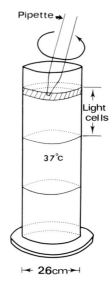

Fig. 1.   The sterile glass column of complete medium used to separate the G_1 cells. A column (2.6 × 19 cm) containing 90 ml of prewarmed culture medium was prepared, and the cells suspended in serum-free medium were deposited carefully on the top of this column with a wide-mouthed pipette. The tip of the pipette touched the surface of medium, which was gently stirred to make a cell layer. The cells were then allowed to settle for 40 minutes at 37°C, after which time the upper third of medium containing light cells were removed.

total cell population was harvested by centrifugation at 500 rpm, then was applied to a column as described above.

The cells thus selected were held in suspension culture over a period of 30 hours, during which their multiplication, their mitotic index, and their incorporation of ^{3}H-labeled thymidine were studied (see Figs. 2 and 3).

IV.   Methods for Characterization of Synchronous Cultures

In order to estimate the number of cells in mitosis, at selected time intervals aliquots of the cells were added to an equal volume of a mixture of ethanol, acetic acid, and distilled water (5:2:3, v/v), allowed to stand at room temperature for at least 30 minutes, recovered by centrifugation, and resuspended in a small volume of crystal violet [0.025% w/v in dilute acetic acid (1% v/v)] (Schindler et al., 1970). The percentage of cells in mitosis (metaphase) was established by counting in a Bürker hemacytometer.

For measuring the rate of DNA synthesis, aliquots (1 ml) of cell suspension were removed from the cultures at 2-hour intervals and incubated with thymidine-methyl-^{3}H (2 μCi/ml, specific activity 18 Ci/m mole, from Radiochemical Centre, Amersham) for 20 minutes at 37°C in hemolysis tubes. At the end of this time the cells were washed twice at 0°C with 5 ml of medium containing cold thymidine (0.6 mM) and precipitated with cold trichloroacetic acid (5% v/v). The precipitate was recovered on Whatman GF/C filters washed with further 5% TCA, and its radioactivity was determined in a Packard scintillation counter after drying.

In order to determine the proportion of labeled cells at the same time intervals, aliquots (1 ml) of cell suspension were removed from culture and labeled with thymidine-^{3}H as described above. After labeling, the cells were washed three times in medium containing cold thymidine (0.6 mM) at 0°C, and one further wash with cold phosphate-buffered saline (pH 7.2), the cells were spread onto gelatinized slides (Ilford), and fixed with acetic acid/ethanol mixture (1:3 v/v). The radioautography procedure was the same as that previously described (Ooka et al., 1973).

Aliquots of cells were counted every 2 hours using either a Bürker hemacytometer or a Coulter counter (Coultronics).

V.   Characterization of the Synchronization

A.   Synchronization by the Combined Method

Figure 2 shows the results obtained from the combined method of selective detachment and natural gravity sedimentation. Less than 10% of the cells are labeled within the first hour of culture, and at the same time the incorporation of thymidine-^{3}H is only 15% of the maximum, which occurred after 10 hours of culture. We can estimate the duration of G_1 phase as about 5–6 hours. The number of cells in the culture remains essentially stable until 12 hours after selection, then a doubling of the number of cells takes place between 12 and 19 hours. The maximum number of cells in metaphase is observed as 16 hours. Furthermore, synchrony is well maintained even during a second cycle. About 50% of the cells were synthesizing DNA at the time corresponding to the second S phase.

B.   Synchronization by Either Gravity Sedimentation or Selective Detachment Alone

Figure 3a shows the results observed with cells synchronized by the gravity sedimentation method (suspension cultured cells simply separated

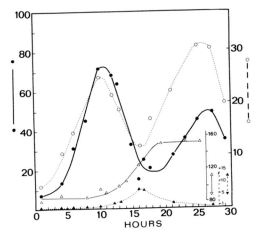

FIG. 2.   Synchronization of KB cells by the combined method of selective detachment from monolayer followed by a natural gravity sedimentation through a column of culture medium. The light cells show a synchronous growth pattern. The following parameters were measured to demonstrate the degree of synchrony: incorporation of ^{3}H-labeled thymidine per milliliter of culture (cpm $\times$ 10^3) O---O; percentage of labeled cells (radioautography), ●——●; number of cells ($\times$ 10^3)/ml, △——△; percentage of mitotic cells (metaphase), ▲---▲. From Ooka and Daillie (1974).

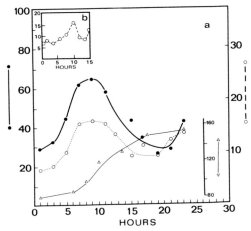

FIG. 3.   Synchronization of KB cells by column method (gravity sedimentation) or by selective detachment method alone. For the column method (a), suspension-cultured cells were harvested by centrifugation, resuspended in serum-free medium, and deposited on the top of a column of medium. For the selective detachment method (b), suspension-cultured cells were seeded in Roux bottles. After 48 hours, the bottles were agitated and the detached cells were recovered by centrifugation. The synchrony of the cell population, one separated by a column and the other detached by agitation, was examined by measuring the following parameters: incorporation of ^{3}H-labeled thymidine per milliliter of culture (cmp $\times$ 10^3), O---O; percentage of cells labeled (radioautography), ●——●; number of cells ($\times$ 10^3)/ml, △——△. From Ooka and Daillie (1974).

on a column of medium). About 10 hours after synchronization, some 60% of the cells are in S phase; however, nearly 30% of the cells are already in S phase within the first hour. The incorporation of thymidine-^{3}H follows the same pattern. The number of cells doubles over the experimental period (30 hours), the major increase occurring between 10 and 15 hours. The above observation implies that this method produces a population of cells that are only moderately synchronized. The overall yield is also rather low (2%). A similar degree of synchronization was achieved by the method of selective detachment alone. Figure 3b shows the incorporation of thymidine-^{3}H in a culture prepared by this method.

It can be pointed out, however, that in both cases a maximum of DNA synthesis occurs at about 10 hours after inoculation. These cell cultures are, however, only partially synchronized, as the incorporation of thymidine-^{3}H at other times was higher than that observed using cells prepared by the combined method.

In conclusion, it appears that a very considerable increase in the synchronization was achieved when the two methods were combined (Fig. 3a), the selectively detached cells being reseparated by gravity sedimentation. The present study clearly shows that the great majority of the cells recovered from the top of a column, after a preliminary selection by ease of detachment from a culture vessel, are in the postmitotic phase, as the population does not begin to increase until at least 12 hours after the beginning of incubation.

The cell volume of the postmitotic cells recovered from the top of the column is about 40% of that of the sedimenting cells. Such an estimation is in good agreement with that for cells obtained from the top of a column of medium by Shall and McClelland (1971) or from a sucrose gradient centrifugation by Schindler et al. (1970).

The overall yield of postmitotic cells with the combined method is about 4% of the original cell population. This agrees well with the observation that, in an asynchronous culture, approximately 4% of the cells will be in mitosis at a given time (Terashima and Tolmach, 1963; Mittermayer et al., 1968) and means that our technique is in fact yielding the maximum number of obtainable naturally synchronized cells.

VI.   Concluding Remarks

We have demonstrated in this chapter that the combined method can provide well-synchronized cells from any cell line which can be grown as a monolayer. The small cells separated from the top of the column of culture medium are clearly in early G_1 phase. It has recently been observed that the results obtained using chemical synchronization methods are sometimes in disagreement with those obtained using natural synchronization (Pica-Mattocia and Attardi, 1972; Ooka, 1976; Comings and Okada, 1973; Painter and Shaefer, 1971). This implies that the avoidance of chemical agents in the preparation of synchronized cells is desirable in all cases. The cells prepared by our method appeared to have undergone little damage and are well suited to structural and/or functional investigation. The method can provide a large number of well-synchronized cells; for instance, using a large column, a synchronous culture containing 1×10^8 cells can be obtained without difficulty. In addition when large amounts of cells are required, the yield of synchronized cells could be increased by storing the detached cells at $0°C$ to block their progression through the cell cycle (Tobey et al., 1967), while the parent culture is reincubated with fresh medium for about 60 minutes when more detachable cells will have appeared and can be harvested and added to pool at $0°C$, which is later brought to $37°C$ (Nagasawa and Dewey, 1972) and separated by gravity sedimentation. It also would be possible to do a preliminary synchronization by the double thymidine block method. The cells would then be allowed to proceed through one synchronous mitosis or, better, two, by which time the daughter cells would have recovered from the thymidine treatment. At a time corresponding to the maximum of early G_1 phase cells, these would be harvested using our combined method.

Our technique has, in fact, permitted the successful study of the evolution of enzyme activities involved in the DNA synthesis, particularly DNA polymerase (Ooka and Daillie, 1975), thymidine, and thymidylate kinases (Ooka, 1976), during the cell cycle.

ACKNOWLEDGMENTS

Deep appreciation is expressed to Professor J. Daillie for his valuable collaboration in this study. I also thank Dr. W. Bernhard, Dr. W. C. Dewey (Colorado State University), Professor R. Sohier, and Dr. T. Greenland for their advice and critical discussion of this method. I acknowledge the skilled technical assistance of Miss A. Larama.

REFERENCES

Bostock, C. J., Prescott, D. M., and Kirkpatrick, J. B. (1971). *Exp. Cell Res.* **68**, 163.
Comings, D. E. (1971). *Exp. Cell Res.* **67**, 441.
Comings, D. E., and Okada, T. A. (1973). *J. Mol. Biol.* **75**, 609.
Joklik, W. K. (1968). Grand Island Biological Company Catalog, p. 69. Grand Island, New York.
Kasten, F. H., Strasser, F. F., and Turner, M. (1965). *Nature (London)* **207**, 161.
Kim, J. H., Gelbard, A. S., and Perez, A. G. (1967). *Cancer Res.* **27**, 1301.
Mittermayer, C., Kaden, P., and Sandritter, W. (1968). *Histochemie* **12**, 67.

Nagasawa, H., and Dewey, W. C. (1972). *J. Cell Physiol.* **80**, 89.

Ooka, T. (1976). In preparation.

Ooka, T., and Daillie, J. (1974). *Exp. Cell Res.* **84**, 219.

Ooka, T., and Daillie, J. (1975). *Biochimie* **57**, 235.

Ooka, T., Girgis, A. M., and Daillie, J. (1973). *Exp. Cell Res.* **81**, 207.

Painter, R. B., and Schaefer, A. W. (1971). *J. Mol. Biol.* **58**, 289.

Pfeiffer, S. E., and Tolmach, L. J. (1967). *Nature (London)* **213**, 139.

Pica-Mattoccia, L., and Attardi, G. (1972). *J. Mol. Biol.* **64**, 465.

Plagemanne, P. G. W., Pickey, D. P., and Erbe, J. (1974). *Exp. Cell Res.* **83**, 303.

Robbins, E., and Marcus, P. I. (1964). *Science* **144**, 1152.

Rueckert, R. R., and Mueller, G. C. (1960). *Cancer Res.* **20**, 1584.

Schindler, R. (1963). *Biochem. Pharmacol.* **12**, 533.

Schindler, R., Ramseier, L., Schaer, J. C., and Grieper, A. (1970). *Exp. Cell Res.* **59**, 90.

Shall, S., and McClelland, A. J. (1971). *Nature (London), New Biol.* **229**, 59.

Sinclair, W. K. (1965). *Science* **150**, 1729.

Sinclair, W. K., and Bishop, D. H. L. (1965). *Nature (London)* **205**, 1272.

Stubblefield, E. R., Klevez, R., and Deaven, L. (1967). *J. Cell Physiol.* **69**, 345.

Terashima, T., and Tolmach, L. J. (1963). *Exp. Cell Res.* **30**, 344.

Tobey, R. A., Anderson, E. C., and Peterson, D. F. (1967). *J. Cell Physiol.* **70**, 63.

Xeros, N. (1962). *Nature (London)* **194**, 683.

Yu, C. K., and Sinclair, W. K. (1968). *J. Cell Physiol.* **72**, 39.

Chapter 27

Hypertonicity and the Synchronization of Mammalian Cells in Mitosis

D. N. WHEATLEY

*Department of Pathology, University of Aberdeen,
Foresterhill, Aberdeen, Scotland*

And all the little oysters stood
And waited in a row.

Tweedledee *in* "Through the Looking Glass"
L. Carroll

The need for synchrony stems from an increasing interest in the basic concept of the cell cycle and all that it implies. Before choosing any particular method of synchronizing cells from a range which has been increasing both in number and ingenuity over the last fifteen to twenty years, the investigator should ask at least the three following questions: What am I synchronizing cells for? Are the synchronized cells representative of their unsynchronized counterparts in the same phase of the cell cycle? And can I get enough this way? If reasonable answers can be given, then the experimenter is already well on the way to using a suitable technique to his greatest advantage. Because this decision of the right technique is of paramount importance, these questions should be examined more closely.

I.   Objectives and Rationale

A.   What Is the Purpose of the Synchrony Technique?

In studying cellular events in the cell cycle, there is much greater safety for more than purely logistic reasons in recording what happens to a hundred cells at one time than to make separate recordings on one hundred individual cells on a hundred occasions. In some studies, for example in the detailed movement of cells in confrontation, there is no overwhelming need to study synchronized cells; nor is there need for synchronization when measurements of a physicochemical or biochemical nature can be made at the individual cell level, as, for example, in the studies of Killander and Zetterberg (1965) and Zetterberg and Killander (1965). But in studies such as the latter, it is not immediately apparent what stage of the cycle cells are in just by looking at the cells unless they have been individually followed by time-lapse cinematography. In synchronous cultures the experimenter sets out to produce a population of cells in a precisely known phase of the cell cycle. Many of the differences that are observed between cells in the various phases of the cell cycle are small, and therefore work at the single-cell level would be tiresome. To overcome this problem, large numbers of cells are regimented by synchrony techniques in order to make such comparisons, and differences between average values of the populations are amenable to statistical analysis. In many cases the demonstration of a statistically significant difference will depend largely on the degree of synchrony achieved in the populations under comparison. This, then, is one of the most important reasons for obtaining a high degree of synchronization.

Ideally, methods should be adopted that will get "*all* the little oysters" to stand in a row, but even with cells the chances of attaining this objective are very low. If we take a completely asynchronous culture, one that shows continual mitotic activity at a steady rate with no oscillatory patterns of division, a certain percentage of cells will be in a given phase at any instance. The actual percentage will depend on the length of this phase in relation to the complete cycle time. If for any reason more or less than expected are present, then a degree of *parasynchrony* exists. The term parasynchrony might truthfully be applied to all our synchrony procedures as the ideal (100% in synchrony) is rarely if ever attained. In day-to-day usage, the terms are used more loosely, and we might take a HeLa S3 cell population growing in suspension culture as an example. The cells divide once every 24 hours and spend 1 hour in mitosis. In a purely asynchronous culture, there should be approximately 4% of the cells in mitosis at any time. Values significantly above or below this value indicate mild parasynchrony. This low-grade fluctuation is often seen when cultures are fed daily by the batch procedure instead of continuously, after short periods of cooling and rewarming, after trypsinization, and after almost any procedure that perturbs cells. Parasynchrony describes the low-grade synchrony obtained by using inappropriate technical procedures or by poor execution of synchrony techniques. But when the large majority of cells are aligned (say 70% or more), cultures are said to be synchronized. Few artificial synchrony procedures achieve a very high degree of synchrony. We have spoken of aligning cells into a particular phase of the cell cycle, but this starting point is of particular importance and must be carefully considered.

When synchronization in the S phase of the cell cycle is required, it is often possible to bring almost every cell in a population into S phase. But S phase could contribute anything from a quarter to a half of the entire interphase time of many cultured mammalian cells. Some of the "synchronized" cells will be as much as half a cycle out of step, yet still in the same phase. But if the *purpose* of this synchronization is to study DNA synthesis without any particular need to relate the findings to the orderly sequence of replication occurring in S phase, synchrony does not need to be strict. If the question concerns initiation of DNA synthesis, cells are best synchronized in late G_1 so that they can be released on an instant into S phase. Such cells would be aligned with respect to a single, important cell cycle event of short duration. Cells synchronized for this purpose will provide an accentuated initiation event in the population, but there is a greater probability of their being disturbed by the synchrony technique, as we shall see in the next section.

It can be readily appreciated that it is best to synchronize cells in the shortest cell cycle phases or at the "interfaces of interphase" (e.g., G_1/S or S/G_2). Mitosis is, for this and other reasons, one of the best periods in which to synchronize most mammalian cells. This can often be quite easily achieved. But to return to the question of purpose, if the investigator wants to study mitotic cells, many synchrony procedures for selecting mitotic cells will probably be suitable. But if G_2 cells are the focus of the investigator's attention, then in most cases mitotic synchronization would not be recommended as the starting point. When two investigators synchronize their watches at 10:00, they may do so in order that they may independently perform a coordinated task at 23:00. They assume that their watches are reasonably accurately constructed and regulated. This is not so easily accomplished with cells. They run at slightly varying rates from one another and only a mean expected time can be given for event B after synchronization at event A. The longer event B is after A, the less precise its predicted time, and the wider the scatter around the mean. In HeLa cells, for example, G_2 is about 21–23 hours after mitosis. Some cells will reach it in 17–18 hours whereas others take more than a day. In brief, synchrony is quickly dissipated by cells, and much can be lost within a single generation cycle. A pure mitotic population would not be nearly so synchronized by the time the next G_2 is reached. Some methods are available that might help to reduce this dissipation of synchrony in cell populations by subcloning (e.g., Nias, 1968).

In its fullest sense, synchrony is not to have the cells or watches correct at the check point alone, but to have them still in harmony at any subsequent time of referral. For this to be successful requires first the ability to "stathmosynchronize" accurately, and second, that the population of cells be homogeneous with respect to their cell cycle kinetics so that they will continue to function at their normal rate after stathmosynchronization. In a word, the cells should have the characteristics of an *ergodic* culture (Engelberg and Hirsch, 1966). Abbo and Pardee (1960) have referred to the two conditions as *synchronized* for the act of bringing cells into line, and *synchronous* for the harmonious behavior of cells thereafter. While this neatly describes these states, the two words may be too easily confused for general use.

Finally, with regard to purpose, we must consider the possibility that the investigator is not specifically interested in the cell *division* cycle. If the division cycle were the master control, then all other cycles would be directly related to it. While this appears to be so for many cellular functions, cases have been made for some constituent parts of a cell having a degree of autonomy in their own cyclical life histories. Hence Mazia (1974a) refers to a chromosomal cycle; the centrioles have another (Robbins, *et al.*, 1968), and mitochondria a separate cycle (Pica-Mattoccia and Attardi, 1972).

B. Are the Synchronized Cells Representative of Their Normal Counterparts?

A stathmosynchrony technique which holds cells at the G_1/S boundary may, on release of an S-phase wave, affect the cell which had just arrived at this boundary least, while another may have been arrested at the boundary for many hours and will be much more affected. The latter may proceed through its subsequent cycle much more slowly than the former, whereas in Engelberg and Hirsch's ergodic culture, one would ideally expect to find both proceeding together at the same rate, and furthermore at a rate that is identical to an unsynchronized control culture. This, as we have seen, is unlikely to occur under certain synchronizing regimes, and this is probably the main reason why techniques inducing synchronization often produce artificial or distorted results. The investigator should be aware of these shortcomings in his chosen experimental procedure and be cautious in assessing the meaningfulness of his results. Typical illustrations are (1) the *unbalanced* growth of cells held from entering S phase where cytoplasmic functions may progress well ahead of nuclear functions (Rueckert and Mueller, 1960; Studinski and Lambert, 1969), (2) the greater setback experienced by "old" as opposed to "young" cells to heat or other shocks used in synchrony (Scherbaum and Zeuthen, 1954), and (3) the presence in populations of synchronized cells obtained by, for example, high specific radioactivity thymidine procedures of damaged, dying, or even "dead" cells which are nevertheless contributing biochemically, even if they are not doing so reproductively (Nias and Fox, 1971).

Any disturbance, however mild, will affect the cells in a synchrony procedure. The aim is to keep it to a minimum or within controllable limits. Cells marshalled into the company of only those in the same stage as themselves enter an abnormal society compared with their asynchronous controls. We have hardly yet considered this question of whether cells in a synchronized population influence one another differently from asynchronous controls.

C. Can I Get Enough?

The development of synchrony procedures is often an experimental approach to a statistical problem; the more we have at any one time doing the same thing, the more accurately we can describe what is done. The essence is for synchrony to be the magnifying glass of cell studies. For many biochemical studies it may be necessary to have say $10^8 - 10^9$ cells for each analysis, and an appropriate synchrony procedure would not only have to satisfy the requirements we have already outlined, but be capable of yielding the required mass of cells. In many cases, these requirements make heavy demands on culture facilities and manpower. The greater the number of technical maneuvers required, the more time elapses during the setting up of a synchronized culture, and time is a particularly scarce commodity if high levels of synchrony are to be achieved.

II.   Techniques

A.   Synchrony by Selection and Induction

Selection synchrony means extracting cells of the right kind from a population, e.g., early G_1 cells because they are smallest or late G_2 cells because they are largest. *Induction* techniques are those which constrain cells into a particular phase of the cell cycle from which they can be released like athletes at the start of a race. The former usually involves a small fraction of the total population whereas the latter often tries to involve most or the whole of the population. Both have many disadvantages and advantages, some of which have already been discussed.

B.   Synchrony by Hypertonic Arrest in Metaphase

Synchrony by hypertonic arrest in metaphase, better known as SHAM, can be described as a simultaneous exploitation of both induction and selection techniques, and must on these grounds be considered in relation to other available methods of collecting mitotic cells. Hypertonicity holds cells up in the middle stage of mitosis, viz. metaphase. Its mitostatic effects has been observed for many years (Ebeling, 1914; Hughes, 1952; Stubblefield and Mueller, 1960), but its application to the field of cell synchronization is recent (Wheatley and Angus, 1973; Wheatley, 1974b). Since cells are held in M phase by hypertonicity, they will pile up just as in any other induction procedure carried out with more conventional agents, such as Colcemid or vinblastine. The distinct advantage of synchronizing monolayer cells in metaphase is that they can be selected out very cleanly to give consistently pure metaphase cultures. This is because metaphase cells generally round up and have a looser attachment to the substrate than interphase cells. A gentle washing action of the medium over these cells will break the few processes by which they usually remain attached. This basic technique was outlined originally by Terasima and Tolmach (1963). Combination treatments have been devised by Stubblefield and Klevecz (1965) and by Romsdahl (1968) where Colcemid was used with a shakeoff synchronization. These technique have been applied to particular cell lines. They work on the delicate balance between a concentration of alkaloid sufficient to arrest (and therefore augment) metaphases for the shakeoff procedure and that from which cells would recover more or less immediately when the drug was removed from the culture. This is not easily accomplished with cultures different from the one for which it was devised, and is probably too fickle to be reliable.

C. Ions and Cell Growth

The basic ion requirements of mammalian cells in culture were established by Eagle (1956). Ions alone do not make up the whole osmotic environment, intracellularly or extracellularly, but they contribute much more significantly than other solutes. Proteins themselves do not greatly contribute to the osmotic pressure as solutes, but do play a fundamental role in modifying the behavior of ions both inside and outside the cell. There is no space here to deal with this important relationship, and the reader is referred to the review by Robinson (1975). For the purposes of this discussion, we will take the gross simplification that the ionic component contributes overwhelmingly to the tonicity of the medium. Only changes of ion concentration, not of other solutes, in media will be considered here.

The ionic environment of cultured mammalian cells is relatively constant since most media still contain the ions specified by Eagle (1956) in approximately the same concentrations as originally stated (see Table I). The total molarity is about 0.15 M (150 mM). The recipe provides an ionic environment which resembles quite closely that of normal body fluids. The large majority of different mammalian cell lines will survive and grow in such a solution if it contains vitamins, amino acids, cofactors, an energy

TABLE I

CONCENTRATION OF PRINCIPAL IONS (mM) IN DIFFERENT SALT SOLUTIONS USED AS BASES FOR CULTURE MEDIA FOR MAMMALIAN CELLS

Ions	Eagle (1956)	Earle's BSS	Hanks' BSS	Suspension medium[a]
Cations				
Na^+	85–115	143.1	141.5	144.4
K^+	1–10	5.4	5.8	5.4
Ca^{2+}	1–2	1.8	1.3	—
Mg^{2+}	0.1–0.4	0.4	0.9	—
	—	150.7	149.5	149.8
Anions				
$HPO_4/H_2PO_4^-$	0.2–2.0	0.9	0.9	1.0
HCO_3^-	"Not essential"	26.2	4.2	26.2
Cl^-	50–100	123.4	144.9	121.8
HSO_4^-	—	—	—	0.8
	—	150.5	149.6	149.8
Total mM for salts	—	150.6	149.5	149.8

[a] After Rueckert and Mueller, 1960.

source, and serum. The early media and almost all subsequent formulations based thereon have been concerned with satisfying one overriding consideration, that cells should proliferate as rapidly as possible in it with the least cell death. This continues to be the objective of many culturists; but it is important that media should be developed with objectives other than proliferation in mind. The effects of hypertonic and hypotonic deviations on cell growth, metabolic function and other cell parameters have been little explored. Our observations stemmed from an analysis of the effects of tonicity on the ability of cells to progress from the end of S phase into mitosis (Wheatley and Angus, 1973). In deliberately collecting the G$_2$ population (see Fig. 1), it was noted that Colcemid cultures exposed to hypertonic medium of about 220–230 mM for 3 hours (i.e., about 20–30 minutes longer

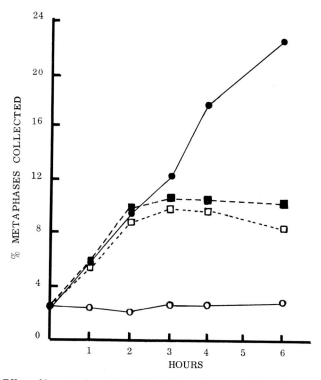

FIG. 1.   Effect of hypertonic medium (227 mM) on the entry of G$_2$ cells into mitosis. The free-running control (● ——— ●) was a culture growing in 149 mM medium to which 0.15 μM Colcemid has been added. Another culture (O——O) had no Colcemid added. To delineate the G$_2$ cohort, a third culture (■---- ■) had 2 mM hydroxyurea added at 0 hour, and therefore only G$_2$ cells were collected by the addition of Colcemid to this 149 mM medium. The last culture (□----□) shows the effect of raising the hypertonicity to 227 mM *without* the addition of Colcemid.

than the average G_2 time) collected as many cells in mitosis as controls in 150–165 mM medium (Table I). It was noted in two such experiments that cultures given only hypertonic medium, and to which no Colcemid should have been added, gave almost identical collections of metaphases as the Colcemid-treated isotonic controls. The similarity was so striking that it was first suspected that Colcemid had been inadvertently added to the former groups. Subsequent experiments proved, however, that hypertonicity alone was an efficient mitostatic agent. This led to defining the optimal conditions for arrest and the changes caused by such treatment in cells. Subsequent studies were principally on the kinetics of cell recovery from hypertonic arrest of metaphase, from which came the development of a practical SHAM procedure for HeLa S3 cells.

D. Raising the Tonicity

Medium of about 150 mM can be raised by adding calculated amounts of concentrated solutions. The use of concentrated balanced salt solutions (e.g., Earle's BSS 20× concentrated stock is a 2.44 M solution), or a simple NaCl solution (e.g., 3 M) gave identical results. Table II sets out a few calculated additions as guidelines for the above solutions.

It is important to note here that much of the development of SHAM dealt with HeLa cells grown in and adapted to 165 mM medium. While

TABLE II

VOLUMES OF PREPARED HIGH IONIC STRENGTH MEDIA REQUIRED
TO ADJUST MEDIUM OF 149 mM TO HIGHER TONICITIES

Volume of culture medium	Vol. of concentrate required (mls)		Final mM
	3 M NaCl	2.44 M BSS[a]	
100	2.80	3.52	227
20 ⎫	0.16	0.20	172
20 ⎬ additive[b]	0.48	0.60	194
20 ⎭	0.80	1.00	216
20 ⎫	0.18	0.22	174
20 ⎬ additive[c]	0.55	0.70	200
20 ⎭	0.95	1.20	227

[a] This is equivalent to a 20× concentrated stock for Earle's BSS.
[b] Additive for triple step increase; the second 20 ml plus supplement is added to the first after 10–15 minutes, and so on. This schedule gives slightly lower tonicities than the next.
[c] As for b, but producing slightly higher tonicities.

raising the molarity to 227 mM gives good collection, it was not always found that 150 mM-adapted cells tolerated this level well, especially if the cultures were "old" before treatment, i.e., beginning to exhaust their medium. In general, therefore, the molarities indicated in this paper are only guidelines. It may often be found that 205–215 mM medium is more suitable, and for this reason several different schedules for adjusting tonicity are given in Table II.

Medium in an established monolayer culture is removed, and debris is filtered or spun away. Using part of this or other preconditioned medium, the appropriate amount of concentrated solution is added to give the required hypertonicity. After thorough mixing, the medium is returned to the cells at its original temperature. The use of the cells' own medium avoids the possibility of a shock being caused by the addition of fresh medium at this time. When concentrated solution is to be added directly to medium in a culture vessel to raise the tonicity, it is usually done in some part of the vessel away from the cells, e.g.,Blake bottles can be turned over and Burler bottles stood on end. There are several ways of reducing the effects of a sudden increase in tonicity in a culture. One is to drip-feed the appropriate amount of concentrated solution into the medium while gently agitating over a period of about half an hour. A slightly different method, which we often employ, is to use three steps, usually of about 25–30 mM tonicity increases each time at intervals of 10 minutes. This has proved to be very satisfactory, and in Table II the amounts of concentrated salt solution required to sequentially raise the tonicity of medium by the addition for three lots of 20 ml of medium is given. After the third step has been added, just over 60 ml of the original medium, now at 227 mM, is present in the culture. Occasionally the pH of medium has altered significantly on addition of concentrated salt solution, and it is advisable, therefore, to adjust the pH of the concentrate to 7.2 before sterilizing it.

There is no single optimal concentration for arrest of mammalian cells in metaphase by higher ionic strength media since some cell lines require higher levels than others (Wheatley and Angus, 1973). As the level rises, the possibility of moving into toxic levels of tonicity is increased and the progression of cells into mitosis is prevented. Thus cell lines requiring very high levels of hypertonicity for arrest, such as the hamster fibroblast line BHK21/C13, are not usually suitable for SHAM.

The tonicity of the culture medium used prior to raising the tonicity must also be considered, and the amount of serum present may modify the effects on metaphase arrest. Clearly all these factors make it difficult to specify effective hypertonic levels for synchrony, and the experimenter must be prepared to find empirically the most appropriate level for the cell system being used.

E.   Arrest of Metaphase Cells

There is clear evidence that in some cell lines, such as HeLa, HEp2, and Chang, hypertonicity arrests cells in metaphase without disturbing their progress into this phase from G_2 (Wheatley and Angus, 1973; Wheatley, 1974a,b). This is of particular interest to devotees of the G_2 phase and is rather unexpected, considering the sensitivity of this same stage to such treatments as X-irradiation (Walters and Petersen, 1968; Sinclair and Morton, 1966). Where cells have been disturbed, and this occurs predominantly in the first 15 minutes after the tonicity has been raised, they are usually discarded. Thus, at the end of the time required to raise the medium to its final hypertonic level, mitotic and interphase cells which are loose, damaged, dying, or dead are discarded and the medium is replaced with more of the originally removed medium, which has also been brought up to that same final tonicity. Collection of metaphases begins from this point.

Cells which are in metaphase when the hypertonicity reaches its final level are prevented from moving on. In this respect hypertonicity acts more swiftly than some of the conventional antimitotic drugs. Exactly how hypertonicity acts in this way is not known, but arrested cells develop spindles; their kinetochore links between chromosomes and spindle fibers are established, and most other features of mitosis look relatively normal (see Wheatley, 1974a; this chapter, Fig. 2). The one change characteristically

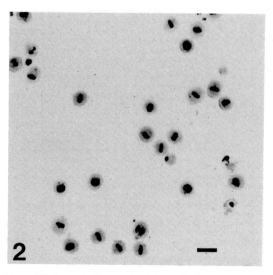

FIG. 2.   Collection of HeLa metaphases after 2 hours in 227 mM medium using a shakeoff procedure. Scale line = 20 μm.

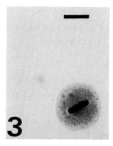

FIG. 3. Higher magnification of a hypertonic (227 mM) metaphase HeLa cell showing tight aggregation of chromosomes as a metaphase plate and clear evidence of a spindle. Scale line = 50 μm.

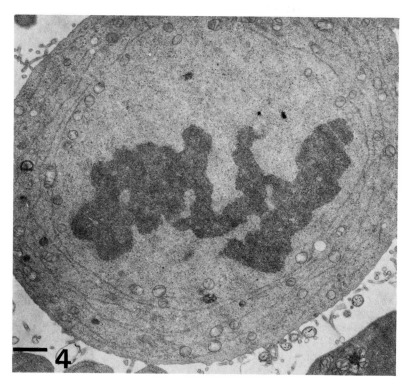

FIG. 4. Electron micrograph of a hypertonic (227 mM) HeLa metaphase showing relatively normal cytoplasmic appearance and the association of the chromosomes in the metaphase plate. Scale line = 1 μm.

found is that chromosomes are condensed and often amassed (Figs. 3 and 4). This occurs early in the arrest, but chromosomes can later be seen to be more discrete from one another. The tentative conclusion at present is that the arrest is more chromosomally mediated than through the spindle, perhaps in a manner akin to the arrest of cells in metaphase by such agents as Lucanthone (Miracil D; D. N. Wheatley and B. Angus, unpublished) or ICRF 159 (Sharpe *et al.*, 1970), which produce excessive stickiness to the chromosomes. There is a distinct difference, however, between SHAM and these antimitotic agents. The amassing and sticking together of chromosomes with drugs does not always stop cells attempting to pass through anaphase and telophase. This can occur to such an extent that the whole of the chromosomal material can be dragged bodily into one of the daughters as the mitotic cell proceeds through telophase, even when the two chromosomal sets had already drawn apart to some extent. It is suggested, therefore, that in SHAM cells the chromosomal changes may reflect changes in stickiness, but these alone might be insufficient to account for the block in metaphase. Perhaps the spindle, while adequately formed, is not properly functional.

F.   Length of Exposure to Hypertonicity

It is relatively simple to add a drug such as colchicine to a culture to arrest cells in metaphase, since cells do not adapt to the treatment. With hypertonicity at the level used in SHAM, HeLa cells adapt to the increased level of ions, and some recover spontaneously in the sense that they will begin moving out of mitosis of their own accord about 3–4 hours after exposure (see Figs. 2 and 10 in Wheatley, 1974b). There are as many cells, however, which do not recover and soon after become pycnotic. While on a long-term basis cells can obviously be made to tolerate and grow in abnormally high tonicities as shown for Ehrlich's ascites tumor cells by Schachtschabel and Foley (1972), on a short-term basis we have found between 3 and 4 hours to be the maximum length of SHAM for collection (Fig. 1), and from which cells neither begin to recover spontaneously nor show undue levels of pycnosis.

G.   Collection of Metaphases

In the conventional shakeoff selection procedure for synchronization (Terasima and Tolmach, 1963), there is good evidence that the combined SHAM procedure assists collection because it tends to loosen further the tenuous attachment of metaphases. Thus only a gentle agitation is required, especially with cells like HeLa. We have found it practical to use rocking tables such as designed by Luckhams Ltd. (Burgess Hill, Sussex, England,

Model SM/RT10) for stacking large monolayer bottles during shakeoff. This allows a standardization of the shakeoff procedure; similar ideas have also been tried with other machines elsewhere (Lindahl and Soreby, 1966; Tobey *et al.*, 1967). For large-scale collection, cells are usually grown in Roller culture. This operates at about 3 rph for growth of cultures and can simply be turned up to about 2 rpm for shakeoff collection. Many machines do not have this range of speed, but it can easily be accomplished by using a by-passable Thyristor control unit, which has the additional advantage of allowing a slow speed of rolling without loss of power.

Cells are usually collected in the second aliquot of their original medium made up to the full hypertonic strength. While it is quite common to shakeoff continuously during the 3 hour hypertonic exposure, it is not recommended because there is also a tendency to loosen interphase cells as well. Therefore, the shakeoff at the higher speed is done for only the final 15 minutes. It is possible to concentrate cells by passing the collected medium from one bottle on to the next just before a brief final shakeoff, but it entails careful removal of the medium from the second bottle, and there is a likelihood of "throwing out the baby with the bath water." The concentration of collected metaphases from any bottle will depend on the volume of medium

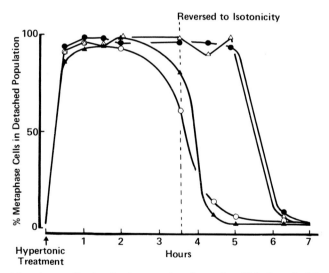

Fig. 5. Metaphase collection for hypertonic cultures in which there had been either a gradual rise in tonicity or a sudden increase to 227 mM. The dashed line indicates reversal to isotonicity. O——O, Metaphase percentage in gradually raised tonicity; ●——●, percentage of mitotic cells including doublets of early interphase; ▲——▲, metaphase percentage for abruptly raised tonicity; △——△ percentage of mitotic cells including early doublets. From Wheatley (1974b), by permission of the Company of Biologists Ltd.

used; it is advisable to keep this to a minimum since recovery involves dilution of this medium back to isotonicity.

Additional procedures for aiding mitotic selection, e.g., use of mild trypsinization (Stubblefield and Klevecz, 1965; Archer and Wheatley, 1971) and use of divalent ion free media (Robbins and Marcus, 1964), are unnecessary with SHAM. Nevertheless we have often employed suspension culture medium free of divalent ions for the procedure, but this is mainly because it is convenient for the further cultivation of the collected cells as suspension cultures.

Monitoring metaphase levels every 30 minutes, we have found the pattern of collection shown in Fig. 5.

H. Yield and Purity

Ideally, a 3-hour collection by hypertonicity should bring a total of about 12% of a HeLa cell population into metaphase. The ideal has not been achieved, but it has been approached in collections of the order of 10% (Wheatley, 1974b). The problem is that, even with hypertonic medium, some metaphases do not detach in a shakeoff collection. Increasing the vigor of shaking not only detaches more of these metaphases, but also more unwanted interphases. In almost all synchrony techniques, purity is only obtained at the expense of yield, and vice versa. The optimal conditions of shakeoff which give the maximum yield and purity would have to be empirically determined for any given experimental operation. Table III shows the purity of collection which is possible with HeLa cells using the proto-

TABLE III

PURITY OF MITOTIC COLLECTIONS OBTAINED FROM HYPERTONIC AND ISOTONIC CULTURES OF HeLa S3 MONOLAYER CULTURES BY THE SHAKEOFF PROCEDURE OF TERASIMA AND TOLMACH (1963).[a]

	% Cells in				
Procedure	Interphase	Prophase	Metaphase	Anaphase	Telophase
Hypertonicity (single step to 227 mM)	2.2	0.0	96.7	0.0	1.2
Hypertonicity (triple step to 227 mM)	4.4	0.2	94.1	0.0	1.3
Colcemid ($5 \times 10^{-7} M$)	1.1	0.3	98.6	0.0	0.0
Isotonic	4.1	0.0	37.0	6.2	52.7

[a] Average values for the first 2 hours of collections (4 samples/group) after discarding the initial 30-minute collection.

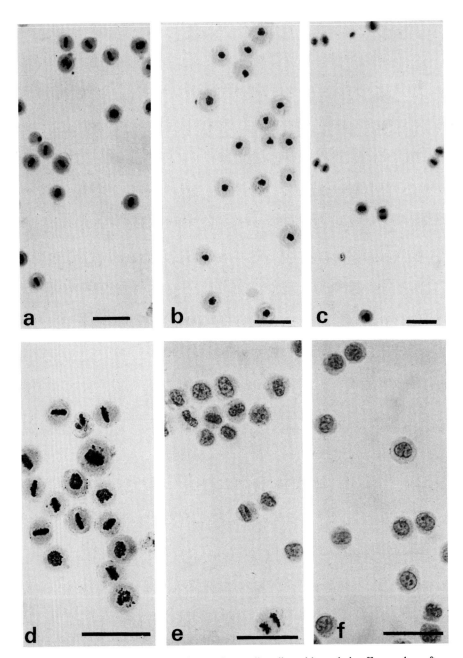

Fig. 6. (a) Hypertonic (227 mM) metaphase cells collected by a shakeoff procedure after 3 hours of treatment. (b) Colcemid-arrested metaphase cells collected after 2 hours exposure by a shakeoff procedure. (c) Isotonic mitotic collection over a 30-minute period. (d) Hypertonic (227 mM) collection of metaphase cells. (e) Cells from culture after 40 minute reversal in isotonic medium. (f) Same culture after 80 minutes in isotonic medium. (a–f): Scale line = 50 μm. Adapted from Wheatley (1974b), by permission of the Company of Biologists Ltd.

cols already described, and Figs. 6a–c show the harvest from hypertonic Colcemid, and isotonic mitotic shakeoffs. In terms of total cell yield, this will depend upon the scale of the operation and the behavior of the mono-layered cells used.

I. Lowering the Tonicity; Recovery

The acid test of any synchrony procedure is that, after strathmosyn-chronization, cells should be recoverable and proceed through the cell cycle normally. One decided advantage of the hypertonicity technique is that nothing need be removed at recovery; it simply requires that water or diluted medium be added to restore the initial tonicity. This is one of the reasons for keeping the concentration of cells in SHAM high at the time of collection. Dilution of the medium must be done slowly with thorough mixing, since at this time it cannot be done away from the cells.

Cells pass smoothly out after SHAM collection through anaphase and telophase (Figs. 6d–f), and into G_1. This is usually seen much more clearly than in conventional Terasima and Tolmach (1963) isotonic collections because the starting point of the latter cells is not as uniform (Fig. 6c). Synchrony of passage through the first cell cycle is not significantly different

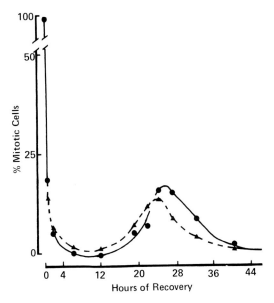

FIG. 7. Mitotic activity in an isotonic mitotic population (▲) of cells detached at 0 hour compared with a hypertonic mitotic population (●) taken after 3 hours in 227 mM medium. At 0 hour the cells were resuspended in normal medium. From Wheatley (1974b), by permission of the Company of Biologists Ltd.

from that achieved by isotonic collected metaphases (Fig. 7), and could probably be improved by preselecting cells for robustness in hypertonicity and a 24-hour cycle. Nevertheless synchrony is largely dissipated within the next generation; this could be due to any of several different factors, such as (1) too much natural variation even in the apparently most homogeneous of cell populations (Engelberg, 1964), (2) unequal division immediately after the starting point (Mazia, 1974), and (3) the probability of a cell proceeding to a further division being governed by a chance event occurring sometime after mitosis in G_1 (Smith and Martin, 1973).

Abnormalities are seen in cells recovering from SHAM. The extent to which this occurs is variable, but in some instances it can be seen that almost every cell in the culture is affected. Microscopically, cells tend to undergo nuclear lobulation several hours after entering G_1. Often this first occurs as an almost perfect reduction of the nucleus into two halves to give cells that can easily be mistaken for binucleates. Details of these changes are given elsewhere (Wheatley, 1974b); the reason for them is unknown. This is one simple demonstration that SHAM cells returned to isotonicity are in some respects disturbed. The fact that the cells arrive at the next mitosis on schedule should not be taken as evidence that the cells went through all their interphase functions either normally or on their individual schedules.

III. Ions, Macromolecular Synthesis, and Cell Division

A.   A Time to Halt

There is one distinct advantage in arresting cells in metaphase. Anabolic activity in mitotic cells is markedly depressed. DNA synthesis is not occurring, RNA synthesis is negligible, and protein synthesis is considerably curtailed (Pederson and Robbins, 1970; Fan and Penman, 1970, 1971). Polysomes break up, and free ribosomes increase. While there is information on the anabolic side, there is a dearth of information on catabolism. Thus, unfortunately it is difficult to assess what overall changes in DNA, RNA, and protein levels occur during normal mitosis. In SHAM cells, hypertonicity will presumably further depress RNA and protein synthesis as it does in asynchronous cultures (Robbins, et al., 1970; Pederson and Robbins, 1970; Wheatley, 1974a). This further suppression of anabolism might in itself be insignificant. The point to emerge, however, is that the mitotic stage of the cell cycle is the one during which the least gene activity is being expressed, and therefore *cells held in this phase are much less likely to become unbalanced*, and their nuclear and cytoplasmic functions should be in phase during their release into G_1.

B. Ions and Mitosis

The possibility that ions are involved in mitosis, perhaps as initiators and almost certainly in secondary phenomena, has been briefly mentioned. Changes, for example in transmembrane potential (Cone, 1969), the zeta potential (Brent and Forrester, 1967), movement of ions between the nucleus and cytoplasm, or between the nucleolus and the rest of the cell (Robbins and Pederson, 1971), may be due primarily or secondarily to ion fluxes. The effect of ions other than Na^+ on mitotic cells has been described elsewhere (Wheatley, 1974a).

C. Chromatin in Hypertonic Situations

Hypertonicity has been claimed to produce a mitosis-like appearance in cells (Rixon and Whitfield, 1962; Robbins *et al.*, 1970). It changes the chromatin pattern in fixed cells to give a prophaselike castellation as seen in the light microscope (Rixon and Whitfield, 1962; Wheatley, 1974a), and in the electron microscope (Robbins *et al.*, 1970) it can mimic a prophaselike condensation pattern. The changes that are reminiscent of chromosome condensation may only be partial chromatin condensation, since discrete structures the size of chromosomes are not seen on the nuclear envelope. Furthermore, these changes are transient and reversible at levels of hypertonicity that can be tolerated from an hour or two (220–260 m*M* for HeLa) as initially noted by Rixon and Whitfield (1962). Higher tonicities induce more severe chromatin condensation (i.e., levels of tonicity about 1.7 times normal), and this is simply a prelude to pycnotic degeneration. There is no evidence that hypertonic treatment of cells promotes their entry into mitosis.

IV. Comment

SHAM is a relatively easy way to augment the basic shakeoff mitotic selection procedure (Terasima and Tolmach, 1963). It works well for HeLa and similar cells. It is based on inhibiting cells at the stage of the cell cycle in which they can best afford to be held without becoming unbalanced and from which they can be reversed by simple dilution. In the initial studies, particular attention was paid, not to the use of this technique in synchronizing cells, but to the physiological reasons why hypertonicity was, within

TABLE IV

SYNCHRONY BY HYPERTONIC ARREST IN METAPHASE: ADVANTAGES AND
DISADVANTAGES AS A TECHNIQUE FOR AUGMENTING SELECTION SYNCHRONY

Advantages	Disadvantages
No new substances introduced to culture	Increased tonicity inevitably leads to increased toxicity
Arrest of metaphase cells is immediate	Arrest limited to a 3–4 hour period
Aids detachment of metaphase cells	Slightly increased tendency for interphase cells to detach if high tonicities used
Purity of stathmosynchronized cells can be checked microscopically	Some contaminations with damaged cells inevitable
Synchrony is more precise (metaphase) than with isotonic collections of mitotic cells	Not suitable for arrest in all cell lines
Arrest occurs when cells are in their least anabolic phase	Significantly suppresses macromolecular in interphase cells moving toward mitosis
Reversed simply by dilution	Reversed cells may develop abnormal lobulations of nuclei
Synchrony through subsequent cycle is as good as for isotonics	Optimal conditions for synchrony have to be determined for each experimental system
Amenable to presynchrony techniques (e.g., Pederson and Robbins, 1971)	
Makes cooling procedures to augment yield, as used in isotonic cultures (Lesser and Brent, 1970), redundant	

limits, a good mitostatic agent. Perhaps when such aspects as catabolism of normal and hypertonic mitotic cells, redistribution of ions in anaphase and telophase and other such problems have been investigated, SHAM can be improved by making it operate to its maximum by physiological brinkmanship. At present the investigator requiring synchronized metaphase cells, especially for studying metaphase → anaphase → telophase → G_1 progression might find SHAM valuable, but its advantages and disadvantages set out in Table IV are best weighed carefully against alternative methods.

ACKNOWLEDGMENTS

This work was largely supported by the Cancer Research Campaign. I am indebted to Dr. Brian Angus and Dr. Stuart Brown for their help with many aspects of this work, as also Mrs. M. Inglis and Miss Ann Mackay.

References

Abbo, F. E., and Pardee, A. B. (1960). *Biochim. Biophys. Acta* **39**, 478.

Archer, F. L., and Wheatley, D. N. (1971). *J. Anat.* **108**, 277.

Brent, T. P., and Forrester, J. A. (1967). *Nature (London)* **215**, 92.

Cone, C. D., Jr. (1969). *Trans. N.Y. Acad. Sci.* [2] **31**, 104.

Eagle, H. (1956). *Arch. Biochem. Biophys.* **61**, 356.

Ebeling, A. H. (1914). *J. Exp. Med.* **20**, 130.

Engelberg, J. (1964). *Exp. Cell Res.* **36**, 647.

Engelberg, J., and Hirsch, H. R. (1966). *In* "Cell Synchrony" (I. L. Cameron and G. M. Padilla, eds.), pp. 14–37. Academic Press, New York.

Fan, H., and Penman, S. (1970). *J. Mol. Biol.* **50**, 655.

Fan, H., and Penman, S. (1971). *J. Mol. Biol.* **59**, 27.

Hughes, A. (1952). *Q. J. Microsc. Sci.* **93**, 207.

Killander, D., and Zetterberg, A. (1965). *Exp. Cell Res.* **38**, 272.

Lesser, B., and Brent, T. P. (1970). *Exp. Cell Res.* **62**, 470.

Lindahl, P. E. and Soreby, L. (1966). *Exp. Cell Res.* **43**, 424.

Mazia, D. (1974a). *In* "Cell Cycle Controls" (G. M. Padilla, I. L. Cameron, and A. Zimmerman, eds.), pp. 265–272. Academic Press, New York.

Mazia, D. (1974b). *Sci. Am.* **230**, 54.

Nias, A. H. W. (1968). *Cell Tissue Kinet.* **2**, 153.

Nias, A. H. W., and Fox, M. (1971). *L. H. Gray Mem. Conf., 3rd, 1971*, pp. 63–108.

Pederson, T., and Robbins, E. (1970). *J. Cell Biol.* **47**, 734.

Pederson, T., and Robbins, E. (1971). *J. Cell Biol.* **49**, 942.

Pica-Mattoccia, L., and Attardi, G. (1972). *J. Mol. Biol.* **64**, 465.

Rixon, R. H., and Whitfield, J. F. (1962). *Exp. Cell Res.* **26**, 591.

Robbins, E., and Marcus, P. (1964). *Science* **144**, 1152.

Robbins, E., and Pederson, T. (1971). *In Vitro* **6**, 323.

Robbins, E., Jentzsch, G., and Micali, A. (1968). *J. Cell Biol.* **36**, 329.

Robbins, E., Pederson, T., and Klein, P. (1970). *J. Cell Biol.* **44**, 400.

Robinson, J. R. (1975). *In* "Pathobiology of Cell Membranes" (B. F. Trump and A. U. Arstila, eds.), pp. 173–189. Academic Press, New York.

Romsdahl, M. M. (1968). *Exp. Cell Res.* **50**, 463.

Rueckert, R. R., and Mueller, G. C. (1960). *Cancer Res.* **20**, 1584.

Schachtschabel, D. O., and Foley, G. E. (1972). *Exp. Cell Res.* **70**, 317.

Scherbaum, O., and Zeuthen, E. (1954). *Exp. Cell Res.* **6**, 221.

Sharpe, H. B. A., Field, E. O., and Hellmann, K. (1970). *Nature (London)* **226**, 524.

Sinclair, W. K., and Morton, R. A. (1966). *Radiat. Res.* **29**, 450.

Smith, J. A., and Martin, L. (1973). *Proc. Natl. Acad. Sci. U.S.A.* **70**, 1263.

Stubblefield, E., and Klevecz, R. (1965). *Exp. Cell Res.* **40**, 660.

Stubblefield, E., and Mueller, G. C. (1960). *Cancer Res.* **20**, 1646.

Studinski, G. P., and Lambert, W. C. (1969). *J. Cell. Physiol.* **73**, 109.

Terasima, T., and Tolmach, L. J. (1963) *Exp. Cell Res.* **30**, 344.

Tobey, R. A., Anderson, E. C., and Petersen, D. F. (1967). *J. Cell. Physiol.* **70**, 63.

Walters, R. A., and Petersen, D. F. (1968). *Biophys. J.* **8**, 1475.

Wheatley, D. N. (1974a). *Exp. Cell Res.* **87**, 219.

Wheatley, D. N. (1974b). *J. Cell Sci.* **15**, 221.

Wheatley, D. N., and Angus, B. (1973). *Experientia* **29**, 1393.

Zetterberg, A., and Killander, D. (1965) *Exp. Cell Res.* **40**, 1.

Chapter 28

Selection of Synchronized Populations of HeLa Cells

ALISON M. BADGER AND SIDNEY R. COOPERBAND

Departments of Microbiology, Surgery, and Medicine, and the Cancer Research Center,
Boston University School of Medicine,
Boston, Massachusetts

I. Introduction and Review of Synchronization Methodology

Cultured cells have an active cell replication cycle, which was first recognized by Howard and Pelc (1953). This cycle consists of pre- and post-DNA synthetic periods (G_1 and G_2), a DNA-synthetic period (S), and mitosis (M). The time period required to complete this cycle varies from cell type to cell type. Techniques that will select populations of cells in one or another part of this cycle are of obvious advantage for investigation of biochemical events and regulatory mechanisms which change with cell growth.

At present there are several methods available for the production and collection of synchronized cell cultures. These methods are almost exclusively aimed at obtaining large numbers of synchronized cells, only a certain percentage of which are actually in synchrony. The methods fall into two major categories: (1) the use of metabolic inhibitors and antagonists to interfere with the cycle or the deprivation of an essential metabolite for a period of time, and (2) the methods that do not require the addition or removal of metabolites and are probably less traumatic for the cells; these are cold shock, recovery of cells in mitosis by virtue of decreased adhesion to surfaces, and velocity sedimentation of cells with different densities.

A. Metabolic Interference

A widely used method for synchronization of large populations of cells is the use of metabolic agents to block either the synthesis of DNA or mitosis. Examples of the chemical agents used for this type of synchronization are thymidine (Xerox, 1962; Puck, 1964), amethopterin (Reukert and Mueller, 1960; Jasinka et al., 1970), 5-fluorodeoxyuridine (Reukert and Mueller, 1960; Littlefield, 1962), Colcemid (Stubblefield et al., 1967), or a combination of cytosine arabinoside and Colcemid (Verbin et al., 1972). Although the use of these chemicals and antimetabolites provides large numbers of synchronized cells they also induce in the population unwanted changes, such as chromosome abnormalities (Colcemid), interference with the normal deoxyribonucleotide metabolism of the cells (thymidine), and general cytotoxicity. We have found that large populations of cells synchronized by the methods described above can survive the treatment, but small populations are adversely affected, and attempts to clone single cells are almost impossible. Another method that is utilized for cell synchronization is the deprivation of an essential metabolite, usually an amino acid, such as leucine (Everhart and Prescott, 1972), isoleucine (Tobey and Ley, 1971), or isoleucine and glutamine (Ley and Tobey, 1970). The removal of serum from cell cultures has also been used for synchronization. These starvation methods, however, must also prove to have a detrimental effect on those cells since the synthesis rate for new proteins and various intracellular enzymes is different for each moiety, and the deprived cell will undoubtedly undergo a period of metabolic instability. It is understandable, therefore, that interpretation of results obtained during the recovery period would be difficult.

B. Cold Shock and Other Physical Methods

Several investigators have used temperature shock treatments for synchronization of cell cultures (Sinclair and Morton, 1963; Newton and Wildy, 1959; Lesser and Brent, 1970). This method does not appear to be as effective as those described in Section I, A, and again there is the question of the undesirable effects of this kind of treatment since here too there are metabolic disturbances which are produced and must be adjusted (Nelson et al., 1971; Nelson and Kruuv, 1972).

Axelrod and McCullouch (1958) were the first to observe that glass-adherent cells in mitosis tend to round up and become more easily detached from the glass. Several workers have put this observation to good use by collecting this synchronized population of cells for experimentation (Terasima and Tolmach, 1963; Robbins and Marcus, 1964; Lindahl and Sorenby, 1966). Recovery is about 1–2% of the total population, but 80–90% of these cells are in metaphase to late telophase. The enormous advantage of this method over the others described is that there has not been interference with the cell's metabolism. Cells are, in fact, in their normal state. The disadvantage of this method is that the cells dislodged in mitosis frequently are in suspension in clumps of various sizes and even filtering the suspension through gauze does not provide a single-cell suspension.

Another method that has been used both alone and in conjunction with the Axelrod and McCullouch procedure described above is that of velocity sedimentation in various density media. Postmitotic cells are smaller and have greater density than at other times, and populations of these cells have been prepared by centrifugation in a sucrose gradient (Sinclair and Bishop, 1965; Schindler et al., 1970), in a colloid silica density gradient (Wolff and Pertoft, 1972), or by simply allowing the cells to descend a column of culture medium by gravity (Shall and McClelland, 1971). Ooka and Daillie (1974) have used the selective detachment technique in combination with natural gravity sedimentation to produce a good yield of well-synchronized cells. Although this procedure produces minimal metabolic alteration of cells, the resulting population is frequently clumped, and obtaining a single-cell suspension is difficult.

During a series of studies on a lymphocyte-produced, growth-regulating factor (Green et al., 1970; Badger et al., 1971, 1974), we found it necessary to determine whether this factor had selective activity during any single portion of the cells' normal replication cycle. We found that the usual methods of producing a synchronous cell population produced too many environmental perturbations for the cells to be in a steady metabolic state. We therefore found it necessary to develop a system for *selection* of a synchronously growing subpopulation of cells from within a culture of cells growing asynchronously, but under steady-state environmental conditions. This procedure allowed us to observe continuously the growth characteristics of single isolated cells and their progeny and determine their individual life-cycle histories. It had the additional advantage of assessing subtle toxicity effects as well as allowing the addition of agents that interfere with cell replication at selective parts of the cell cycle. This is not a procedure for preparing large numbers of cells, or for examining biochemical events in relatively large populations which may contain a variable portion of the population out of synchrony.

II.   Methods and Materials

HeLa cells (obtained from Flow Laboratories, Rockville, Maryland) were maintained as a continuous cell line by accepted techniques (Lennette and Schmidt, 1964). Cells were grown in Minimal Essential Medium with Earle's salts (MEM–Earle's) with 10% heat-inactivated (56°C for 30 minutes) fetal calf serum (FCS) and 50 units of penicillin and 50 units of streptomycin per milliliter. A barely confluent sheet of HeLa cells was used to prepare the cell suspension. The growth medium was decanted from the cell sheet, and the surface was rinsed once with warm saline. A volume of 0.25% trypsin or trypsin–EDTA (Gibco) just necessary to cover the cells was then added (trypsin–EDTA is more likely to assure a single cell suspension). The bottle was then incubated, undisturbed, at 37°C for 10–15 minutes. As soon as the cells were released from the surface of the bottle, the trypsin activity was halted by adding MEM–Earle's with 20% FCS, and the cells were pipetted up and down several times using a 5-ml pipette to assure a single-cell suspension. The suspension was then centrifuged for 5 minutes at 20 g at room temperature, the supernatant medium was discarded, and the cells were suspended in MEM–Earle's with 10% FCS. All procedures from this point to the incubator are carried out at room temperature. After mixing thoroughly by pipetting, the suspension was examined microscopically in a hemacytometer chamber; an adequate suspension was one that contained less than 0.5% of the cells in clumps of 2, 3, or more. If the suspension contained too many clumps of cells, it was discarded. Two thousand cells in 0.2 ml of medium were pipetted into milk dilution bottles (12 × 4 × 4 cm), and the final volume was brought to 10 ml with MEM–Earle's with 10% FCS. For experimentation purposes, this final volume was adjusted to 9 ml so that additives in volumes of not more than 1 ml could be added after the synchronized cells were selected. Bottles were then placed in a CO_2 incubator (5%) at 37°C immediately, and allowed to incubate for 1–1.5 hours. At the end of this time the bottles were removed from the incubator and examined by inverting them and scanning the surface with a dissection microscope at 30–80 ×. Any 2-cell colonies that could be found were assumed to have originated from a cell division that occurred after the essentially single-celled suspension had been added to the bottle. These 2-cell colonies were numbered and marked for future identification by circling them with a fine felt-tipped black marking pen. It is also advisable to prepare a chart and mark the position of the colony in the circle so that there will be no confusion with other single cells that may have been included in the circle and may divide later. The bottles were then replaced in the incubator. The elapsed time from the initial trypsinization until this final incubation is usually about 3 hours. The cells may be removed from the incubator, the bottles inverted, and the identified colonies examined at various time intervals for cell growth. It is advisable to treat the bottles gently at all times, since cells in mitosis tend to detach from the glass more easily than nondividing cells.

III. Results and Discussion

When 2-cell HeLa colonies are selected in the manner described, 75% or more of them divide synchronously into 4-cell colonies (within an inclusive time period of 3–4 hours—the time necessary to establish the procedure) approximately 20 hours after their selection. This time period may vary depending upon the general metabolic health of the cells at the beginning of the experiment and the rapidity with which the preparation and selection of cells is carried out. Figure 1 shows the collective data for 20 clones in the same bottle and their distribution as 2-, 3-, and 4-cell colonies when the bottle was examined 19 hours (Fig. 1A) and 23 hours (Fig. 1B) after cells were first added to the bottle (taking into account that it takes

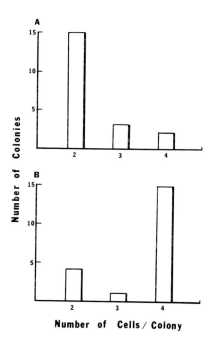

FIG. 1.   Examination of the fate of 20 selected 2-cell colonies 19 hours (A) and 23 hours (B) after attachment to the glass (Badger and Cooperband, 1973).

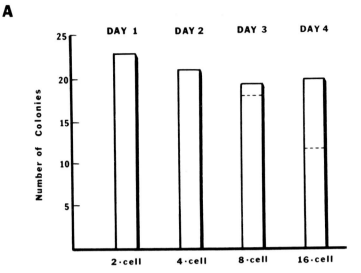

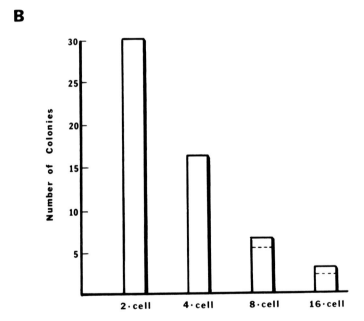

FIG. 2. (A) Synchronous cultures: On day 1, 23 2-cell colonies were selected within 3 hours of mitosis, and these were followed for 3 days. On days 2, 3, and 4 the number of colonies remaining in synchrony was determined—i.e., the number of colonies containing 4, 8, and 16 cells. (B) Asynchronous cultures: On day 1, 30 2-cell colonies were selected 15–18 hours after attachment to the glass and again followed for 3 days. In both histograms that area above the dashed line includes the 7–8-cell clones on day 3 and the 14-, 15-, and 16-cell clones on day 4 (Badger et al., 1974).

approximately 3 hours to complete the procedures, this would be 16 and 20 hours after selection). In this experiment 15 of the 20 colonies were in synchrony and divided within 4 hours of each other. Figure 2 demonstrates the length of time that the selected cells remain in synchronous cell division compared with asynchronous cultures. In Fig. 2A it can be seen that 21/23 of the 2-cell colonies were in absolute synchrony on day 2. By day 3, 18/23 of the colonies were in absolute synchrony, and by day 4 this number had dropped to 12/23. When, however, instead of enumerating only the 8- and 16-cell colonies as being in synchrony, we take into account the single cell that may have failed to divide and count now the colonies that contained either 7 or 8 cells (on day 3) and the 14-, 15- and 16-cell colonies (on day 4), the number of clones in synchrony becomes 19/23 and 20/23, respectively. This is shown by the extended portion of the histogram above the dotted line. In order to demonstrate that this is truly a synchronized population of cells, a control group of 30 2-cell clones was selected 15–18 hours after attachment of single cells to the glass. These colonies are in all stages of the cell cycle, and it can be seen that there is no synchronous division at all (Fig. 2B). Of the 30 colonies originally selected, only 16 are in the 4-cell stage on day 2, 7 of them are at 8-cell stage on day 3, and only 4 are at the 16-cell stage on day 4.

This technique offers almost complete cell synchrony for 72 hours after the cells have been selected. These cells have not been subjected to the harsh treatments that most of the other known procedures have to utilize, and therefore it is possible to observe very small numbers of cells growing in clonal growth.

REFERENCES

Axelrad, A. A., and McCullouch, E. (1958). *Stain Technol.* 33, 67–71.
Badger, A. M., and Cooperband, S. R. (1973). *Proc. Soc. Exp. Biol. Med.* 142, 1240–1242.
Badger, A. M., Cooperband, S. R., and Green, J. A. (1971). *J. Immunol.* 107, 1259–1267.
Badger, A. M., Cooperband, S. R., and Green, J. A. (1974). *Cell. Immunol.* 13, 335–346.
Everhart, L. P., and Prescott, D. M. (1972). *Exp. Cell Res.* 75, 170–174.
Green, J. A., Cooperband, S. R., Rutstein, J. A., and Kibrick, S. (1970). *J. Immunol.* 105, 48–54.
Howard, A., and Pelc, S. R. (1953). *Heredity, Suppl.* 1, 261.
Jasinka, J., Steffen, J. A., and Michalowksi, A. (1970). *Exp. Cell. Res.* 61, 333–341.
Lennette, E. H., and Schmidt, N. J., eds. (1964). "Diagnostic Procedure for Viral and Rickettsial Infections," 3rd ed. Am. Public Health Assoc., New York.
Lesser, B., and Brent, T. P. (1970). *Exp. Cell. Res.* 62, 470–473.
Ley, K. D., and Tobey, R. A. (1970). *J. Cell Biol.* 47, 453–459.
Lindahl, P. E., and Sorenby, L. (1966). *Exp. Cell Res.* 43, 424–434.
Littlefield, J. W. (1962). *Exp. Cell Res.* 26, 318–326.
Nelson, R. J., and Kruuv, J. (1972). *Exp. Cell Res.* 70, 417–422.
Nelson, R. J., Kruuv, J., Koch, C. J., and Frey, H. E. (1971). *Exp. Cell Res.* 68, 247–252.

Newton, A. A., and Wildy, P. (1959). *Exp. Cell Res.* **16**, 624–635.

Ooka, T., and Daillie, J. (1974). *Exp. Cell Res.* **84**, 219–222.

Puck, T. T. (1964). *Science* **144**, 565–566.

Reukert, R. R., and Mueller, G. C. (1960). *Cancer Res.* **20**, 1584–1591.

Robbins, E., and Marcus, P. I. (1964). *Science* **144**, 1152–1153.

Schindler, R., Ramseier, L., Schaer, J. C., and Grieper, A. (1970). *Exp. Cell Res.* **59**, 90–96.

Shall, S., and McClelland, A. J. (1971). *Nature (London), New Biol.* **229**, 59–61.

Sinclair, R., and Bishop, D. H. L. (1965). *Nature (London)* **205**, 1272–1273.

Sinclair, W. K., and Morton, R. A. (1963). *Nature (London)* **199**, 1158–1160.

Stubblefield, E., Klevecz, R., and Olaven, L. (1967). *J. Cell Physiol.* **69**, 345–353.

Terasima, T., and Tolmach, L. J. (1963). *Exp. Cell Res.* **30**, 344–362.

Tobey, R. A., and Ley, K. D. (1971). *Cancer Res.* **31**, 46–51.

Verbin, R. S., Diluiso, G., Liang, H., and Farber, E. (1972). *Cancer Res.* **32**, 1489–1495.

Wolff, D. A., and Pertoft, H. (1972). *J. Cell Biol.* **55**, 579–585.

Xeros, N. (1962). *Nature (London)* **194**, 682–683.

Chapter 29

The Preparation and Characterization of Intact Isolated Parenchymal Cells from Rat Liver

DOUGLAS R. LaBRECQUE AND ROGER B. HOWARD

*Department of Internal Medicine, Liver Study Unit, Yale University, New Haven, Connecticut;
and Michael Reese Hospital and Medical Center, Chicago, Illinois*

I. Introduction

The concept of utilizing isolated cell suspensions as a tool to study metabolic events in the liver *in vitro* has been attractive for many years. The advantages of such a system would be manyfold. First, the cells would maintain their precise inner architecture, so important to the orderly sequence of cellular reactions, along with the permeability barrier of the plasma membrane, both of which are disrupted in tissue homogenates. Second, added substrates, oxygen, and hormones would have ready access to all the cells at equal concentrations, removing the problems of diffusion that occur in liver slices. Third, a better perspective of cellular functions might be ascertained by reducing metabolic abnormalities produced in the damaged cells at the periphery of a liver slice or in poorly perfused areas of the isolated, cyclically perfused liver. Fourth, in contrast to the isolated perfused liver, the number of experimental variables would be limited only by the number of cells obtained, and perturbations could be evaluated simultaneously with the cells in identical condition. Fifth, and most important, a single cell type could be isolated permitting precise determination of the response to the addition of substrates or hormones without the confusion of determining whether the parameter measured is a function of one or several of the distinct cell types present in the normal liver.

Numerous attempts to obtain isolated parenchymal cells from adult rodent liver have been made over the past few decades. The early methods involved mechanical treatment of the tissue, alone or in combination with enzymic digestion, or perfusion of the intact liver with chelating agents such as EDTA and citrate. Many of these techniques have been reviewed by Schreiber and Schreiber (1973). It was not until 1967, however, that structurally and metabolically intact isolated parenchymal cells from adult rat liver could be obtained in sufficient yield to permit further characterization (Howard et al., 1967). This was achieved by perfusing the liver with, and incubating the sliced tissue in, a calcium-free medium containing collagenase and hyaluronidase under physiological conditions. This method, with more recent modifications (Howard et al., 1973) is described here. In addition, the evaluation of the structural and metabolic integrity of these cells and problems encountered in using such a preparation are briefly discussed.

II.   Preparation of Cells

A.   Enzyme Solution

The enzyme solution employed is 0.05% collagenase and 0.1% hyaluronidase dissolved in calcium-free Minimum Essential Medium of Eagle (1959) based on Hanks' solution. Types I and II crude collagenase from Worthington Biochemical Corporation (Freehold, New Jersey) are the most effective in this particular procedure. Pure collagenase does not work well, probably owing to the lack of small amounts of contaminant proteases, which aid in the dispersion of the liver slices. Individual lots of collagenase vary in activity, and it is recommended that a small sample be tested before ordering a large quantity of the same batch. Hyaluronidase (type I) is purchased from Sigma Chemical Company. The enzymes are dissolved in ice-cold medium immediately before use. Since the addition of hyaluronidase acidifies the medium, it is necessary to readjust the pH of the enzyme solution to 7.4. Phenol red is added to the medium as an indicator of pH during later incubation.

B. Technique

An entire liver is removed from a 200–300-gm anesthetized rat, rinsed in ice-cold wash medium [calcium-free Hanks' solution (Hanks and Wallace, 1949)], and then perfused with 15–50 ml of the enzyme solution until the liver is blanched. Perfusion is carried out by gently injecting the solution into the portal vein or each of the primary lobular branches with a 10- or 20-ml syringe. To ensure complete perfusion, the tip of the needle may be subsequently placed at the extremities of the lobes. Each lobe is thinly sliced by hand utilizing a Stadie-Riggs tissue slicer at its tightest setting to procure slices 0.3–0.4 mm in thickness. The first and last slices of each lobe are discarded since they contain capsular tissue. The slices are placed into two or three 250-ml Erlenmeyer flasks containing 10 ml of enzyme solution. All glassware is siliconized. Each flask should eventually contain 2–3 gm of tissue. All the above steps are carried out in a cold room at 4°C as rapidly as possible. With practice, this part of the procedure can be accomplished in 15–20 minutes.

The flasks are then incubated in a water bath at 37°C and shaken at a rate of 90 oscillations per minute; 100% oxygen, which has been bubbled through water, is constantly blown onto the slices. After 25–35 minutes of incubation, 0.1 ml of 1.41% calcium chloride solution is added to each flask so as to replace that previously omitted. The slices are then incubated for a further 20 minutes after the addition of calcium for a total incubation time of 45–55 minutes.

It is necessary to monitor the pH of the flask contents during the entire procedure. As the slices disperse, the medium becomes acid. Therefore, drops of 0.1 N NaOH are added until the color of phenol red indicates that the contents are approximately at pH 7.4. These adjustments are generally made after 20 and 40 minutes of incubation. The amount added depends on the gas flow and the weight and nutritional state of the tissue present.

At the end of incubation, approximately 10 ml of ice-cold wash medium are added to each flask. The contents are first filtered through nylon stocking into a small beaker so as to remove undispersed material. The filtrate is then carefully poured into another small beaker through a nylon mesh of approximately 60 μm pore size so as to remove clumps of cells. This mesh can be purchased from Henry Simon Ltd, Cheadle Heath, Stockport, England, or Tobler, Ernst and Traber, Inc. of Elmsford, New York. Filtration is aided throughout by the addition of ice-cold wash medium. Undispersed material is not forced through the meshes. The final volume of filtrate suspension approximates 80 ml. This is carefully poured down the sides of two 50-ml centrifuge tubes, and the cells are sedimented at 50 g for 1 minute. The cloudy supernatants are removed with a pipette. The cells are then washed twice in 10 ml of ice-cold wash medium. This is accomplished by gently blowing the medium onto the pellet with a Pasteur pipette two or three times. After each washing, the cells are sedimented at approximately 20 g for 1 minute. The final pellet of cells is resuspended in the chosen medium of incubation and used immediately. Storage of the cells at 4°C is not recommended since they will take up water, calcium, and sodium ions while losing potassium and magnesium ions (van Rossum, 1970; Baur *et al.*, 1975). The resulting cell suspension consists almost exclusively of parenchymal cells without contamination by cellular debris.

C.   Modifications

Higher yields of cells may be obtained by the following modifications: (1) The enzyme solution is adjusted to pH 7.1 before use and is then readjusted to approximately pH 7.4 after 15 minutes of incubation with the slices. (2) Ten minutes after the addition of calcium, the contents of the flasks are gently drawn in and out of a wide-bore 10-ml pipette two or three times. Timing of this mechanical stress is important. If carried out too soon, the plasma membrane of a high percentage of the cells will be torn as a result of insufficient intercellular loosening. In addition, it is important that the flask contents be incubated for a further 10 minutes after the mechanical treatment in order for the enzymes to digest any damaged cells produced.

D.   Difficulties

Perfusion and slicing of the liver require practice. Experience in teaching this technique has shown that about ten attempts are necessary before the expected yields are obtained. Extensive studies in our laboratories have verified the importance of carefully adhering to each step of the procedure as outlined above. Modifications of the procedure can result in damaged cells and low yields. Many investigators had difficulty in reproducing the methods of Howard *et al.* (1967) and Howard and Pesch (1968) because they deviated from the described procedure. Thus, Crisp and Pogson (1972) had calcium present during the entire procedure. Hommes *et al.* (1970) added EDTA to the enzymes, and Gallai-Hatchard and Gray (1971) minced the liver instead of slicing. The cells prepared by Jezyk and Liberti (1969) respired at a low rate. This undoubtedly was the result of incubating the slices in air instead of oxygen (see LaBrecque *et al.*, 1973). In order to eliminate some of these harmful modifications, Howard *et al.* (1973) outlined the rationale for each step in the procedure.

E.   Yield of Cells

Four to eight million cells per gram wet weight of liver slices are obtained by the method of Howard *et al.* (1973). This is increased to 11–12 million cells per gram wet weight by utilizing the two modifications described above, so that approximately 100 million cells can be obtained from one liver. This provides enough cells for most metabolic experiments since only 1–3 million cells per incubation flask are necessary. If more cells are required, two livers may be used. Alternatively, one of the methods described in the next section should be employed.

Cell counts are determined with a hemacytometer. One million cells contain, on the average, 2.3 mg of protein. Cell yields can be estimated by other methods, such as dry weight or protein content.

F.   Other Techniques

A considerably greater yield of cells may be obtained by using a more elaborate procedure, such as that described by Berry (1974). This is an updated version of the method introduced by Berry and Friend (1969) in which the enzymes are continuously recirculated through the liver *in vivo* before dispersing the cells *in vitro*. Numerous variations of this method have been described, since most laboratories adopt their own minor modifications. Two methods that differ significantly are those by Seglen (1973) and Wagle and Ingebretson (1975).

Cells produced by the method of Berry (1974) are comparable to the cells obtained by the method of Howard *et al.* (1973) in most respects, but the procedure of Berry yields 5–10 times more cells. However, a special and costly apparatus is required. In addition, skill in cannulating blood vessels and maintaining adequate perfusion and oxygenation of the liver are necessary, and the technique often requires two persons.

III.   Characterization of Isolated Cells

There is a certain art in preparing isolated cells, and it should not be assumed that the method selected automatically produces suspensions of intact cells. It is necessary to carefully characterize the cells before initiating further studies, and it is imperative that the cells obtained from every preparation be observed under the light microscope.

A. Structural Integrity

1. LIGHT MICROSCOPY

In contrast to damaged cells, which are irregularly shaped, intact liver parenchymal cells appear as spherical bodies with a well-defined, sharply refractile plasma membrane under the light microscope. Cells with an intact

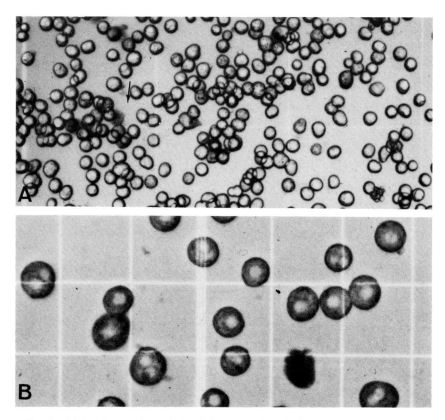

FIG. 1. Light micrographs of isolated liver parenchymal cells suspended in Hanks' solution containing 0.2% trypan blue. (A) Immediately after isolation. Arrow points to a damaged cell which has taken up the dye and reveals a darkly stained nucleus and lightly stained cytoplasm. × 150. (B) Incubated for 90 minutes. A severely damaged cell is diffusely stained. All other cells are viable. × 375.

plasma membrane exclude the vital stain trypan blue (Fig. 1). A cell with any sign of dye uptake is considered nonviable. The presence of albumin or serum in the medium during the determination of cell viability should be avoided since trypan blue has a greater affinity for extracellular than for intracellular protein (Phillips, 1973) and the number of damaged cells is underestimated.

In general, 90–98% of the cells obtained by the method described herein exclude trypan blue. This is improved to 95–98% if the two modifications are incorporated. However, exclusion of vital stains does not preclude intracellular damage (LaBrecque et al., 1973; Baur et al., 1975), and other measurements of cellular integrity are necessary in order to evaluate the quality of the cells more completely.

2. Electron Microscopy

The fine structure of the cells observed by electron microscopy provides a better assessment of structural integrity. Several electron micrographs of intact hepatocytes have been published (Howard et al., 1967, 1973; Berry and Friend, 1969; Drochmans et al., 1975). Figure 2 shows part of a typical liver parenchymal cell obtained by the method of Howard et al. (1973). It is seen that the intracellular organelles are well preserved and glycogen is abundant. In contrast, cells prepared by mechanical methods have a torn plasma membrane and considerable alteration of the intracellular organelles (Berry and Simpson, 1962; Howard et al., 1967).

3. Retention of Soluble Enzymes

Since cells prepared by mechanical means have a torn plasma membrane, it is not surprising that most of the soluble enzymes leak out (Berry, 1962; Takeda et al., 1964; Prydz and Jonsen, 1964). In contrast, cells prepared by enzymic methods retain their soluble enzymes (Berry and Friend, 1969; Crisp and Pogson, 1972). The percentage of soluble enzymes retained corresponds with the percentage of cells excluding trypan blue (Ontko, 1972; Baur et al., 1975).

B. Metabolic Integrity

1. Respiratory Activity

The measurement of oxygen uptake by cells utilizing their own endogenous substrates when suspended in a physiological saline medium, such as Krebs–Ringer phosphate solution (DeLuca and Cohen, 1964), is a good index of metabolic as well as structural integrity. Cells prepared by mechanical means do not respire when incubated in the above medium (Berry, 1962;

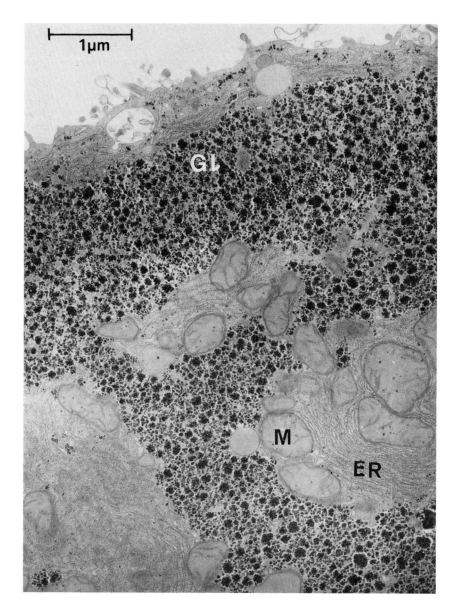

FIG. 2. Electron micrograph of isolated liver parenchymal cell. Mitochondria (M) are surrounded by granular endoplasmic reticulum (ER) arranged in parallel cisternae. Glycogen particles (GI) are abundant. × 22,000. Part of this figure has been published at higher magnification (Howard *et al.*, 1973).

Iype and Bhargava, 1965). This is the result of mitochondrial damage as well as the leakage of the soluble enzymes necessary for producing fuel to drive the tricarboxylic acid cycle. Low rates of endogenous oxygen uptake can be observed in such cells when they are incubated in media that normally support the respiration of isolated mitochondria. Addition of calcium, however, abolishes respiration (Berry, 1962; Iype and Bhargava, 1965).

In contrast, hepatocytes prepared by enzymic techniques consume oxygen in Krebs–Ringer phosphate solution, and the addition of calcium stimulates consumption (Howard et al., 1973). Rates of endogenous respiration can be correlated with structural integrity. Thus, the progressive improvement in the quality of liver cell suspensions made by Howard and his co-workers has resulted in a simultaneous increase in the rate of endogenous respiration (Howard et al., 1967, 1973; Howard and Pesch, 1968) so that the $Q_{O_2}^{protein}$ of 13 attained is greater than that for liver slices (Kratzing, 1961). Several respiratory characteristics of the cells have been described by Howard and his co-workers above and by Berry (1974).

2. BIOSYNTHESIS

Intact liver cells have been shown to synthesize a variety of complex molecules at rates comparable with the perfused liver. There have been reports on the synthesis of lipids (Jezyk and Liberti, 1969; Wright and Green, 1971; Lipson et al., 1972), glucose (Berry and Friend, 1969; Johnson et al., 1972; Ingebretson and Wagle, 1972; Garrison and Haynes, 1973), urea (Krebs et al., 1974), and specific proteins such as fatty acid synthetase (Burton et al., 1969), albumin (Weigand et al., 1971), and fibrinogen (Crane and Miller, 1974; Jeejeebhoy et al., 1975). Other metabolic characteristics reported are the esterification of fatty acids (Ontko, 1972; Homey and Margolis, 1973) and the response to hormones (Johnson et al., 1972; Wagle and Ingebretson, 1973; Garrison and Haynes, 1973; Pilkis et al., 1975).

Caution is advised in evaluating the quality of cell suspensions based on the ability to synthesize complex molecules. This particularly applies when the incorporation of radioisotopes is involved (see LaBrecque et al., 1973; Schreiber and Schreiber, 1973). One of the more reliable and easy methods of assessing the metabolic integrity of cell suspensions is the chemical or enzymic determination of glucose produced from precursors such as lactate or pyruvate (Garrison and Haynes, 1973).

3. RETENTION OF CHARACTERISTICS INDUCED *IN VIVO*

Isolated liver parenchymal cells retain characteristics induced *in vivo*. An example of this is the synthesis of DNA, which is markedly stimulated in the liver of partially hepatectomized rats. Table I shows that tritiated thymidine is incorporated into DNA at a very low rate in cells prepared from the livers

TABLE I

INCORPORATION OF THYMIDINE-^{3}H INTO DNA OF ISOLATED
PARENCHYMAL CELLS FROM NORMAL AND REGENERATING LIVER[a]

Source of livers	Specific activity (dpm $\times$ 10^{-2}/mg DNA)
Sham-operated rats	62
34% Hepatectomized rats	1774
68% Hepatectomized rats	8170

[a] Three to six million cells were incubated in 3 ml of Krebs–Ringer phosphate solution containing 20 mM pyruvate and 2 μCi of ^{3}H-labeled thymidine (6.7 Ci/mmole) for 2 hours. Cell viability fell from approximately 90% to 80% during this period. Results are the average of duplicate flasks from three cell suspensions, each prepared from livers of the three animal states 23 hours after surgery. Analytical methods have been described elsewhere (LaBrecque and Pesch, 1975). Specific activity of controls, performed by adding isotope at the end of incubation, has been subtracted.

of sham-operated animals, but at much higher rates in cells obtained from the livers of hepatectomized rats 23 hours after surgery. It is seen that the incorporation of tritiated thymidine into DNA increases with the degree of hepatectomy in a manner similar to that found *in vivo* (Bucher, 1967).

Other investigators have also obtained isolated liver cells that retain characteristics induced *in vivo*. Werner and Berry (1974), for example, compared the metabolism of cells from normal and hyperthyroid rats. Another approach has been to treat rats with phenobarbitol and then separate the altered parenchymal cells by isopycnic centrifugation on Ficoll density gradients (Wanson *et al.*, 1975). The isolated cells themselves can be altered *in vitro* so that changes in the metabolism of isolated organelles are observed. Such experiments have been carried out with isolated mitochondria from control and glucagon-treated cells (Garrison and Haynes, 1975).

IV. Culture of Isolated Cells

The basic method of cell preparation is easily adapted to the provision of cells for culture by merely carrying out all procedures with sterile equipment and solutions. There are several reports of isolated parenchymal cells from normal and regenerating adult rat liver being maintained in nonproliferating monolayer culture on plastic or collagen-coated vessels for 5–10 days (Pesch *et al.*, 1968; Iype, 1971; Bissell *et al.*, 1973; Bonney, 1974; Bonney *et al.*, 1974). During this time, the cells aggregate in cords and the plates of polygonal cells retain typical parenchymal cell morphology and metabolism. The cultures are not overwhelmed by rapidly dividing fibroblasts as has occurred with liver cultures in the past, since an essentially pure suspension of parenchymal cells is the starting material.

The failure to propagate these cells in culture should not be considered a defect of the cells, since the parenchymal cells of normal adult liver *in vivo* have extremely low rates of cell division (Bucher, 1967). Until the factors responsible for cell division during liver regeneration are elucidated and can be incorporated into the culture procedure, it is not expected that parenchymal cells from adult liver will divide and retain their normal characteristics *in vitro*. One of us has described a hepatotrophic factor extractable from normal weanling or regenerating adult rat liver which stimulates incorporation of tritiated thymidine into DNA of the liver of 34% hepatectomized adult rats (LaBrecque and Pesch, 1975). When added to cultured liver parenchymal cells, it stimulates incorporation of tritiated thymidine into DNA in a dose-dependent manner (D. R. LaBrecque and N. R. Bachur, unpublished data). It should be noted that as a measure of the quality of cells obtained by a particular method, the ability to culture the cells has its limitations, since partially damaged cells have the ability to repair that damage during culture (Pariza *et al.*, 1975).

V. Potential Errors in Evaluating Data

The metabolism of isolated liver parenchymal cells, in addition to being a function of the structural intactness of the cells, is also a function of the incubation medium utilized. Table II shows that cells from the same preparation synthesize protein at different rates in the two media employed. The addition of exogenous substrate enhances the synthesis of both protein and albumin. Other effects of the extracellular medium on the metabolism of isolated cells have been reported (Howard *et al.*, 1973; Schreiber and Schreiber, 1973). Therefore, the nature of the media employed should always be considered when comparing rates of metabolism in isolated cells prepared by different techniques or in other liver preparations.

Another point to consider while studying the metabolism of isolated cells is the significant loss in cell viability when the pH of the medium deviates from the range of 7.0 to 7.4. The addition of some substrates produces changes in pH of the incubation medium (Howard *et al.*, 1973). Consequent

TABLE II

EFFECT OF MEDIUM AND PYRUVATE ON INCORPORATION OF
LEUCINE-^{14}C INTO PROTEIN OF LIVER CELLS[a]

Medium	Specific activity (dpm/mg protein)
Krebs–Ringer phosphate	2677 ± 247
+ 20 mM Pyruvate	3894 ± 244
Krebs–Ringer bicarbonate	4341 ± 93
+ 20 mM Pyruvate	6074 ± 538

[a] Four to five million cells were incubated in 3 ml of medium containing 0.2 μCi of ^{14}C-labeled leucine (2.56 Ci/mmole) for 2 hours. Cell viability fell from approximately 95% to 85% during this period. Results are the mean average ± SEM of duplicate flasks from three cell preparations. Specific activity of controls, performed by adding isotope at the end of incubation, have been subtracted. Similar effects are seen on the synthesis of albumin.

reduction in the number of viable cells can occur. Since the rate of metabolism will be a function of the number of living cells present, it is possible that an observed effect of an added substrate on a metabolic event may be due solely to differences in the number of living cells. Therefore, it is strongly recommended that the pH of the medium and cell viability be determined at the end of every experiment for each variable tested.

VI. Conclusion

The technique for obtaining intact isolated liver parenchymal cells is relatively simple. However, the whole procedure requires care and precision. We would recommend that investigators new to the field obtain further insight into the mechanisms and critical factors involved before commencing preparation of these cells (see Howard *et al.*, 1967, 1973; Berry and Friend, 1969; LaBrecque *et al.*, 1973; Drochmans *et al.*, 1975). The major aim should be to obtain cells of high structural and metabolic integrity in sufficient number to permit further study rather than high yields of partially damaged cells. We have described, therefore, several parameters for measuring cellular integrity.

It is important to realize that the cells are living and should not be treated as though they are just another chemical or enzyme. Therefore, conditions during subsequent incubation should be provided to keep the cells alive. We stress again that monitoring the pH of the extracellular medium and of cell viability during or at the end of incubation is necessary in order to adequately interpret the data obtained from many experiments.

Rat liver was the source of the isolated cells described here. The method has also been applied to mouse (Curtis *et al.*, 1968) and chicken liver (Goodridge, 1975). Presumably it should also be adaptable for use with the liver from other animals.

ACKNOWLEDGMENTS

Most of the work described in this manuscript was performed in the laboratory of Dr. LeRoy A. Pesch at the Stanford University School of Medicine; the School of Medicine, State University of New York at Buffalo; and Michael Reese Hospital and Medical Center, Chicago. We are extremely appreciative not only of his support and interest, but also for his allowing us individuality and freedom in our research efforts. Funds were obtained from Grant AM-10680 from the National Institute of Arthritis and Metabolic Diseases, United States Public Health Service, and the Medical Research Institute Council of Michael Reese Hospital and Medical Center. A portion of this work was carried out in the laboratory of Dr. Nicholas R. Bachur, Baltimore Cancer Research Center, and we gratefully acknowledge his interest, support, and valuable criticisms.

REFERENCES

Baur, H., Kasperek, S., and Pfaff, E. (1975). *Hoppe-Seyler's Z. Physiol. Chem.* **356**, 827–838.
Berry, M. N. (1962). *J. Cell Biol.* **15**, 1–8.
Berry, M. N. (1974). *Methods Enzymol.* **32**, 625–632.
Berry, M. N., and Friend, D. S. (1969). *J. Cell Biol.* **43**, 506–520.
Berry, M. N., and Simpson, F. O. (1962). *J. Cell Biol.* **15**, 9–17.
Bissell, D. M., Hammaker, L. E., and Meyer, U. A. (1973). *J. Cell Biol.* **59**, 722–734.
Bonney, R. J. (1974). *In Vitro* **10**, 130–142.
Bonney, R. J., Becker, J. E., Walker, P. R., and Potter, V. R. (1974). *In Vitro* **9**, 399–413.
Bucher, N. L. R. (1967). *N. Engl. J. Med.* **277**, 686–696 and 738–744.
Burton, D. N., Collins, J. M., and Porter, J. W. (1969). *J. Biol. Chem.* **244**, 1076–1077.
Crane, L. J., and Miller, D. L. (1974). *Biochem. Biophys. Res. Commun.* **60**, 1269–1277
Crisp, D. M., and Pogson, C. I. (1972). *Biochem. J.* **126**, 1009–1023.
Curtis, J. C., Lawner, P., and Minasian, D. (1968). *J. Cell. Biol.* **39**, 30a–31a (abstr.).
DeLuca, H. F., and Cohen, P. P. (1964). *In* "Manometric Techniques" W. W. Umbreit, R. H. Burris, and J. F. Stauffer, eds.), pp. 131–133. Burgess, Minneapolis, Minnesota.
Drochmans, P., Wanson, J.-C., and Mosselmans, R. (1975). *J. Cell Biol.* **66**, 1–22.
Eagle, H. (1959). *Science* **130**, 432–437.
Gallai-Hatchard, J. J., and Gray, G. M. (1971). *J. Cell Sci.* **8**, 73–86.
Garrison, J. C., and Haynes, R. C., Jr. (1973). *J. Biol. Chem.* **248**, 5333–5343.
Garrison, J. C., and Haynes, R. C., Jr. (1975). *J. Biol. Chem.* **250**, 2769–2777.
Goodridge, A. G. (1975). *Fed. Proc., Fed. Am. Soc. Exp. Biol.* **34**, 117–123.
Hanks, J. H., and Wallace, R. E. (1949). *Proc. Soc. Exp. Biol. Med.* **71**, 196–200.
Homey, C. J., and Margolis, S. (1973). *J. Lipid Res* **14**, 678–687.
Hommes, F. A., Draisma, M. I., and Molenaar, I. (1970). *Biochim. Biophys. Acta* **222**, 361–371.
Howard, R. B., and Pesch, L. A. (1968). *J. Biol. Chem.* **243**, 3105–3109.

Howard, R. B., Christensen, A. K., Gibbs, F. A., and Pesch, L. A. (1967). *J. Cell Biol.* **35**, 675–684.

Howard, R. B., Lee, J. C., and Pesch, L. A. (1973). *J. Cell Biol.* **57**, 642–658.

Ingebretson, W. R., Jr., and Wagle, S. R. (1972). *Biochem. Biophys. Res. Commun.* **47**, 403–410.

Iype, P. T. (1971). *J. Cell. Physiol.* **78**, 281–288.

Iype, P. T., and Bhargava, P. M. (1965). *Biochem. J.* **94**, 284–288.

Jeejeebhoy, K. N., Ho, J., Greenberg, G. R., Phillips, M. J., Bruce-Robertson, A., and Sodtke, U. (1975). *Biochem. J.* **146**, 141–155.

Jezyk, P. F., and Liberti, J. P. (1969). *Arch. Biochem. Biophys.* **134**, 442–449.

Johnson, M. E. M., Das, N. M., Butcher, F. R., and Fain, J. N. (1972). *J. Biol. Chem.* **247**, 3229–3235.

Kratzing, C. C. (1961). *In* "Biochemists' Handbook" (C. Long, ed.), pp. 795–810. Van Nostrand-Reinhold, Princeton, New Jersey.

Krebs, H. A., Cornell, N. W., Lund, P., and Hems, R. (1974). *In* "Regulation of Hepatic Metabolism" (F. Lundquist and N. Tygstrup, eds.), pp. 726–750. Academic Press, New York.

LaBrecque, D. R., and Pesch, L. A. (1975). *J. Physiol. (London)* **248**, 273–284.

LaBrecque, D. R., Bachur, N. R., Peterson, J. A., and Howard, R. B. (1973). *J. Cell. Physiol.* **82**, 397–400.

Lipson, L. G., Capuzzi, D. M., and Margolis, S. (1972). *J. Cell Sci.* **10**, 167–179.

Ontko, J. A. (1972). *J. Biol. Chem.* **247**, 1788–1800.

Pariza, M. W., Yager, J. D., Jr., Goldfarb, S., Gurr, J. A., Yanagi, S., Grossman, S. H., Becker, J. E., Barber, T. A., and Potter, V. R. (1975). *In* "Gene Expression and Carcinogenesis in Cultured Liver" (L. E. Gerschenson and E. B. Thompson, eds.), pp. 137–167. Academic Press, New York.

Pesch, L. A., Peterson, J. A., and Howard, R. B. (1968). *J. Clin. Invest.* **47**, 78a (abstr.).

Phillips, H. J. (1973). *In* "Tissue Culture: Methods and Applications" (P. F. Kruse, Jr. and M. K. Patterson, Jr., eds.), pp. 406–408. Academic Press, New York.

Pilkis, S. J., Claus, T. H., Johnson, R. A., and Park, C. R. (1975). *J. Biol. Chem.* **250**, 6328–6336.

Prydz, H., and Jonsen, J. (1964). *Scand. J. Clin. Lab. Invest.* **16**, 300–306.

Schreiber, G., and Schreiber, M. (1973). *Sub. -Cell. Biochem.* **2**, 307–353.

Seglen, P. O. (1973). *Exp. Cell Res.* **82**, 391–398.

Takeda, Y., Ichihara, A., Tanioka, H., and Inoue, H. (1964). *J. Biol. Chem.* **239**, 3590–3596.

van Rossum, G. D. V. (1970). *J. Gen. Physiol.* **55**, 18–32.

Wagle, S. R., and Ingebretson, W. R., Jr. (1973). *Biochem. Biophys. Res. Commun.* **52**, 125–129.

Wagle, S. R., and Ingebretson, W. R., Jr. (1975). *Methods Enzymol.* **35**, 579–594.

Wanson, J. -C., Drochmans, P., May, C., Penasse, W., and Popowski, A. (1975). *J. Cell Biol.* **66**, 23–41.

Weigand, K., Müller, M., Urban, J., and Schreiber, G. (1971). *Exp. Cell Res.* **67**, 27–32.

Werner, H. V., and Berry, M. N. (1974). *Eur. J. Biochem.* **42**, 315–324.

Wright, J. D., and Green, C. (1971). *Biochem. J.* **123**, 837–844.

Chapter 30

Selective Cultivation of Mammalian Epithelial Cells

ROBERT B. OWENS[1]

*Cell Culture Laboratory, School of Public Health,
University of California,
Berkeley, California*

I. Introduction

Epithelial cells are often recognizable in primary cell cultures, but are usually overgrown in later subcultures by the more rapidly dividing fibroblastic cells from connective tissue. Although there are a number of long-established epithelial cell lines listed by the American Type Culture Collection (Shannon, 1972), most of these are now recognized to be sublines of HeLa cells (Jones *et al.*, 1971), which have contaminated and overgrown

[1] This research was supported by Public Health Service Contract E 73-2001-NO1-CP-3-3237 from the National Cancer Institute.

other cell lines in many laboratories (Gartler, 1968; Nelson-Rees *et al.*, 1974). The establishment of non-HeLa epithelial cell lines from human tissues is a rare event. Several studies in which cultures were attempted from 100 or more epithelial tumors reported success rates of only 5% or less (Dobrynin, 1963; Fogh and Allen, 1966; Feller *et al.*, 1972; Giard *et al.*, 1973). In the case of human brain tumors, however, Westermark *et al.* (1973) obtained permanent epithelial cell lines from 12 of 28 (43%) grade III–IV gliomas. Normal human epithelial cells, capable of serial subculture through varying numbers of generations, have been cultured from brain (Pontén and Macintyre, 1968), mesothelium (Castor and Naylor, 1969), and endothelium (Lewis *et al.*, 1973).

In recent years, a few reports have described methods for selective culture of epithelial cells from normal or neoplastic tissues of rats, mice, or hamsters (Yasumura *et al.*, 1966; Coon, 1968; Iype, 1971; Lasfargues and Moore, 1971; Waymouth *et al.*, 1971; Gerschenson *et al.*, 1972; Williams *et al.*, 1971, Douglas and Kaighn, 1974). Some of these methods do not completely eliminate fibroblasts, while others employ complex technical procedures (e.g., organ perfusion, cloning, alternate tissue culture and animal passages, mechanical destruction of undesired cell types) which limit their general usefulness. For a more complete review of methodology and currently available epithelial cell lines, see articles by Rafferty (1975) and Fogh and Trempe (1975).

By combining the collagenase digestion method of Lasfargues and Moore (1971) with brief trypsinization to selectively remove fibroblasts from primary cultures, Owens (1974) was able to grow epithelial cell strains (Federoff, 1967) almost routinely from mouse tissues, including mammary carcinomas (Owens and Hackett, 1972) and normal liver, ovary, skin, and mammary gland (Owens *et al.*, 1974). This method, which is relatively simple and reproducible, has recently been used, with minor modification, to isolate epithelial cell strains from human carcinomas and normal tissues with a success rate of about 40% (Owens *et al.*, 1975). A detailed description of the method is presented below.

II. Primary Culture Method

A. Growth Medium

For growth of mouse epithelial cells, we use the high glucose version of Dulbecco's modified Eagle's medium (DMEM) made by Gibco (Grand Island Biological Company, Grand Island, New York). This medium contains 4.5 gm of glucose per liter and no sodium pyruvate. It is available by special order either as the instant powder (catalog No. H-21HG) or liquid medium (catalog No. 196G). Most commercial suppliers now list a revised formulation which contains only 1 gm of glucose and 42 mg of pyruvate per liter. We have found this revised formula definitely inferior to the high-glucose formula for growth of epithelial cells.

The medium is supplemented with 10 μg of insulin per milliliter (Calbiochem, San Diego, California) and with 10% fetal calf serum (Gibco) from carefully selected lots. We confirm the observation by Kaighn (1974) that many lots of fetal calf serum are toxic for epithelial cells, even though adequate for growth of fibroblasts. We test each lot against a good reference serum for its ability to promote maximum growth rate and saturation density in a very serum-sensitive mouse mammary epithelial cell strain. (NMuMg; see Owens et al., 1974). Replicate cultures are seeded with 1×10^4 cells/cm^2 and fed twice a week for 14 days. Viable cells, which exclude trypan blue dye, are trypsinized from two flasks every other day and counted in a hemocytometer. Saturation density is taken as the value where three successive harvests show no increase in cell number.

B. Processing of Fresh Tissues

Tissue samples are collected aseptically in cold Leibovitz's L-15 medium (Gibco) containing 10% fetal calf serum, penicillin (100 units/ml), streptomycin (100 μg/ml), and Fungizone (5 μg/ml). This medium is self-buffering and provides a stable pH during transport of the tissue from surgery to the tissue culture laboratory. If the tissue is minced to fragments smaller than 1 cm^3 and stored at 4° C, many cells will remain viable for 48 hours. Best results, however, are obtained if the tissue is dispersed for cell culture within several hours after surgery.

The tissue is poured into a sterile plastic cell culture dish (100 × 20 mm, Falcon Plastics, Oxnard, California), and the collecting fluid is removed. After any fat, muscle, or investing membranes are cut away, the tissue is minced to about 2-mm cubes using two sterile scalpels (Bard-Parker, No. 20) pulled in opposite directions in a scissors-like action. This is done most easily in a plastic dish because the cutting scores the plastic in a grid pattern, which helps to anchor each piece of tissue as it is minced.

C. Tissue Dissociation in Collagenase

Portions of this mince (1 or 2 scalpel bladefuls) are transferred to each of several plastic culture flasks (Falcon, 75 cm^2) and dispersed in 15 ml of complete DMEM containing 1 mg/of collagenase per milliliter (Worthington Biochemical Corp., Freehold, New Jersey, catalog No. 4194). The medium is adjusted to pH 7.0–7.2 by addition of 10% CO_2 to the air in each flask before tightening the cap. This can be accomplished by placing the flasks with loose caps in a CO_2 incubator or gassing chamber for several hours. We prefer to gas each flask immediately by passing a gentle flow of 100% CO_2 (hospital grade) through a sterile, cotton-plugged pipette into the air (not the medium) inside the flask. A flow gauge on the CO_2 tank is necessary to maintain a standard flow rate. A little practice will determine how many seconds of flow are needed to adjust the pH to an acceptable range as indicated by a salmon-pink color of the medium. The color of the medium is checked 15 minutes after gassing and sealing the flask. If the medium is red to purple, more CO_2 is added; if the medium is yellow, some of the air in the flask is removed through a sterile pipette attached to a vacuum source. DMEM contains 3.7 gm/of sodium bicarbonate per liter and must be equilibrated with CO_2 soon after being placed in the culture flask in order to avoid a rapid rise in pH, which can be tolerated by only a few fast-growing, established cell lines.

The tissue mince is incubated at 37° C for 18–24 hours in the collagenase solution, then the suspended fragments are drawn in and out of a 10-ml plastic pipette (Falcon) several times to complete the dispersion. If the fragments are too large to pass through the orifice of the pipette, the tip of the pipette can be cut off with a hot scalpel to enlarge the opening. This process will usually disperse connective tissue to single cells while leaving many clusters of epithelial cells undispersed (Fig. 1). The flask is placed upright for several minutes to allow settling of the cell clusters, then slowly tilted so that the fluid approaches the neck of the flask along a top:side angle. About 90% of the collagenase solution is withdrawn slowly (1–2 minutes) by lightly touching the fluid surface with the flame-bent tip of a Pasteur pipette

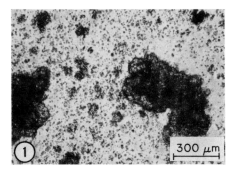

Fig. 1. Undissociated epithelial cell clusters after 24 hours' digestion in collagenase medium. Live culture, × 100.

attached to a suction flask. Most visible cells clusters are left behind by this skimming process while single cells and debris are swept away.

D.   Seeding of Primary Cultures

The cell clusters are diluted and gathered in fresh growth medium containing the same antibiotics used in the collecting medium (Section II,B) but without collagenase. Aliquots are distributed to fresh culture flasks (Falcon, 25 cm² or 75 cm²) providing 2 or 3 clusters per square centimeter. Growth medium is renewed twice a week (7- or 20-ml/flask), with care not to discard unattached cell clusters, and the pH is carefully adjusted at each feeding. Antibiotics are discontinued after the second subculture, except in one reserve flask. When planted in medium without collagenase and incubated undisturbed for several days, most of the epithelial cell clusters will attach and begin to produce monolayer outgrowths of epithelial cells, which often exceed 1 cm in diameter within 10 days (Fig. 2). Unfortunately, the few fibroblastic cells remaining after the settling and skimming procedure grow much faster than the epithelial-cell islands. Nevertheless, a large proportion of the confluent primary cell sheet usually consists of scattered epithelial islands of various sizes (Fig. 3).

E.   Selective Detachment of Fibroblasts

At this point, the cell sheet is rinsed once with STV, a solution of trypsin (0.05%) and Versene (0.02%) in calcium–magnesium-free saline (Gibco, catalog No. 530); 2 ml of fresh STV is added and gently washed across the cells for about 1 minute while the flask is warmed gently by occasional passes through a soft flame. When viewed through the flask held up to fluorescent

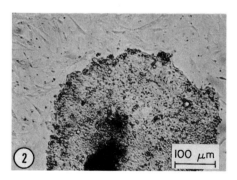

FIG. 2.   Epithelial cell island among fibroblasts after 7 days' growth. Live culture, × 100.

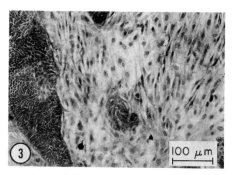

FIG. 3. Confluent monolayer of fibroblasts and epithelial islands from mouse liver after 10 days' growth. May–Grünwald–Giemsa stain, × 100.

light, the fibroblastic cell areas are seen to retract and begin detachment more rapidly than do the epithelial cell islands. The culture is quickly rinsed several times with growth medium to stop the enzymic dispersion and remove the lossened cells. When the rapidly detached cells are centrifuged and replated in growth medium, uniform fibroblast-like cultures are produced (Fig. 4).

If properly timed, the brief STV treatment leaves the epithelial cell islands relatively undisturbed. These islands are again incubated with growth medium. This procedure sometimes has to be repeated several times at weekly intervals before all fibroblasts are eliminated and a uniform, confluent monolayer of epithelial cells is formed (Figs. 5–8). If the primary epithelial cell islands are widely separated after removal of fibroblasts with STV, it is sometimes necessary to feed them with one-third to one-half volume of 24-hour conditioned medium from a culture of the autologous fibroblasts to stimulate further outgrowth of the epithelial cells.

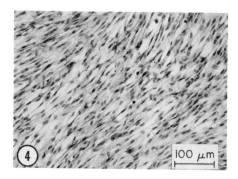

FIG. 4. Morphologically uniform fibroblastic cells removed from mixed primary culture of mouse liver by selective trypsinization. May–Grünwald–Giemsa stain, × 100.

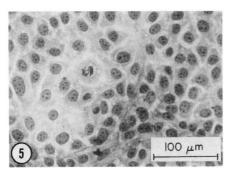

FIG. 5.   Morphologically uniform epithelial cells remaining in primary culture of normal mouse liver after removal of fibroblasts by selective trypsinization. May–Grünwald–Giemsa stain, × 175.

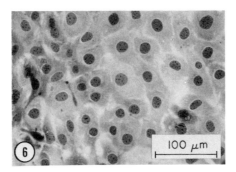

FIG. 6.   Epithelial cells isolated from normal mouse ovary by selective trypsinization. May–Grünwald–Giemsa stain, × 175.

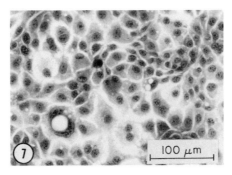

FIG. 7.   Epithelial cells isolated from normal mouse mammary gland by selective trypsinization. May–Grünwald–Giemsa stain, × 175.

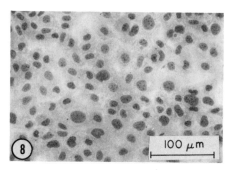

FIG. 8.   Epithelial cells isolated from normal mouse skin by selective trypsinization. May–Grünwald–Giemsa stain, × 175.

F.   Subculture of Epithelial Cells

The epithelial cell sheet can usually be dispersed to single cells by 5–10 minutes' exposure to STV at 37° C and divided among 2 or 3 new flasks to begin serial subcultures. A few epithelial cell strains attach very tightly to the plastic substrate and to each other, requiring long exposure to STV and vigorous pipetting to disperse them. Often only about 10% of the cells remain viable after such treatment. We have found that Pronase (Calbiochem, San Diego, California, grade B, No. 53702), at 0.25% in normal saline, will release these cells rapidly from the substrate in large sheets. These can be gently pipetted to produce smaller cell aggregates, most of which will reattach and grow after they are washed twice to remove the enzyme.

III.   Epithelial Cell Strains from Mouse Tissues

Using the method described above (Section II) we attempted cultures from 16 epithelial tissues of normal adult, weanling, or fetal mice. From these, 9 epithelial cell strains (56%) were isolated (Owens et al., 1974). The tissues of origin include skin (1), ovary (4), liver (3), and mammary gland (1).

After many subcultures, all these strains continue to display a mosaic or cobblestone-like growth pattern in which individual cells assume the polygonal or cuboidal shapes typical of epithelial cells in culture. No elongated spindle-shaped cells growing in the parallel arrangement typical of

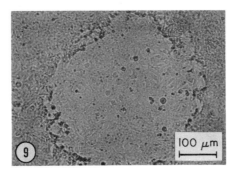

FIG. 9.   Blister-like hemicyst raised above dense monolayer of mouse liver cells. Live culture, × 100.

fibroblasts have appeared in these cultures since completion of the selective detachment process.

In addition to their epithelial morphology, 6 of the 9 strains develop unique, blister-like structures when dense monolayers are maintained for several weeks without subculture. These structures, variously described as hemicysts (Leighton *et al.*, 1969), domes (McGrath, 1969), or secretory vesicles (Auersperg, 1969), seem to form by secretion of fluid beneath the cell sheet. This suggests that these cultured cells have retained differentiated secretory functions typical of glandular epithelium *in vivo*. These hemicysts (Fig. 9) could still be induced in the mouse liver strains after 40 subcultures.

The well known tendency of normal mouse fibroblasts to spontaneously "transform" in culture to aneuploid, neoplastic cells is also evident in these epithelial strains, several of which contained fewer than 50% diploid cells after 24 or fewer subcultures (personal communication from Walter A. Nelson-Rees). All of 5 strains tested by inoculation into newborn isogeneic mice produced some type of tumorous growth, ranging from benign, differentiated cystadenomas to undifferentiated sarcomas. The normal liver and mammary gland cells produced benign cystic structures lined with columnar or cuboidal epithelium and filled with up to 8 ml of fluid (Fig. 10). Some tumors also contained areas of well-differentiated adenocarcinoma (Fig. 11).

When examined by electron microscopy, the cell strains derived from mammary gland, liver, and ovary show many ultrastructural features typical of secretory epithelium (Sandborn, 1970). Cells in close apposition had interdigitating microvilli and were held together by desmosomes (Fig. 12) or intermediate junctions. Tonofibrils, well-developed Golgi zones, and abundant polyribosomes were seen in the cytoplasm (personal communication from E. L. Springer).

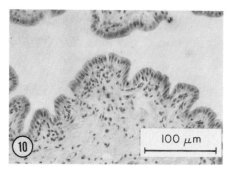

FIG. 10. Cystic tumor produced by mouse liver cells (NMuli) inoculated into inter-capular fat of newborn isogeneic mouse. Hematoxylin and eosin (H & E) stain, × 100.

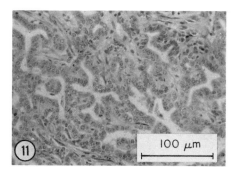

FIG. 11. Well-differentiated adenocarcinoma produced by mouse liver cells (Mm15Li) inoculated into interscapular fat of newborn isogeneic mouse. H and E stain, × 185.

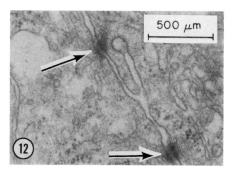

FIG. 12. Desmosomal junctions (arrows) between epithelial cells of a strain (Hs700T) isolated from human metastatic adenocarcinoma. Electron micrograph, × 34,000. Courtesy of E. L. Springer and Nancy Robertson.

IV.   Epithelial Cell Strains from Human Tissues

A.   Modification of the Method

The method we use for growth of human epithelial cells is identical to the method for mice (Section II) with two exceptions: (1) the growth medium is further supplemented by addition of Eagle's nonessential amino acids (Gibco No. 114), 10 ml/liter, and (2) the fetal calf serum is tested for toxicity (Section II,A) on a low-passage frozen stock of epithelial cells (FHs 74 Int) isolated in this laboratory from human fetal intestine.

B.   Results with Human Cells

Mixed outgrowth of epithelial and fibroblastic cells was obtained from 50 (83%) of 60 carcinoma tissues dissociated in collagenase (Section II,C). From these 50 cultures, 19 cell strains (38%) of epithelial morphology were isolated by selective detachment of fibroblasts with STV (Section II,E). Most of the successful cultures were obtained from metastatic carcinomas and from epithelia of the intestinal and urinary tracts (Owens et al., 1975).

In addition to the 19 carcinoma strains, we have isolated human epithelial cell strains from 3 normal fetal intestines, 1 normal adult skin, and "normal" tissue adjacent to carcinomas in 1 adult bladder and 5 adult kidneys. Representative light micrographs of some of these strains are shown in Figs. 13–16. Desmosomes have been observed in 10 of 18 carcinoma strains and 5 of 9 normal strains (personal communication from E. L. Springer). The population doubling times of all the human epithelial strains are slow, compared to normal human fibroblasts; ranging from 3 to 20 days. Several of

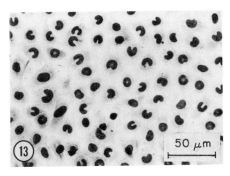

FIG. 13.   Epithelial cell strain (FHs677 Int) isolated from human fetal intestine. May–Grünwald–Giemsa stain, × 225.

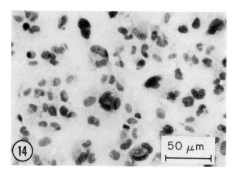

FIG. 14. Epithelial cell strain (Hs761T) isolated from human kidney carcinoma. May–Grünwald–Giemsa stain, × 225.

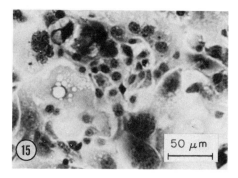

FIG. 15. Epithelial cell strain (Hs766T) isolated from human pancreas carcinoma. May–Grünwald–Giemsa stain, × 225.

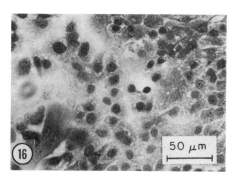

FIG. 16. Epithelial cell strain (Hs700T) isolated from human adenocarcinoma metastasis. May–Grünwald–Giemsa stain, × 225.

these strains have produced histotypical tumors when inoculated into immunosuppressed or nude mice (personal communications from H. S. Smith and P. Arnstein).

V. Discussion

Using the methods described in this chapter, we have isolated 38 cell strains (9 mouse and 29 human) which have maintained an epithelium-like morphology and growth pattern through 6 to 45 subcultures. Morphologically, these cells are clearly distinguisable from autologous fibroblasts grown under the same conditions. This morphological stability and the finding of specialized intercellular junctions (desmosomes) in most of these strains argue strongly for their epithelial origin. In addition, many of these cells have produced histotypical tumors when inoculated into isogeneic or immunosuppressed mice.

We cannot exclude the possibility that some of these strains are derived from endothelial or mesothelial elements, which may be present in any tissue sample. Although these mesodermal derivatives are generally classified as simple, squamous epithelia, they may differ extensively from the epithelia of ectodermal or endodermal origin, which most often give rise to carcinomas. Resolution of this question must await the identification of tissue-specific markers for each type of epithelium (Franks and Cooper, 1972; Lewis et al., 1973; Gimbrone et al., 1974).

While the slow population doubling time of human epithelial cells may be an inherent property, it might also indicate that tissue-specific nutrients or hormonal factors are not provided by the growth media in common use. In our early attempts to grow epithelial cells, we tested a number of commercially available media, including Eagle's MEM, RPMI 1640, Ham's F12, and the low glucose version of Dulbecco's modified MEM (all from Gibco). All were supplemented with 10% fetal calf serum and 10 μg of insulin per milliliter. Epithelial cells did not survive beyond the second passage in any of these media, all of which were adequate for growth of fibroblasts. We believe that the increased amounts of amino acids, vitamins, and glucose, as found in the high-glucose version of Dulbecco's modified MEM, are important for survival and growth of mamalian epithelial cells in vitro. Addition of insulin (above that in the serum supplement) probably facilitates uptake of amino acids and glucose (Griffiths, 1972). Several epithelial cell lines recently established in other laboratories were grown in media containing greater than the usual amounts of amino acids, vitamins, and glucose

(Coon and Weiss, 1969; Williams *et al.*, 1971; Giard *et al.*, 1973; Kaighn, 1973; Douglas and Kaighn, 1974; Kaighn and Babcock, 1975).

In our studies, as in most other attempts to grow epithelial cells, the method for tissue dissociation and the choice of growth medium, supplements, and culture conditions were determined empirically. While we believe that our method is relatively simple, reproducible, and efficient, we are aware that much remains to be learned about specific nutritional requirements of differentiated cells in primary and low-passage cultures. For excellent discussions of the present status and future prospects in the field of primary cell culture, the reader is referred to the proceedings of a symposium on "New Horizons for Tissue Culture in Cancer Research" convened by Sandford *et al.* (1974) and to a review by Ham (1974).

References

Auersperg, N. (1969). *J. Natl. Cancer Inst.* **43**, 151–173.
Castor, C. W., and Naylor, B. (1969). *Lab. Invest.* **20**, 437–443.
Coon, H. G. (1968). *J. Cell Biol.* **39**, 29a.
Coon, H. G., and Weiss, M. C. (1969). *Wistar Inst. Symp. Monogr.* **9**, 83–96.
Dobrynin, Y. V. (1963). *J. Natl. Cancer Inst.* **31**, 1173–1195.
Douglas, W. H. J., and Kaighn, M. E. (1974). *In Vitro* **10**, 230–237.
Federoff, S. (1967). *J. Natl. Cancer Inst.* **38**, 607–611.
Feller, W. F., Stewart, S. E., and Kantor, J. (1972). *J. Natl. Cancer Inst.* **48**, 1117–1120.
Fogh, J., and Allen, B. (1966). *In Vitro* **2**, 125.
Fogh, J., and Trempe, G. (1975). *In* "Human Tumor Cells *in Vitro*" (J. Fogh, ed.), pp. 115–159. Plenum, New York.
Franks, L. M., and Cooper, T. W. (1972). *Int. J. Cancer* **9**, 19–29.
Gartler, S. M. (1968). *Nature (London)* **217**, 710.
Gerschenson, L. E., Okigaki, T., Andersson, M., Molson, J., and Davidson, M. B. (1972). *Exp. Cell Res.* **71**, 49–58.
Giard, D. J., Aaronson, S. A., Todaro, G. J., Arnstein, P., Kersey, J. H., Dosik, H., and Parks, W. P. (1973). *J. Natl. Cancer Inst.* **51**, 1417–1423.
Grimbrone, M. A., Jr., Cotran, R. S., and Folkman, J. (1974). *J. Cell Biol.* **60**, 673–684.
Griffiths, J. B. (1972). *Exp. Cell Res.* **75**, 47–56.
Ham, R. G. (1974). *In Vitro* **10**, 119–129.
Iype, P. T. (1971). *J. Cell. Physiol.* **78**, 281–288.
Jones, H. W., McKusick, V. A., Harper, P. S., and Wuu, K. D. (1971). *Obstet. Gynecol.* **38**, 945–949.
Kaighn, M. E. (1973). *In* "Tissue Culture: Methods and Applications" (P. F. Kruse, Jr. and M. K. Patterson, Jr., eds.), pp. 54–58. Academic Press, New York.
Kaighn, M. E. (1974). *In Vitro, Monogr.* **3**, 21.
Kaighn, M. E., and Babcock, M. S. (1975). *Cancer Chemother. Rep., Part 1* **59**, 59–63.
Lasfargues, E. Y., and Moore, D. H. (1971). *In Vitro* **7**, 21–25.
Leighton, J., Brada, Z., Estes, L. W., and Justh, G. (1969) *Science* **163**, 472, 473.
Lewis, L. J., Hoak, J. C., Maca, R. D., and Fry, G. L. (1973). *Science* **181**, 453–454.
McGrath, C. M. (1969). *J. Natl. Cancer Inst.* **47**, 151–173.

Nelson-Rees, W. A., Flandermeyer, R. R., and Hawthorne, P. K. (1974). *Science* **184**, 1093–1096.

Owens, R. B. (1974). *J. Natl. Cancer Inst.* **52**, 1375–1378.

Owens, R. B., and Hackett, A. J. (1972). *J. Natl. Cancer Inst.* **49**, 1321–1332.

Owens, R. B., Smith, H. S., and Hackett, A. J. (1974). *J. Nat. Cancer Inst.* **53**, 261–269.

Owens, R. B., Smith, H. S., Nelson-Rees, W. A., and Springer, E. L. (1975). *J. Natl. Cancer Inst.* **56**, 843–849.

Pontén, J., and Macintyre, E. (1968). *Acta Pathol. Microbiol. Scand.* **74**, 465–486.

Rafferty, K. A. (1975). *Adv. Cancer Res.* **21**, 249–272.

Sandborn, E. B. (1970). "Cells and Tissues by Light and Electron Microscopy," Vol. 1. Academic Press, New York.

Sandford, K. K., Gantt, R. R., and Taylor, W. G. (1974). *J. Natl. Cancer Inst.* **53**, 1437–1464.

Shannon, J. E., ed. (1972). "Registry of Animal Cell Lines." Am. Type Cult. Collect., Rockville, Maryland.

Waymouth, C., Chen, H. W., and Wood, B. G. (1971). *In Vitro* **6**, 371.

Westermark, B., Pontén, J., and Hugosson, R. (1973). *Acta Pathol. Microbiol. Scand., Sect. A* **81**, 791.

Williams, G. M., Weisburger, E. K., and Weisburger, J. H. (1971). *Exp. Cell Res.* **69**, 106–112.

Yasumura, Y., Tashjian, A. H., Jr., and Sato, G. H. (1966). *Science* **154**, 1186–1189.

Chapter 31

Primary and Long-Term Culture of Adult Rat Liver Epithelial Cells

GARY M. WILLIAMS

Naylor Dana Institute for Disease Prevention,
American Health Foundation,
Valhalla, New York

I. Introduction

The methods for cell culture of liver cells have improved considerably in recent years. The use of continuous perfusion to obtain high recoveries of rat liver cells (Seglen, 1973; Williams and Gunn, 1974; Laishes and Williams, 1976a) has made it possible to initiate functionally active short-term primary cultures of hepatocytes (Bissell *et al.*, 1973; Bonney *et al.*, 1974; Laishes and Williams, 1976a) which give rise to long-term epithelial cultures (Williams and Gunn, 1974). Study of the behavior of representative hepatocytes in cultures initiated from high-yield perfusion dissociates now permits for the first time assessment of the ability of hepatocytes to adapt to culture and to proliferate (Williams and Gunn, 1974; Bonney *et al.*, 1974; Laishes and Williams, 1976a).

II.   Methods

A.   Perfusion and Dissociation

In situ perfusion was performed with a peristaltic pump (Williams and Gunn, 1974; Laishes and Williams, 1976a) using a modification of the method of Seglen (1973). A wash-out perfusion through the portal vein with 0.5 mM EGTA in Ca^{2+}-, Mg^{2+}-free Hanks' balanced salt solution was conducted for 1.5 minutes at 8 ml per minute, after which the flow rate was increased to 40 ml per minute and allowed to run to waste for 2.5 minutes. During this period, a return cannula was secured in the suprahepatic inferior vena cava. The enzyme solution, 0.02–0.05% collagenase (Sigma type I) in Williams' medium E (WE, Williams and Gunn, 1974), was perfused for 9.5 minutes. In order to maintain the perfusion at 37° C, all solutions were kept at 38° C and a 40-W bulb was positioned 6 cm above the liver, which was covered with sterile gauze. After perfusion, the liver was removed into 50 ml of cold WE in a petri dish on ice and trimmed. The porta hepatis was grasped with a pair of forceps, and the capsule of the liver was opened at numerous points on both surfaces by radial cuts with small scissors. Cells were detached by combing gently with an aluminum comb and shaking off loose cells. A satisfactory perfusion permitted detachment of almost all the brownish liver parenchyma, leaving about 0.5 gm of white connective tissue. The cells were transferred to a 50-ml centrifuge tube by pipetting gently with a 10-ml pipette. They were sedimented at 100 g for 5 minutes at 10°C and resuspend in WE on ice.

B.   Culturing

Williams' medium E was formulated to contain every essential nutrient for the rat. It provides good support for the growth of rat liver epithelial cells (Williams *et al.*, 1971; Williams and Gunn, 1974), as well as many other lines, and maintains a higher level of cell function than does a minimal medium (Gunn *et al.*, 1976). Cells inoculated in WE with 10% fetal bovine serum (FBS) yielded the greatest attachment efficiency. One hour was allowed for attachment, after which the cultures were carefully washed and refed.

Medium was changed daily for the first week and twice weekly thereafter. For initiated epithelial lines, 5% FBS provided almost as good growth support as 10% (Williams and Gunn, 1974).

III. Results

A. Perfusion and Dissociation

Perfusion with 0.02% collagenase resulted in no destruction of hepatocytes while producing an average recovery of 329 × 10⁶ cells per 100 gm body weight with about 57% viability (Laishes and Williams, 1976a). In addition to damaged cells, free nuclei in the dissociate equaled about 20% of the total yield and were a good indication of the trauma involved in detachment of cells and, therefore, the efficacy of the perfusion. Increasing the collagenase concentration to 0.05% raised the average recovery to 344 × 10⁶

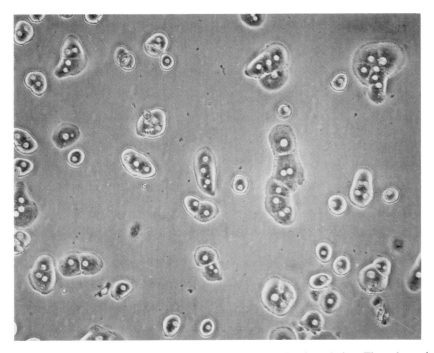

Fig. 1. Primary culture of rat liver cells at 24 hours after inoculation. The polygonal hepatocytes are arranged in cords and clusters. Many cells are binucleate. Phase contrast, × 100.

cells per 100 gm body weight with 63% viability. However, these conditions produced considerable disruption of the liver, which made the determination of selective destruction to regions of hepatocytes impossible. Repeated washing and sedimentation at 50 g markedly decreased free nuclei in the pellet, but did not alter the cell number.

B. Primary Cultures

Inoculation of 5 × 10⁶ viable cells per 75 cm² T-flask (66,000 cells/cm²) permitted 51% ± 5 attachment, which was higher than that observed at greater densities (Laishes and Williams, 1976a), probably owing to the relatively greater surface area available for attachment. It is important that allowing only 1 hour for attachment resulted in attachment of virtually only viable cells.

The percentage of viable cells that attached was increased by the addition of 2 m units of insulin per milliliter (Laishes and Williams, 1976a). Thus,

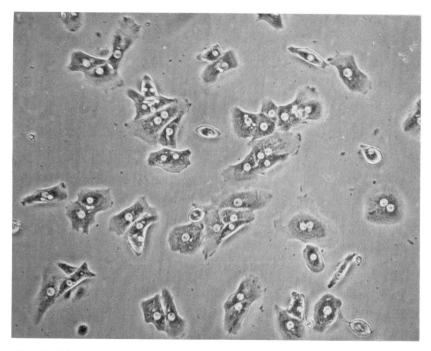

Fig. 2.   Primary culture of rat liver cells at 48 hours after inoculation. The hepatocytes have begun to lose their polygonal shape and to assume a spread-out elongated shape. Phase contrast, × 100.

by inoculating 15×10^6 viable cells per flask in the presence of insulin, it was possible to obtain completely confluent flasks.

The cultures were at least 90% composed of uniform polygonal cells (Fig. 1), many of which were binucleate. These cells arranged themselves in cords and clusters. The cultures possessed a variety of liver-specific functions such as phosphoenolpyruvate carboxykinase, tyrosine aminotransferase, pyruvate kinase, and aldolase B (Laishes and Williams, 1976a).

In regular WE or with added insulin, cells began to die and detach between 24 and 48 hours after inoculation. This was accompanied by a change in morphology from the polygonal cell shape to an elongated one (Fig. 2). By 3 days, about 10% of the original inoculum remained attached (Laishes and Williams, 1976a). This rapid loss of cells was not due to irreparable damage inflicted by dissociation since it was greatly retarded by dexamethasone (Laishes and Williams, 1976b). Addition of 1 μM dexamethasone after attachment resulted in 100% survival for 3 days followed by a slow decline. The change in morphology was also delayed by dexamethasone (Fig. 3).

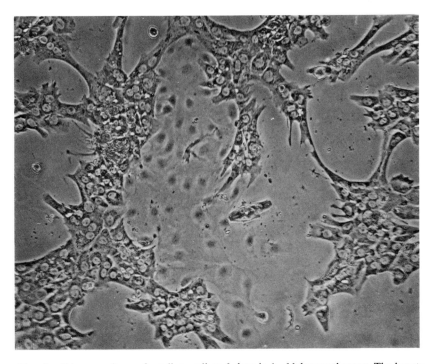

Fig. 3. Primary culture of rat liver cell at 6 days in 1 μM dexamethasone. The hepatocytes have survived longer in the presence of dexamethasone and are arranged in compact clusters. In the center, a colony of clear epithelial-type cells with pale cytoplasms is growing. Phase contrast, $\times 100$.

C. Long-Term Cultures

After the initial attrition to 10% of the original population in the first week, at least 50% of primary cultures showed growth of epithelial-type cells by the end of the second week. These epithelial-type cells had pale translucent cytoplasms (Fig. 3). In addition, proliferating fibroblasts were frequently present. The latter occasionally grew to a substantial extent but often had limited proliferative capability, as did the clear epithelial-type cells. By 3–4 weeks, small islands of epithelial cells with dense cytoplasms developed. These were the cells which eventually gave rise to lines (Fig. 4), but it has not been determined whether they were derived from the clear epithelial cells or arose from a different precursor. The failure of adjacent colonies to blend suggests that these two epithelial types were of different origin.

When a culture contained a mixture of cell types, several procedures were employed to obtain epithelial cultures. Undesirable cells were mechanically destroyed in the primary culture by scraping with a flamed platinum wire (Williams *et al.*, 1971). Also, several extensions were made of the original

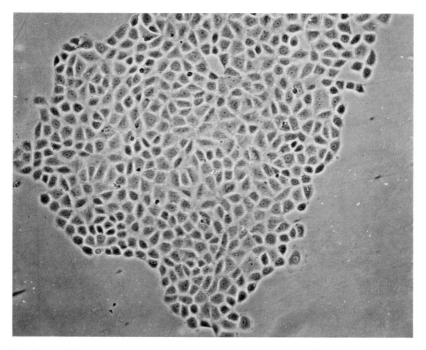

FIG. 4.   An island of cells from a line (ARL 12) of rat liver epithelial cells have denser cytoplasms than those in Fig. 3. They are polygonal and arranged in compact mosaic-like sheets. The periphery of the island has a smooth contour. Phase contrast, × 100.

observation that fibroblasts adhere more readily than epithelial cells to the culture surface (Williams *et al.*, 1971). At subculture, incomplete detachment by short intervals of trypsinization was monitored under the phase microscope to selectively recover epithelial cells which detached more easily than fibroblasts. Also, at subculture, short intervals of attachment (i.e., 30 to 60 minutes) followed by removal and culture of the unattached cells, depleted the final inoculum of the more readily attaching fibroblasts. Perhaps the most effective approach was to maintain the cultures in WE with only 5% FBS subculturing at low density (5 × 10³ cells/cm²). Under these conditions, fibroblasts grew very poorly.

When the principal objective was the establishment of epithelial lines rather than high density primary cultures, several procedures were of help. The first utilized the observation that fibroblasts attach more rapidly than epithelial cells (Williams *et al.*, 1971). Accordingly, the dissociate was plated for 20 minutes after which the medium with unattached cells was withdrawn and plated in a second flask. After 20 minutes, the medium was, again, withdrawn and plated for a third time for 4 hours. With both trypsin-dissociated newborn rat liver and perfusion-dissociated adult rat liver, there was depletion of fibroblasts from the final plating, which permitted easier initiation of epithelial lines. A second approach recently attempted successfully was to inoculate the primary culture at low density (2000–20,000 cells/cm²). In one case, an epithelial clone was obtained without further manipulation. If fibroblasts survived these selection techniques, they were eliminated by the methods detailed above.

IV. Discussion

Primary hepatocyte cultures were initiated by procedures producing high and nonselective recovery, which is necessary to approach the ideal of completely representative liver cultures. The present yields, which were comparable in cell number to those of Seglen (1973), exceeded those obtained by other methods.

The primary cultures were mainly functioning hepatocytes and represented 14% of the liver (Laishes and Williams, 1976a). The addition of dexamethasone resulted in the longest survival of primary hepatocyte cultures reported. This effect clearly indicates that it may be possible to develop conditions to permit prolonged maintenance of hepatocytes.

The initiation of long-term cultures was reasonably reliable, but thus far has not become easier as recovery conditions have improved. The cultures still represented a minority of the cells of adult liver. The lines were morphologically similar to those initiated from newborn rat liver (Williams *et al.*, 1971), suggesting that the latter were not simply a primitive cell type. They possessed sufficient properties of liver (Williams, 1975) to suggest that they contained hepatocytes. It has been found that nutrient (Gunn *et al.*, 1976) and culture (Williams *et al.*, 1973) conditions are important to phenotypic expression of liver epithelial cells, and thus, it remains to be determined how much liver function can be maintained in such lines. On the other hand, these lines may be derived from a special population of liver cells capable of continuous proliferation. The study of factors affecting the fate of representative primary cultures will yield insight into what fraction of hepatocytes can adapt to sustained proliferation.

ACKNOWLEDGMENTS

This investigation was supported by an Alexander Rolston Peacock Memorial Grant for Cancer Research (BC-133A) from the American Cancer Society. The collaboration of Dr. Martyn Gunn, Fels Research Institute, and Dr. Brian A. Laishes, Research Fellow of the National Cancer Institute of Canada, is gratefully acknowledged.

REFERENCES

Bissell, D. M., Hammaker, L., and Meyer, U. A. (1973). *J. Cell Biol.* **59**, 722.
Bonney, R. J., Becker, J. E., Walker, P. R., and Potter, V. R. (1974). *In Vitro* **9**, 399.
Gunn, J. M., Shinozuka, H., and Williams, G. M. (1976). *J. Cell. Physiol.* **87**, 79.
Laishes, B. A., and Williams, G. M. (1976a). *In Vitro* (in press).
Laishes, B. A., and Williams, G. M. (1976b). *In Vitro* (in press).
Seglen, P. O. (1973). *Exp. Cell Res.* **76**, 25.
Williams, G. M. (1975). *In* "Gene Expression and Carcinogenesis in Cultured Liver" (L. E. Gerschenson and E. B. Thompson, eds.), pp. 480–487. Academic Press, New York.
Williams, G. M., and Gunn, J. M. (1974). *Exp. Cell Res.* **89**, 139.
Williams, G. M., Weisburger, E. K., and Weisburger, J. H. (1971). *Exp. Cell Res.* **69**, 106.
Williams, G. M., Stromberg, K., and Kroes, R. (1973). *Lab. Invest.* **29**, 293.

Chapter 32

Isolation of Rat Peritoneal Mast Cells in High Yield and Purity

P. H. COOPER AND D. R. STANWORTH

Department of Experimental Pathology,
Medical School, University of Birmingham,
Birmingham, England

I. Introduction

Mast cells, or cells with similar histochemical characteristics, are widely distributed phylogenetically in the animal kingdom. In higher vertebrates, tissue mast cells have been found to be numerous in cat, dog, goat, mouse, rat, guinea pig, bat, and calf (Keller, 1966). They are generally found in tissues often exposed to trauma, especially in loose connective tissue, around blood vessels, in the interstitium of the myocardium, among fat cells, in peritoneum, in most organ capsules, and in the thymus. In mammals, mast cells generally have an ovoid or elongated shape; rat mast cells in suspension have a mean diameter of 12.6 μm and a spherical nucleus 5 μm in diameter. The cell is tightly packed with histamine-containing membrane-bound granules, each cell containing on average about 500 granules. In the cyto-

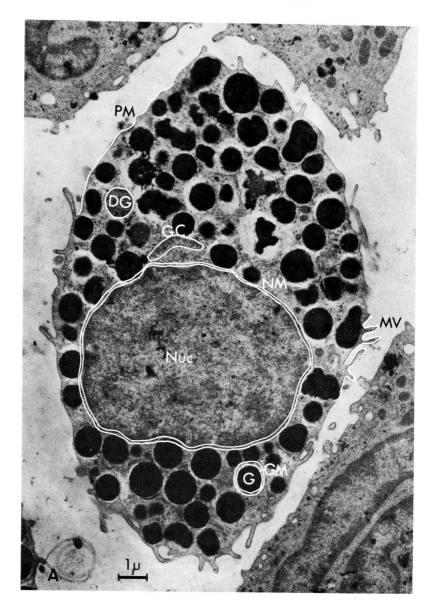

FIG. 1.   Electron micrographs of isolated rat peritoneal mast cells showing the histamine-containing granules (G) surrounded by granular membrane (GM). Partially discharged granules (DG) are also shown. The nucleus (Nuc), nuclear membrane (NM), Golgi complex

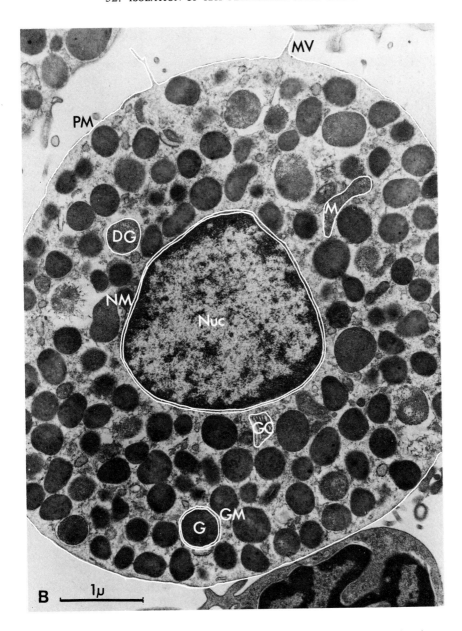

(GC), and mitochondria (M) are indicated. Microvilli (MV) are present on the plasma membrane (PM). (A) × 7500; (B) × 22,500. Photographs supplied by courtesy of Drs. P. M. Mann and C. J. Morris.

plasm the Golgi complex is well developed, but only a few mitochondria and a sparse endoplasmic reticulum are found. The plasma membrane is characterized by the presence of small villous processes (Fig. 1).

The different types of histamine release initiated by antigen–antibody reaction or nonimmunological agents acting directly on or near to the surface of the rat mast cell have been described and studied most successfully by *in vitro* techniques (Goth, 1967; Jasani *et al.*, 1973; Schachter, 1973). Suspensions of mast cells can be obtained by irrigation of the rat peritoneal cavity, but, as such washings only contain 2–5% mast cells, purification is necessary in order to make sure that the primary interaction is between the histamine-liberating agent and the mast cell. Furthermore, studies on the biochemistry and enzymology of the mast cell require the use of purified preparations (Cooper and Stanworth, 1975).

Several techniques have been described for the isolation of rat mast cells. Sucrose-gelatin and sucrose gradient centrifugation have been used, but the cells lose more than 70% of their histamine on isolation and their ability to respond to histamine-releasing agents such as compound 48/80 and antigens is reduced. Morphological changes also occur (Padawer and Gordon, 1955; Glick *et al.*, 1956; Absoe-Hansen and Glick, 1958; Johnson and Moran, 1966). These changes are probably due to the exposure of the cells to the hypertonic solutions during the density gradient centrifugation. Albumin and Ficoll have also been used as isolation media with some success, but, under the conditions used, cells thus prepared also lose some of their responsiveness to antigens (Johnson and Moran, 1966). In the technique described below, a high yield of mast cells is obtained reproducibly with little or no loss in response to the histamine-releasing capacity of liberators.

II. Procedures

A. Materials

1. ANIMALS

Wistar rats, body weight 250–450 gm, of either sex are suitable.

2. MATERIALS

Phosphate-buffered (0.01 M, pH 7.3) saline tablets (Dulbecco A) and mineral salts solution (Dulbecco B) can be obtained from Oxoid, London. N-2-hydroxyethylpiperazine-N'-2-ethane sulfonic acid (HEPES), histamine, o-phthalaldehyde, bovine serum albumin type F, ovalbumin, and ATP-Tris salt can be purchased from Sigma (London) Chemical Co. Ltd. Ficoll (MW

400,000) is obtainable from Pharmacia, Uppsala, Sweden. Its pretreatment by dialysis was found to be unnecessary for use in mast cell isolation. Compound 48/80 is available from Wellcome Reagents Ltd.; Synacthen (ACTH 1-24 polypeptide) has been obtained from Ciba, Horsham, and calcium ionophore A23187 through the courtesy of E. Lilley and Co.

B. Isolation and Purification of Mast Cells

Plastic disposable apparatus is used throughout, and any glass vessels used for storing solutions are washed with a 0.1% solution of the chemical detergent RBS25 (Chemical Concentrates Ltd., London) and rinsed thoroughly. The rats are killed by placing them in an ether-saturated chamber for 2–3 minutes. This was found to be superior to other methods, which tended to cause internal bleeding into the peritoneal cavity. The animals are quickly removed, and the abdominal fur is cleaned with 70% alcohol. The abdominal wall is exposed by removal of a piece of fur 2 inches square. A small incision is made, just large enough to admit the passage of a 10-ml plastic tube, the end of which has been perforated several times with a hot needle. The tube is inserted deep into the abdomen. The rat is held vertically by the tail, and 20 ml of sterile phosphate-buffered saline are poured through the tube. The rat is then laid horizontally, and, by holding the tube free of the abdominal contents, is massaged for 90 seconds. The rat is again held vertically by the tail and the peritoneal fluid is removed by inserting a 10-ml plastic pipette into the perforated tube. The fluid is transferred to 10-ml plastic tubes kept in ice. All subsequent operations are carried out at 4°C. The tubes are centrifuged for 1 minute (including acceleration) at 750 g in an MSE (Measuring and Scientific Equipment Ltd., London) Major refrigerated centrifuge using an angle-head rotor (No. 6882). The tubes are then removed, and the supernatants are decanted. The cells are resuspended in complete Dulbecco buffer (20 ml of phosphate-buffered saline + 0.1 ml of mineral salts solution + 0.1 ml of 0.1 M glucose) and, using a 5-ml plastic pipette, are carefully layered in 3-ml portions (equivalent to the yield from three rats) over 2 ml of 30% Ficoll in complete Dulbecco buffer supplemented with 1% bovine serum albumin (type F). Cellulose nitrate tubes (2 × 0.5 inches) are used for centrifugation (5 minutes at 150 g) in a swing-out head in an MSE Minor bench centrifuge in the cold room. The mast cells can be seen as a white precipitate in the Ficoll layer, underneath the pink interface, which contains erythrocytes and macrophages. The tubes are then placed in a tube cutter (Randolph and Ryan, 1954) (Fig. 2) and sliced 2 mm below the interface. The contaminating layers are thoroughly washed from the tube-cutter chamber with phosphate-buffered saline, and the remainder of the tube contents comprising Ficoll and mast cells recovered. The cells are then

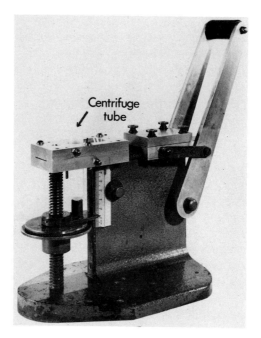

FIG. 2.   Mechanical tube slicer.

washed into a 5-ml test tube (1 tube per sample) using complete Dulbecco buffer supplemented with 0.5% bovine serum albumin, type F. The tubes are made up to 5 ml with the same medium, inverted gently to mix the Ficoll with the buffer, and centrifuged at 750 g for 1 minute (including acceleration) in an MSE Major centrifuge. The supernatant is carefully removed, and the white pellet of cells is resuspended in a small volume of Dulbecco/0.5% bovine serum albumin mixture. The tube contents are made up to 5 ml and recentrifuged for 1 minute at 750 g. This step is repeated twice, and the final preparation of mast cells is resuspended to the desired volume using Dulbecco buffer/0.5% bovine serum albumin. The cells can be stored in this medium at 4°C for up to 2 hours, but they should be used immediately if possible.

MODIFICATIONS TO METHOD

1. When isolating sensitized cells, it is found that prolonged exposure to cold causes some desensitization. For this reason, isolations are performed at room temperature and the cells are used immediately.

2. The technique has also been used successfully by substituting 20 mM HEPES buffer (pH 7.3, 0.9% NaCl) for phosphate-buffered saline (Cooper and Stanworth, 1975).

C.   Other Methods

1.   STAINING AND COUNTING

Mast cell suspension (0.1 ml) is added to Dulbecco/0.5% bovine serum albumin medium (0.85 ml) and preincubated at 37°C for 2 minutes; 0.1% toluidine blue (50 μl) is added, and the incubation is continued for another 2 minutes. The cells are then counted in an Improved Neubauer counting chamber.

2.   HISTAMINE RELEASE STUDIES

In a typical experiment the liberator or antigen is added to complete Dulbecco buffer/0.5% bovine serum albumin in a 5-ml plastic test tube to a final volume of 0.9 ml. After preincubation at 37°C for 5 minutes, the reaction is started by addition of 0.1 ml of mast cell suspension also preincubated at 37°C. After 5 minutes incubation (15 minutes for ATP) at 37°C, the reaction is terminated by immersion in ice for 5 minutes. The tubes are then centrifuged at 750 g for 5 minutes at 4°C. The supernatants are separated from the sediments and placed in 5-ml plastic tubes. To the supernatants 10% trichloroacetic acid (1 ml) is added, and to the sediments 5% trichloroacetic acid (2 ml). The tubes are shaken, using a vortex mixer, then centrifuged at 750 g for 10 minutes at 4°C and stored at this temperature until ready for histamine assay.

3.   LACTATE DEHYDROGENASE-RELEASE STUDIES

To determine the amount of lactate dehydrogenase released when mast cells are challenged, the procedure, as described in Section 2 above, is modified. After separation of supernatant and sediment, Dulbecco medium (1 ml) is added to the pellets, which are then homogenized using an MSE 100W Ultrasonicator at 8 μm for 10 seconds at 4°C. Under these conditions more than 95% of the cells are disrupted. Samples of 50 μl are then taken from both supernatant and pellet samples for lactate dehydrogenase assay by the method of King as described by Varley (1967), the volumes being scaled down by a factor of 4 to give the necessary sensitivity. The incubation time is 30 minutes. From these results a value for the percentage of lactate dehydrogenase released can be calculated. Trichloroacetic acid (10%, 1 ml) is added to each of the supernatants and sediments, and after centrifugation, histamine assay is performed.

4.   HISTAMINE ASSAY

Histamine is determined by a manual (Shore et al., 1959) or automated (Evans et al., 1973) spectrofluorimetric procedure.

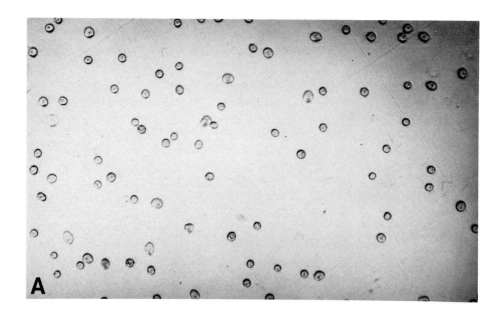

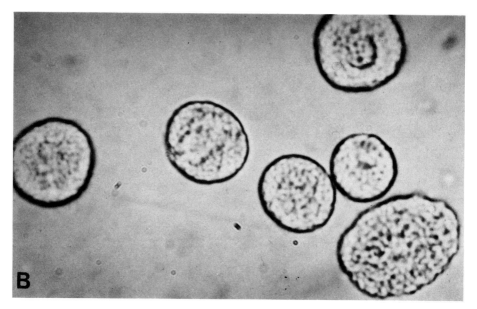

FIG. 3. Microscopic appearance of a purified mast preparation. (A) × 100. (B) × 1000.
Phase contrast.

III. Characteristics of Purified Mast Cell Preparation

A. Morphology

The microscopic appearance of a mast cell preparation obtained in the manner described is shown in Fig. 3. The cells are intact, and the granules can be seen. The characteristic metachromasic staining of the granules with toluidine blue can be demonstrated, and the cells are shown to be viable by the trypan blue exclusion test.

B. Yield and Purity of Preparation

In Table I the results are divided according to the sex of the rats. This is because female rats consistently give more erythrocyte contamination in the original peritoneal washings than do male rats. In a badly contaminated preparation, it is sometimes necessary to repurify the cells using a second Ficoll gradient. Pure cells are obtained, but the response to antigen-induced release is reduced. This can be avoided by initial dilution of such preparations with buffer. The variability of yield is probably due to slight differences in the positioning of the cellulose nitrate tube in the tube cutter. Purity of the preparation is reproducibly greater than 95%, and nonspecific release of histamine is low. In six preparations the recovery of histamine was 78.1 ± 11.3%, which correlates well with the percentage yield of mast cells as shown in Table I.

C. Histamine Release Studies

Figure 4 shows dose-response curves of histamine release induced immunologically by ovalbumin (using mast cells from presensitized rats) and artificial liberators, such as compound 48/80, Synacthen (ACTH 1-24 polypeptide), ATP (adenosine triphosphate), and calcium ionophore A23187. These liberators have been shown to release histamine selectively; that is,

TABLE I

DEGREE OF PURITY AND YIELDS OF MAST CELLS RECOVERED IN REPEATED ISOLATIONS[a]

| Number of preparations | Mast cells (%) | | Yield (%) | Nonspecific release of histamine (% of total available) |
	Original	Final		
12 (male)	2.95 ± 1.79	97.1 ± 2.3	74.1 ± 12.8	0.83 ± 1.28
18 (female)	1.34 ± 0.61	96.9 ± 2.7	72.2 ± 17.0	0.68 ± 1.51

[a] Results are quoted as means ± standard deviation.

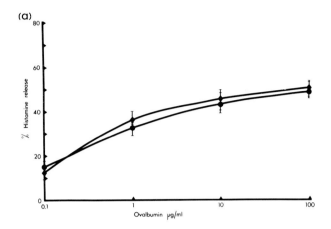

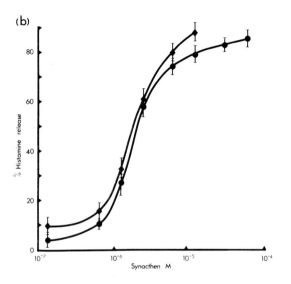

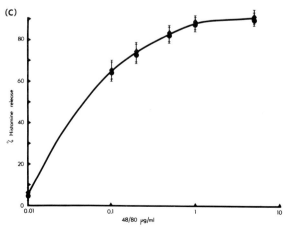

other soluble constituents of the cell are not released at the same time (Goth, 1967; Krüger and Diamant, 1967; Foreman *et al.*, 1973; Jasani *et al.*, 1973; Schachter, 1973). It can be seen that there is little difference between the response shown by the mast cells to the various liberators before and after fractionation, indicating that the procedure has not altered the structural integrity of the cell. To test the selectivity and fragility of the purified cells, lactate dehydrogenase release was measured concomitantly

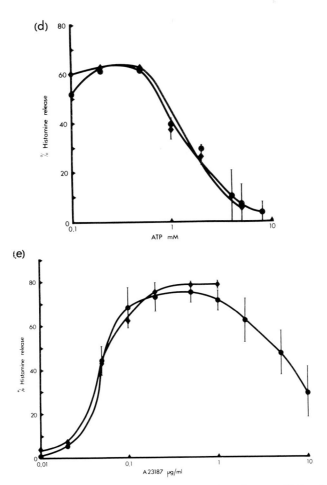

FIG. 4. Log (dose)-response curves for histamine release from purified and unpurified mast cells on challenge with (a) ovalbumin (using sensitized rats); (b) Synacthen; (c) compound 48/80; (d) adenosine triphosphate (ATP); and (e) calcium ionophore A23187. Each curve is the mean of three experiments. Vertical bars indicate ± SD. ●, Pure mast cells; ◆, unpurified mast cells.

TABLE II

RELEASE OF LACTATE DEHYDROGENASE AND HISTAMINE FROM PURIFIED
MAST CELLS ISOLATED FROM SENSITIZED RATS ON CHALLENGE WITH
ANTIGEN (OVALBUMIN) OR ARTIFICIAL LIBERATORS[a]

Liberator	Dose	Histamine release (%)	Lactate dehydrogenase release (%)
Ovalbumin	100 μg/ml	33.6 $\pm$ 5.9	5.4 $\pm$ 1.5
Synacthen	2.8 $\times$ 10^{-6} M	52.4 $\pm$ 6.6	5.4 $\pm$ 0.6
Compound 48/80	0.2 μg/ml	71.3 $\pm$ 11.0	5.1 $\pm$ 2.4
ATP	1 mM	42.5 $\pm$ 11.0	6.8 $\pm$ 1.5
A23187	0.1 μg/ml	73.1 $\pm$ 8.9	6.1 $\pm$ 3.3
Control	—	1.5 $\pm$ 1.1	5.0 $\pm$ 1.1

[a] Results are quoted as mean $\pm$ standard deviation of four experiments.

with histamine release. Table II shows that the cells retain 95% of their cytoplasmic contents even when a high percentage of the histamine is released.

IV.   Discussion

A purified preparation of peritoneal mast cells presents an ideal source of material for pharmacological and biochemical studies of the mechanisms of histamine release. One approach, which we have been pursuing recently in our own laboratory, assumes that the cell-triggering process, in which the antibody plays a central role, can be short-circuited by artificial liberators (such as basic polypeptides), which supposedly act directly upon the cell activation sites, and also that structurally similar activating sites are produced in cell-bound antibody that has been cross-linked by artificial means or by preaggregation in free solution (Stanworth, 1973). Direct kinetic studies performed using basic polypeptides have produced results consistent with the idea that these liberators behave like immunological triggers in exerting their effect selectively through a substrate–enzyme type of reaction (Jasani et al., 1973). We are also now beginning to look at well-defined "tailor-made" synthetic peptides, since our findings indicate that similar structures might be represented within that part of the Fc region of the sensitizing antibody which is intimately concerned with the cell-triggering process.

Another approach has been concerned with the biochemical investigation of the processes leading to histamine release. The rapidity of the exocytotic process, together with the apparent localization of the initial stimulus at the cell surface, is consistent with a possible role of cyclic nucleotides. While conclusive data are not available, there is indirect evidence that indicates an ability of cyclic adenosine 3′,5′-cyclic monophosphate to modulate the response; and recent experiments from other laboratories using purified mast cells have reinforced this theory (Sullivan *et al.*, 1975). In common with other systems that store secretory product within granules, there appear to be the same common denominators in the histamine release process, namely, the requirement for energy and calcium. Furthermore, our present knowledge of the properties of enzymes comprising such energy-requiring systems would suggest the involvement of some kind of ATPase. Subcellular fractionation studies on purified mast cell homogenates have shown that a calcium-stimulated ATPase is present at the external surface of the plasma membrane, and further studies on the properties of this enzyme compared to requirements for histamine release suggest that it may be involved in the release process, possibly acting in the role of a contractile protein (Cooper and Stanworth, 1975).

A third line of investigation involves the use of methods such as immunoadsorption and affinity chromatography techniques for the isolation of membrane fractions retaining the antibody-binding or cell-activation sites. However, because of the small amounts of material available, characterization of such membrane fragments is difficult. Work is now under way in an attempt to overcome these problems. It will prove rewarding to pay further attention to isolated mast cell granules in direct tests to obtain evidence of the possibility of a "second messenger" responsible for conveying the effect of the initial stimulus on the target cell surface to the sites of storage of histamine and other mediators of hypersensitivity.

ACKNOWLEDGMENTS

This work was supported by a grant from the Medical Research Council.

REFERENCES

Absoe-Hansen, G., and Glick, D. (1958). *Proc. Soc. Exp. Biol. Med.* **98**, 458–461.
Cooper, P. H., and Stanworth, D. R. (1976). *Biochem. J.* **156**, 691–700.
Evans, D. R., Lewis, J. A., and Thomson, R. S. (1973). *Life Sci.* **12**, 327–336.
Foreman, J. C., Mongar, J. L., and Gomperts, B. D. (1973). *Nature (London)* **245**, 249–251.
Glick, D., Bonting, S. L., and Den Boer, D. (1956). *Proc. Soc. Exp. Biol. Med.* **92**, 357–359.
Goth, A. (1967). *Adv. Pharmacol.* **5**, 47–78.
Jasani, B., Stanworth, D. R., Mackler, B., and Kreil, G. (1973). *Int. Arch. Allergy Appl. Immunol.* **45**, 74–81.
Johnson, A. R., and Moran, N. C. (1966). *Proc. Soc. Exp. Biol. Med.* **123**, 886–888.
Keller, R. (1966). "The Tissue Mast Cells in Immune Reactions." Am. Elsev., New York.
Krüger, P., and Diamant, B. (1967). *Acta Physiol. Scand.* **71**, 291–302.

Padawer, I., and Gorden, A. S. (1955). *Proc. Soc. Exp. Biol. Med.* **88**, 29–31.

Randolph, M. L., and Ryan, R. R. (1954). *Science* **112**, 528.

Schachter, M., ed. (1973). "Histamines and Antihistamines," Int. Encycl. Pharmacol. Ther. Vol. 1, Sect. 74. Pergamon, Oxford.

Shore, P. A., Halter, A., and Cohn, V. H. (1959). *J. Pharmacol. Exp. Ther.* **127**, 182–186.

Stanworth, D. R. (1973). *Front. Biol.* **28**, 290–328.

Sullivan, T. J., Parker, K. L., Stenson, W., and Parker, C. W. (1975). *J. Immunol.* **114**, 1473–1485.

Varley, H. (1967). "Practical Clinical Biochemistry, "4th ed., pp. 279–270. Wiley (Interscience), New York.

Chapter 33

Monocyte Preparation from Blood

C. R. W. RAYNER[1]

*St. George's Hospital and Westminster Hospital,
London, England*

I. Introduction

The study of monocyte functions, until recent years, has, broadly speaking, been aimed in two directions. First, to examine the possible role of the monocyte as a fibroblast precursor, and second, to examine its role in resistance to infection, e.g., monocyte to macrophage transformation, and most recently its role in the immune response. The most sophisticated methods of studying these problems have been based on either cross-circulation experiments in animals or on the use of skin windows in either normal or immunosuppressed patients or animals. Such methods of study may have certain inherent weaknesses; for example, Ross *et al.* (1970) in a cross-circulation model could not demonstrate monocyte-to-fibroblast conversion in wounds in recipient animals where monocytes had been derived only from the donor animal. However, as the recipient animals had been subjected to total body irradiation in order to suppress their own cellular responses, it could be argued that their monocyte-activating system had been suppressed. Allgower *et al.* (1974) in a similar cross-circulation experi-

[1] *Present address*: Department of Plastic Surgery, Withington Hospital, Manchester, M20 8LR, England.

ment in rabbits used sex chromatin as a cellular marker. Monocytes from female donors to male recipients could not be shown to transform into fibroblasts. If the possible immunological problems of this model are ignored, it could still be argued that this model had simply failed to demonstrate one of the "reserve" functions of the monocyte. Rebuck et al. (1971) studied monocyte functions in immunosuppressed patients. Once again the normal cellular interactions may be impaired and conclusions concerning *normal* monocyte function invalid.

It can be seen, therefore, that monocytes require study from two main viewpoints: first, their normal function in their normal cellular environment, i.e., *in vivo*; and second, their "reserve" functions, e.g., possible ability to convert to fibroblasts or to endothelial cells (Milward and Rayner, 1976), which may be more easily studied *in vitro*, together with the factors controlling monocyte coalescence, etc. Recent studies on monocyte function seem to suggest a closer role in wound repair and in the immune response than has previously been apparent. Thus, it is accepted that monocytes may transform into macrophages and that macrophages in turn play an important role in fibrin degradation and fibroblast orientation in healing wounds. Douglas and Goldberg (1972) have described receptors on monocytes for immunoglobulin and complement, and Snyderman et al. (1973) have reported a defective chemotactic response in monocytes in chronic candidiasis. Thus, it is possible to postulate the existence of an immune response-stimulated, cell-based, "servo" mechanism controlling fibrosis. The experimental study of such a complex mechanism might be made easier if the cells were available in a pure vital suspension. This should allow more direct methods of study to be employed than were previously possible, especially if the cells can be "marked," e.g., with a radioactive label.

II.   The Identification of Monocytes

While there is little difficulty in identifying "normal" monocytes on routine blood film, it may be difficult to distinguish the smaller cells in this series from the larger cells in the lymphocyte series. Rebuck et al. (1971), using a "skin window" technique in humans, claimed that lymphocytes transformed into monocytes in experimental wounds. This conclusion was based on direct observation of coverslip preparations. Rayner (1976), however, using chromium-[31]-labeled and tritiated thymidine-labeled thoracic duct lymphocytes injected into experimental wounds, was unable to confirm this. Whitelaw (1966) has been able, by means of peroxidase staining (by Rytomaa's method), to show that some cells which would morphologically be regarded as lymphocytes are, in fact, monocytes. Lymphocytes are never peroxidase positive in peripheral blood. Furthermore, the presence of monocyte precursors, i.e., promonocytes, in the bone marrow would seem to indicate that other cells are unlikely to be monocyte precursors as well. Thus, it is reasonable to regard the monocyte, for experimental purposes, as a peroxidase-positive mononuclear cell found in peripheral blood. Provided that only peroxidase-positive cells are studied, the question whether or not it has been derived from other cells in the blood can probably be ignored. In fact, it appears unlikely that such a possibility exists. The use of pure suspensions of monocytes fulfilling the above criteria would overcome the problem of identifying cells on the identification "borderline."

III.   Methods of Isolating Monocytes

A.   History

In 1935, Ungar and Wilson stimulated a chemical peritonitis in guinea-pigs, obtained a cell suspension from the abdominal cavity, and incubated this with India ink and lithium carmine and then injected the labeled cells (80% mononuclear) into other animals. Their conclusion that monocytes transform into macrophages has subsequently been confirmed by more refined techniques. Since that time, many attempts have been made to obtain various cell fractions from peripheral blood with high yields and purity. Pertoft *et al.* (1968) described a method of separating various white cell fractions by centrifuging blood on colloidal silica. These authors used tritiated thymidine incorporation to demonstrate lymphocyte viability after separation, with good results. Unfortunately the monocyte yield was low (22%). Bennett and Cohn (1966) also reported a method of separating monoctes from large quantities of human blood (150 ml). Their method [also used by Holm and Hammarström (1973) in a modified form with a gelatin gradient] employed a final stage involving cell adhesion to plastic materials with subsequent difficulties in detaching the cells. The final concentration involved lymphocyte contamination of the order of 5–20%. These methods are not suitable for obtaining a pure monocyte fraction with large numbers of cells from small amounts of blood, e.g., as obtained from experimental animals. They are, however, at the present time the best methods available to obtain samples from human blood.

Boyum (1968) described a technique combining sedimentation and centrifugation of whole blood across a sucrose (Ficoll)/sodium metrizoate gradient. With this method Boyum noted difficulty in obtaining a lymphocyte preparation uncontaminated by monocytes; variation in the osmolarity of the blood solution resulted in further loss of lymphocytes, but not of monocytes. With Boyum's method the gradient medium causes red cell clumping which leads to rapid red cell precipitation through the medium as well as a gradient to separate cells of different densities. It would seem from Boyum's work that a very large monocyte load in the blood under separation should result in a greater number of these cells remaining with the lymphocytes at the end of the separation procedure.

In 1949, Stanley described a method for obtaining a lipid extract from the bacterial organism *Listeria monocytogenes*. Girard and Murray (1954) gave a detailed description of a method using chloroform extraction and filtration to obtain large quantities of this monocyte-producing agent and showed that it could be used to produce a sustained monocytosis. Tadayon *et al.* (1969) used a similar technique but managed to obtain improved yields by using "smooth" colonies of *Listeria monocytogenes* in the primary culture of the organism. The monocyte-producing agent is obtained as a dry powder. These authors give a detailed account of the factors affecting the manufacture and ultimate yield of the monocyte-producing agents. Rebuck and Crowley (1955) had used the whole organism to stimulate a monocytosis in rabbits with experimental wounds. Although successful, the method had been difficult to establish and obviously could not provide an exact cell "dose" to the wound.

B. A Method for Obtaining a Pure Suspension of Monocytes

This method (Rayner, 1975a) attempts to combine the advantages of a stimulated monocytosis with Boyum's refined cell separation technique.

1. ANIMALS

Adult Sprague-Dawley rats. Six animals were injected with chloroform-soluble (4°C) lipid extract of *Listeria monocytogenes*. This substance is responsible for the monocytosis associated with infection with this organism (Tadayon *et al.*, 1969). The extract is supplied in powder form and reconstituted in normal saline by sonification.

Each animal received 6 mg intraperitoneally daily for 3 days. A brisk monocytosis was noted after 36 hours (see Table I). Eighteen hours after the third injection (i.e., on the fourth day) blood was obtained by cardiac puncture using heparin anticoagulant to give a final heparin concentration of 10 units/ml. This was then diluted one part blood to one part normal saline. Four milliliters of the blood/saline mixture were then layered carefully over 3 ml of a Ficoll/Metrizoate gradient (Lymphoprep, Nyegaard Ltd) in a 21-ml polythene tube. This was then spun at 400 g to the interface for 20 minutes on an MSE Major centrifuge at room temperature.

TABLE I

RANGES OF WHITE CELL COUNTS PER CUBIC MILLILITER
OF WHOLE BLOOD OF NORMAL RATS AND OF RATS
STIMULATED WITH *Listeria* MONOCYTOGENES EXTRACT

Cells	Normal animals (6)	*Listeria*-stimulated animals (9)
Total	6,100–19,200	28,500–56,800
Monocytes	352–726	21,000–39,000
Lymphocytes	4,030–12,800	4,300–11,900
Granulocytes	1,720–5,700	3,200– 5,900

[a] Blood was taken 90 hours ± 2 after first injection of extract.

The buffy layer containing the cells at the blood/gradient interface was removed with half the gradient fluid (after most of the plasma layer had been discarded). The cells were washed once in Earle's Balanced Salt Solution (BSS) and resuspended in 5 ml of BSS. One drop of rat antilymphocyte serum and fresh complement were added, and the suspension was incubated for 1 hour. (Rat antilymphocyte serum was raised on rabbits by three injections of 1×10^8 rat spleen cells.) The cell pellet was then washed twice and resuspended in either 1 ml of BSS or 1 ml of TC 199 culture medium. A cell count was performed and the mixture was diluted to the required cell concentration. Twelve separations were performed. Slides of the final cell suspension were stained with hemotoxylin and eosin and tested for the peroxidase reaction (Rytomaa, 1962).

2. METHOD OF LABELING WITH THYMIDINE-³H

In each of three additional animals, 6 hours after each injection of *Listeria monocytogenes* extract, 250 mCi of high specific activity thymidine-³H was injected intraperitoneally. Monocytes from the blood of these animals were then obtained as described above. Smears were made, and the thymidine-labeled cells were counted by autoradiography, three grains per nucleus or more being taken as indicating a labeled cell, there being less than one grain per nucleus background.

3. RESULTS

The percentage yield obtained from blood of unstimulated animals (12–37.6%) is lower than from stimulated animals (80–82%) (Table II). The difference may be explained by the regular, absolute loss of cells due to entrapment of monocytes by precipitating red cells through the lymphopreparation. This is a function of the red cell mass as well as the mono-

TABLE II

NUMBERS OF MONOCYTES IN WHOLE BLOOD IN NORMAL AND
Listeria-STIMULATED ANIMALS AND THE YIELDS OBTAINED BY
SEPARATION

	Normal animals (6)	*Listeria*-stimulated animals (9)
Monocytes per ml whole blood	352,000–726,000	2.1×10^6–3.9×10^6
Total monocyte yield per ml blood	37,000–273,000	1.9×10^6–3.2×10^6
Percentage yield from 1 ml whole blood	12–37.6	80–82

cyte load. Another factor may be the greater number of large monocytes in the stimulated animals; smaller cells of the series (more in evidence in the unstimulated animal) may exhibit the tendency of rat lymphocytes to follow red cells through the medium. This is possibly a species peculiarity and requires comparison with other experimental animals. The final suspension may be so diluted as to give any cell concentration required for convenient injection according to the experimental circumstances.

Slides from the final suspension showed the typical rat monocyte appearance. There were no cells that could definitely be identified as lymphocytes, and 96% of the cells were peroxidase positive.

4. THE TRYPAN BLUE EXCLUSION TEST

An average of two cells out of two hundred examined on each separation were stained. Six separate separations were performed on blood from the animals given thymidine-^{3}H. There was no detectable difference in total yield or viability between animals receiving thymidine-^{3}H and those that did not. The final cell suspension contained more than 95% labeled cells. No animal developed lymphopenia in this series.

IV. Discussion and Some Unsolved Problems Concerning This Method of Separation

Lymphoprep is a commercially produced version of Boyum's gradient mixture and is prepared specifically to separate human lymphocytes from whole blood. Boyum suggested that the rat monocyte has a greater tendency to stay in the buffy layer then was the case in other species. While it is certain that *Listeria monocytogenes* has the same monocytosis-producing effect in other experimental animals (mice, rabbits, and guinea pigs), further investigation is required to see whether the method of obtaining suspensions is as effective as in the rat. The method is of particular value where syngeneic animals are used for cell transfer experiments (rats and mice), as problems of immune rejection can be circumvented. In view of the dangers of neutropenia associated with *Listeria* inoculation, this method is not suitable for experiments in humans. It may, however, prove to be of value for research studies in patients suffering from monocytoses from other causes.

The short-term viability of the monocytes appears to be unaffected by the addition of antilymphocyte serum, and studies in healing wounds (Rayner, 1975b) suggest that reinjected cells are viable for at least 3 days *in vivo*. It is probable that the presence of antimacrophage activity in antilymphocyte serum implies a degree of antimonocyte activity, though the trypan blue test in this series indicates that the great majority of cells were viable after addition of antilymphocyte serum. Jakobsen (1973) showed that the amount of antimacrophage activity varied greatly according to the method of raising the antilymphocyte serum. According to Jakobsen, the three-pulse/spleen cell method of raising the serum, as used in this series, results in the least antimacrophage activity.

The asynchronous injection of thymidine-^{3}H affords a simple method of obtaining labeled cells. Van Furth *et al.* (1970) demonstrated that 48 hours after a single injection of thymidine-^{3}H, 59% of peripheral blood monocytes were labeled. He has also demonstrated the great rapidity with which promonocytes take up thymidine-^{3}H. Two hours after a single dose of thymidine-^{3}H was injected, 68.7% of promonocytes were labeled. Whitelaw (1966), using a technique of intermittent thymidine-^{3}H injection over a 4-day period, found that more than 95% of blood monocytes were labeled. The high percentage of thymidine-^{3}H-labeled cells in their series may, therefore, be explained as the effect of intermittent injection of thymidine-^{3}H coupled with the stimulus of *Listeria* extract to acute monocyte production. The cells produced under such a stimulus may be expected to retain some of the labeling characteristics of their precursors.

It is unclear from our studies whether there is such a cell as a peroxidase-negative monocyte present in normal peripheral blood. There are certainly a few peroxidase-negative monocytic cells present in the final separation after *Listeria* stimulation. Another possibility is that the numerous large monocytes in peripheral blood after 3 days of stimulation with *Listeria monocytogenes* extract represent a "younger" type of cell tending morphologically toward the promonocyte series. (Whitelaw suggested that monocytes have an intravascular life-span of about 3.1 days.) It is probable that any monocytes in the blood stream prior to *Listeria* stimulation would have left the blood stream after 3 days, and that their replacements would be the result of *Listeria* stimulation, i.e., a large type of monocyte. Van Furth *et al.* (1970) have pointed out the marked peroxidase-positive staining properties of the promonocytes. Thus, it may be that the *Listeria*-stimulated monocyte appearing in an acute response to this stimulation in peripheral blood still retains some of the promonocyte staining properties. This may explain the very high percentage of peroxidase-positive cells in the final suspension.

Are the monocytes produced by *Listeria* stimulation normal? Morphologically, they are large cells with horseshoe nuclei indistinguishable from normal large monocytes. In their peroxidase activity and thymidine-^{3}H labeling properties, these cells correspond to the promonocyte of bone marrow. It would be expected, therefore, that these cells should be functionally similar to the promonocyte series.

ACKNOWLEDGMENT

I am grateful to the publishers of *Experimental Cell Research* for permission to reproduce details from "Monocyte Separation from Whole Blood, A Method for Experimental Animals" (Rayner, 1975a).

REFERENCES

Allgower, M., Brennwald, J., Kallenberger, A., Moor, R., and Perren S. M. (1975). *Proc. Int. Symp. Wound Healing 1974*, pp. 16–20.

Bennett, W. E., and Cohn, Z. A. (1966). *J. Exp. Med.* **123**, 145–160.

Boyum, A. (1968). *Scand. J. Lab. Invest.* **21**, Suppl. 97, 77–89.

Douglas, S. D., and Goldberg L. S. (1972). *Vox Sang.* **23**, 214–217.

Girard, K. F., and Murray, E. G. D. (1954). *Can. J. Biochem. Physiol.* **32**, 1–13.

Holm, G., and Hammarström, S. (1973). *Clin. Exp. Immunol.* **13**, 29–43.

Jakobsen, A. (1973). *Acta Pathol. Microbiol. Scand., Sect. B* **81**, 353–358.

Milward, T., and Rayner, C. R. W. (1976). *Plast. Reconstr. Surg.* (in press).

Pertoft, H., Bäck, O., and Lindahl-Kiessling, K. (1968). *Exp. Cell Res.* **50**, 355–368.

Rayner, C. R. W. (1975a). *Exp. Cell Res.* **90**, 436–439.

Rayner, C. R. W. (1975b). *Proc. Int. Symp. Wound Healing, 1974*, pp. 244–246.

Rayner. C. R. W. (1976). M. S. Thesis (submitted London University).

Rebuck, J. W., and Crowley, J. H. (1955). *Ann. N.Y. Acad. Sci.* **59**, 757–794.

Rebuck, J. W., Le Sher, D. A., Perkins, M. J., and Leal, F. (1971). *Immunopathol. Inflammation, Proc. Symp., 1970* Excerpta Med. Found. Int. Congr. Ser. No. 229, pp. 79–97.

Ross, R., Newton, B., Tyler, E., and Tyler, R. (1970). *J. Cell Biol.* **44**, 645.

Rytomaa, T. (1962). *Blood* **19**, 439–442.

Snyderman, R., Altman, L. C., Frankel, A., and Blase, R. M. (1973). *Ann. Intern. Med.* **78**, 509–513.

Stanley, N. F. (1949). *Aust. J. Exp. Biol. Med. Sci.* **27**, 123–131.

Tadayon, R. A., Carroll, K. K., and Murray, R. G. E. (1969). *Can. J. Microbiol.* **15**, 421–428.

Ungar, J., and Wilson, G. R. (1935). *Am. J. Pathol.* **11**, 681–691.

Van Furth, R., Hirsch, J. G., and Fedorko, M. E. (1970). *J. Exp. Med.* **132**, 794.

Whitelaw, D. M. (1966). *Blood* **28**, 455–464.

Chapter 34

Differentiation and Proliferation of Hemopoietic Cells in Culture

T. M. DEXTER AND N. G. TESTA

Paterson Laboratories, Christie Hospital, and Holt
Radium Institute, Manchester, England

I. Introduction

In the last decade several clonal techniques have been developed for studying hemopoietic cells *in vitro*. Bradley and Metcalf (1966) and Pluznik and Sachs (1966) independently reported that mouse bone marrow cells would form colonies of granulocytes and macrophages when plated in a soft gel medium, provided that a suitable source of colony-stimulating factor was present. These granulocyte colony-forming cells (or CFU_c) have since been detected in a variety of species, including man (Pike and Robinson, 1970).

More recently, it has been shown that under suitable conditions and using appropriate stimulating factors, erythroid (Stephenson *et al.*, 1971), megakaryocytic (Metcalf *et al.*, 1975a), T-lymphocyte (Schredni *et al.*, 1975), and B-lymphocyte (Metcalf *et al.*, 1975b) colony formation also occurs under *in vitro* conditions.

From such cultures a great deal of useful information has accumulated on the nature of the colony-forming cells and the characteristics of the stimulating factors. However, in these systems, proliferation is maintained usually for only a matter of days and the pluripotent, self-maintaining hemopoietic stem cells (CFU_s) (Till and McCulloch, 1961) rapidly disappear. Consequently such systems are obviously unsuitable for studying those factors which control the proliferation and differentiation of stem cells. However, in a liquid culture system reported recently (Dexter and Lajtha, 1974), it was shown that proliferation of stem cells can occur for several months, concomitantly with production of CFU_c and formation of granulocytes. Using this "long-term" culture in combination with the clonal techniques described previously, it may be possible to investigate some of the control mechanisms regulating proliferation of stem cells and their differentiation into committed precursor cells, as well as the changes occurring during leukemic transformation.

This chapter will be restricted to methods for the study of granulocytic systems and will describe both short-term cultures for the assay of granulocyte precursor cells and long-term liquid cultures for the maintenance of stem cells.

II.   Stem Cell Assay

Cell suspensions (either from normal bone marrow or from cultured cells) were tested for the presence of pluripotent hemopioetic stem cells (CFU_s) using the technique of Till and McCulloch (1961). Briefly, this involves injecting the cell suspensions into mice whose hemopoietic systems have been ablated by potentially lethal X-irradiation. At 8–10 days after injection of the cells the mice are killed, the spleens are removed, and the number of nodules are counted. Each nodule represents the progeny of a single stem cell. These *in vivo* colonies may be predominantly erythroid, granulocytic, or megacaryocytic, but mixed colonies containing both erythroid and granulocytic cells also occur (particularly at later times). For a review of this technique the reader is referred to Metcalf and Moore (1971).

III. CFU_c Assay

In this section are discussed technical aspects of the *in vitro* assay used in our laboratory, which result in granulocytic or monocytic colony formation when cells from hemopoietic tissues are cultured in soft agar in the presence of colony-stimulating activity (CSA).

A.   Materials

1.   PREPARATIONS OF CELL SUSPENSIONS

($C_3H \times AKR$) or (C57B1 × DBA) F_1 hybrid mice 8–12 weeks old are used as cell donors. The mice are killed by cervical dislocation, then the femora are removed and freed from muscle, and both ends of the bone are cut with scissors and discarded. The marrow cells are obtained by repeated flushing with a known volume of Fischer's culture medium using a syringe attached to a 23-gauge needle inserted in the bone cavity. Single-cell suspensions are prepared by passing the suspensions through successively finer gauge needles. Usually cells from 3 femora are pooled in a volume of 15 ml of medium, and the cell concentration is determined in a hemacyto-meter using standard hematological techniques. With this method about 2×10^7 cells per femur are obtained (range in 32 recent experiments was 1.4 to 2.7×10^7). The cell suspensions are kept in culture tubes placed on crushed ice and are plated as soon as possible, usually within an hour. This time can be increased if necessary to about 3 hours without any deleterious effect on colony formation.

2.   CULTURE MEDIUM

Fischer's medium (Fischer and Sartorelli, 1964) is purchased from Gibco Bio Cult Ltd., Paisley, Scotland, as 10 × concentrated medium, and can be stored at 4°C for periods up to 3 months.

Single-strength medium is prepared weekly. For a liter of medium. 100 ml of 10 × concentrated stock are added to 865 ml of sterile double-distilled water, to which 1430 mg of sodium bicarbonate in 32.5 ml of 4.4% solution (Oxoid-Lab), 500,000 units of benzil penicillin (Beecham Research Lab.) and 50 mg of streptomycin sulfate (Glaxo Lab. Ltd.) dissolved in 2.5 ml of sterile water has been added. The medium is stored at 4°C and is used within 5 days.

Other culture media are also suitable, and a number are used in different laboratories, e.g., Eagle's (Bradley and Metcalf, 1966), McCoy's (Hahn and Kersen-Bax, 1971), MRL 1066 (Worton *et al.*, 1969), and α-medium (Iscove, 1971).

3. SUPPLEMENTATION OF THE MEDIUM

Some additional nutrients have been used to supplement the medium for bone marrow cultures, e.g., a four fold concentration of amino acids and vitamins (Pluznik and Sachs, 1966), 10% trypticase soy broth (Bradley and Metcalf, 1966), or 2% bovine serum albumin (Worton et al., 1969). In our hands, addition of excess amino acids and vitamins, l-glutamine or bovine serum albumin did not result in better colony growth when added to freshly prepared Fischer's medium.

Preparations of trypticase soy broth or bovine serum albumin may vary in the degree of purity and may have toxic or inhibitory effect (Worton et al., 1969; Metcalf, 1971a). Washed red blood cells or red cell lysates have a marked enhancing effect on colony growth, increasing the incidence by a factor of 5 to 10 (Bradley et al., 1971). However, as this effect is more marked in some subpopulations of cells with colony-forming ability (Testa et al., 1973; Bradley, 1973), the addition of erythrocytes to bone marrow cultures has to be evaluated carefully when considering the experimental design. l-Asparagine (20 μg/ml) is essential for colony growth and ought to be present in the culture medium (Worton et al., 1969). (It is present in Fischer's medium.)

4. SERUM

Horse serum is purchased from Gibco Bio Cult Ltd., Paisley, Scotland or from Flow Laboratories, Scotland, and can be stored at $-20°C$ for up to 6 months. Longer storage periods may result in loss of ability to support colony growth, although some batches were stored for up to 18 months without apparent change in their quality. Serum is thawed as needed, and any remaining thawed serum can be kept at 4°C for about 5 days.

TABLE I

INFLUENCE OF THE CONCENTRATION OF HORSE SERUM
ON THE NUMBER OF IN VITRO COLONIES

		Concentration (%)			
Batch	Source	10	20	30	40
1	Gibco	12^a	63	120	—
2	Gibco	0	65	73	—
3	Gibco	—	48	69	106
4	Flow	—	165	—	180
5	Flow	100	140	—	160
6	Flow	86	150	160	—

[a] CFU_c per 10^5 bone marrow cells.

There is variation both in the ability of sera to support colony growth and in the concentration that gives optimal colony numbers (Table I). This makes it necessary to pretest each serum batch before it can be used for experiments.

Fetal calf serum, either alone or mixed with horse serum, is used in a number of laboratories with good results (Bradley and Metcalf, 1966; Metcalf, 1971a). We have assayed several batches of fetal calf serum over a period of 4 years, and with one exception they did not support any colony growth in the soft agar system. Even batches of fetal calf serum that were adequate for erythroid colony growth from mouse or human bone marrow cells (Stephenson et al., 1971; Testa and Dexter, 1975), or for granulocytic colony growth from human bone marrow cells (Pike and Robinson, 1970) failed to support the growth of colonies of granulocytes and macrophages derived from mouse bone marrow cells.

5. Colony-Stimulating Activity (CSA)

In vitro colony formation depends on the continuous presence of CSA in the cultures. A convenient source of CSA is medium which has been conditioned by the growth of some cell types (Worton et al., 1969; Pluznik and Sachs, 1966; Bradley and Sumner, 1968; Stow, 1969). For the preparation of conditioned medium (CM) from mouse embryo cells, embryos between 15 and 18 days of age are removed from the uterus and killed by cervical dislocation. After washing the embryos in sterile saline, the tissues are minced finely with scissors for about 15 minutes, then incubated at 37°C for 1 hour with 5 ml per embryo of 0.25% solution of trypsin (Bacto-Trypsin, Difco). After trypsinization the cells are washed twice in Fischer's medium and are resuspended in medium plus 10% horse serum at a concentration of 4×10^6 cells per milliliter. Aliquots, 10 ml, are dispensed into medical flat bottles. After gassing with 5% CO_2 in air, the bottles are closed with screw caps and the cultures are incubated at 37°C. A confluent monolayer of cells adhering to the glass surface is formed after 3–4 days, and the medium becomes very acid. Every 7 days the supernatant (CM) is harvested and replaced with 10 ml of fresh Fischer's medium plus 10% serum. The time of harvesting is chosen for convenience and can be shortened to 3–4 days if necessary. The CM is centrifuged for 5 minutes at 1000 g to separate cell debris and is stored at −20°C until used.

The embryo cell monolayer produces active CM for a long period of time, sometimes as long as one year (Table II). It is not necessary to trypsinize or subculture the cells; the simplicity of the procedure allows the preparation of large amounts of CM, which remain active for at least 2 years when stored at low temperature (Table III).

Heart CM has also been found to be a convenient source of CSA (Metcalf

TABLE II

Colony-Stimulating Activity in Conditioned
Medium (CM) from Mouse Embryo Cell
Cultures of Different Ages

Age of monolayer (weeks)	$CFU_c/10^5$ bone marrow cells
1–19[a]	65
20–29	61
30–38	48

[a] CM was harvested weekly and pooled before the CFU_c assay.

TABLE III

Colony-Stimulating Activity in Mouse
Embryo Conditioned Medium (CM) after
Storage at $-20°C$

Batch	Storage (months)	$CFU_c/10^5$ bone marrow cells
A	0	50 (2)[a]
	10	46 (3)
	23	55 (5)
B	0–1	65 (30)[b]
	4	63 (5)
	22	59 (5)
C	0–1	61 (30)[b]
	5	60 (5)

[a] In parentheses, number of replicate plates.
[b] Mean of 6 determinations.

et al., 1975c). Heart tissue from adult mice is minced coarsely with scissors and suspended in Fischer's medium with 10% horse serum (2 hearts in 10 ml). Harvesting and storing is done as described above for mouse embryo cells CM. Cultures of mouse heart have produced active CM for 15 weeks to date (Table IV).

Conditioned medium may have an inhibitory effect on colony growth when used at high concentrations (usually more than 25%). It has been reported that heat treatment (30–90 minutes at 60°C) and dialysis enhance the activity of CM (Bradley and Sumner, 1968). We have been unable to improve CM by either procedure. On the contrary, both procedures resulted in loss of CSA in CM from different sources (Table V). There is a sigmoid dose–response relationship between the number of colonies developed and the con-

TABLE IV
COLONY-STIMULATING ACTIVITY IN MOUSE
HEART CONDITIONED MEDIUM (CM)

Age of the culture (weeks)	Culture A[a]	Culture B
1	160[b]	62[b]
2	142	65
3	145	—
4	151	—
5–15	—	86[c]

[a] Different batches of horse serum were used for the CFU_c assays in A and B.
[b] All batches of CM were assayed at 15%. Results are expressed as $CFU_c/10^5$ bone marrow cells.
[c] Mean of 11 individual assays.

TABLE V
EFFECT OF HEAT TREATMENT OR DIALYSIS ON THE
COLONY-STIMULATING ACTIVITY OF CONDITIONED MEDIUM (CM)
FROM VARIOUS SOURCES

CM[a]	Treatment		
	None	Heat 60° C, 1 hr	Dialysis, 48 hr
Mouse embryo[b]	95[d]	65	0
Mouse embryo[c]	33	15	0
Mouse bone[e]	65	12	15
Mouse lung[f]	43	0(56° C, 30 min)	—

[a] Assayed at optimum concentration.
[b] Cells were cultured in Fischer's medium plus 15% horse serum.
[c] Cells were cultured in Fischer's medium without serum.
[d] CFU_c per 10^5 bone marrow cells.
[e] Prepared according to Chan and Metcalf (1972).
[f] Prepared according to Sheridan and Metcalf (1973).

centrations of CSA (Bradley and Sumner, 1968). Plateau numbers are usually reached at concentrations of 5–15% CM. Before experimental use, each batch of CM is tested to determine plateau concentration for colony growth.

6. AGAR

Bacto-Agar (Difco) is mixed with double-distilled water (5% w/v) and heated to 90°C in a water bath for 1 hour. The preparation is dispensed into glass bottles, sterilized by autoclaving at 15 psi during 15 minutes, and stored at 4°C until used. When needed for plating, it is melted in a boiling water bath.

Purified agar gives similar results to Bacto-agar, but some batches have been found to be less efficient in supporting colony growth. Methyl cellulose (0.8%) or hydrolyzed starch (7%) can be used instead of agar (Pluznik and Sachs, 1966).

7. PETRI DISHES

Sterilin (ordinary grade) or Nunc (tissue culture or ordinary grade) plastic petri dishes 5 cm in diameter are used. These have to be pretested, since some batches have been found to be toxic and unable to support colony growth.

B. Establishing Cultures

1. PLATING

CM to give an adequate concentration of CSA is added to Fischer's medium supplemented with appropriate (optimal) concentrations of horse serum. The required number of hemopoietic cells suspended in Fischer's medium are added and mixed gently by pipetting before warming to 37°C in a water bath. Molten agar cooled to approximately 45°C is then added to give a final concentration of 0.3%, and 3-ml aliquots are pipetted into petri dishes.

Usually 3 replicate cultures are plated for each cell suspension. The cultures are allowed to gel at room temperature for about 10 minutes before being placed in an incubator.

Normal femoral bone marrow or fetal liver cells are plated at 2×10^5 cells per dish; spleen cells or nucleated peripheral blood cells at 10^6 cells per dish. There is a linear dose–response relationship within the range that gives countable colonies (up to about 5×10^5 bone marrow or fetal liver cells). In the case of splenic or peripheral blood cells (where the incidence of colonies is lower), there may be inhibition of growth if more than 5×10^6 cells per dish are plated.

2. INCUBATION

The cultures are incubated at 37°C for 7 days in a humidified incubator with a continuous flow of 5% CO_2 in air (which maintains the pH at approximately 7.4) and a quick flushing system with pure CO_2, which operates when the incubator door opens. Alternatively, airtight metal boxes with 2 valves for gas inflow and outflow can be used. A water-saturated atmosphere is obtained by placing in the box a few open petri dishes filled with water. Gassing with 5% CO_2 in air for 5 minutes at the beginning of the culture is enough to maintain an adequate pH for 10 days in a sealed box. This system has the advantage that the cultures remain undisturbed during the whole incubation period.

C.   Characteristics of the Cultures

The colonies are composed of up to 3000 differentiating and apparently normal granulocytic cells and/or mononuclear cells (Bradley and Metcalf, 1966; Pluznik and Sachs, 1966). Granulocytic cells in various degrees of differentiation (from myeloblasts to segmented granulocytes) form the majority of the population in young colonies, but as incubation progresses the mononuclear cells become more numerous, and eventually are the only cell type found (Bradley and Metcalf, 1966; Bradley and Sumner, 1968). The mononuclear cells have been characterized as macrophages by their morphology, their ability to phagocytic carbon particles (Ichikawa et al., 1966) and also on the basis of glass adherence and the possession of receptors for immunoglobulins (Cline et al., 1972). It has been shown that mixed colonies with granulocytes and macrophages can originate from single cells (Moore et al., 1972) and also that granulocytic cells can become macrophages in culture (Metcalf, 1971b). However, it is not known whether the appearance of macrophages in the colonies reflects the true potential for differentiation of the CFU_c in vivo, or is an artifact induced by the culture.

The number of colonies developing in vitro from a given number of hemopoietic cells depends on the concentration of CSA and on the quality of the culture medium and serum used. The plating efficiency of the CFU_c cannot be determined, since a pure population of the cells has not yet been isolated. However, there is some experimental evidence indicating that this culture system detects a large proportion (at least 66%) of the CFU_c (Moore et al., 1972). The incidence of CFU_c in adult mouse bone marrow is 0.5–2 per 10^3 cells, in spleen 3–50 per 10^6 cells (with a large variation in different strains of mice) and 10–50 per 10^6 nucleated blood cells (Metcalf and Moore, 1971).

Usually 40–50 cells is considered to be the minimum size for a colony growing in soft agar. Smaller cell aggregates (clusters), which also arise from proliferation of single (probably more mature) cells, and which present a differentiation pattern similar to that of colonies, are also found. The ratio of clusters to colonies may be as high as 10:1 in bone marrow (Metcalf, 1969).

D.   General Comments on the Agar Culture System

The conditions for proliferation and differentiation of granulocytic cells appear to be adequate in the colonies growing in soft agar: there is no evidence of cell death and the cell cycle times are similar to those found for granulocytic cells in vivo (Testa and Lord, 1973). However, cycling starts slowing by

5 days, and after 7–10 days of culture cell proliferation practically ceases. Experimentally, the soft agar cultures are adequate to monitor a population of granulocytic progenitor cells (CFU_c), which are close to the stem cells (CFU_s) and on which specific regulatory mechanisms (CSA) may act, inducing them to proliferate and differentiate in culture. These cultures are also adequate to investigate humoral factors or other cell populations which may influence granulopoiesis. They are not suitable, however, for the study of CFU_s since they do not survive in significant numbers for more than 3–5 days in the cultures in the conditions of the CFU_c assay, and they do not respond to modifications of the culture which influence the CFU_c (Table VI). This indicates that, at least *in vitro*, each cell population is regulated by independent mechanisms. Experimental work on CFU_c has recently been reviewed by Metcalf and Moore (1971).

Several studies have been carried out on normal and leukemic human granulocytic colonies (Pike and Robinson, 1970; Duttera *et al.*, 1972; Iscove *et al.*, 1971; Robinson *et al.*, 1971). It has been shown that leukemic cells proliferate in agar with a variety of growth patterns which permit, in the case of acute leukemias, the identification of subgroups with different probabilities of remission (Moore *et al.*, 1974). This kind of information supports the concept that agar cultures are a useful specialized technique in the study of normal and abnormal hemopoiesis.

TABLE VI

EFFECT OF VARIATIONS OF CULTURE CONDITIONS ON THE
NUMBER OF CFU_s AND CFU_c DETERMINED AFTER 48 OR 72
HOURS OF CULTURE IN AGAR

Modifications[a]	CFU_s[b]	CFU_c[b]
Culture of regenerating bone marrow (3–4 days after 450 rad X-rays	Increase	No change
Addition of washed rat or mouse RBC (0.1–0.2 ml/ml of culture)	Increase	No change
Addition of lethally irradiated spleen cells	Decrease	No change
Absence of CM	No change	Decrease
Replacement of horse serum by fetal calf serum[c]	Decrease	No change

[a] Basic cultures are of normal bone marrow cells in Fischer's medium plus horse serum and Conditioned medium (CM).

[b] 30–50% of the initial number of CFU_s and 200–400% of the initial number of CFU_c are detected after 48–78 hr of culture (Testa and Lajtha, 1973).

[c] Batch is adequate for *in vitro* colony growth.

TABLE VII

ABILITY OF VARIOUS SERA TO SUPPORT PROLIFERATION OF
CFU$_s$ IN VITRO IN LONG-TERM LIQUID CULTURES

Type of sera	Source	Number of batches tested	Number supporting growth of CFU$_s$ in vitro
Horse	Gibco	7	0
Fetal calf	Gibco	8	0
Bobby calf	Gibco	2	0
Fetal calf	Flow	5	0
Horse	Flow	12	6(a)

^{a} Optimal concentrations were between 20 and 30%.

IV. Liquid Cultures

A. Materials

1. SERUM

Several types of sera from various sources have been tested (Table VII) at concentrations ranging from 10 to 40%. Only horse serum obtained from Flow Laboratories (Scotland) supported the long-term maintenance of CFU$_s$—the optimal serum concentration normally being between 20 and 30%. It is of interest that there is no correlation between the ability of different batches of sera to support the growth of CFU$_c$ and their ability to promote the proliferation of CFU$_s$, since all the batches of Flow Laboratories horse sera tested stimulated the development of CFU$_c$ from mouse bone marrow, whereas only 50% supported the growth of CFU$_s$ in the liquid long-term culture system to be described. Similarly, Gibco horse serum has been found consistently to support the growth of mouse CFU$_c$ (Table I) while being uniformly negative for the long-term maintenance of CFU$_s$ in liquid culture.

2. CULTURE MEDIUM

Fischer's medium (Gibco or Flow Laboratories) has been used in all experiments. L-Glutamine is essential and has to be added as per the manufacturers instructions. The medium is supplemented with sodium bicarbonate, penicillin, and streptomycin sulfate as described in Section III,A,2).

3. CULTURE BOTTLES

Screw-capped flat soda-glass bottles (10 cm × 5 cm × 2.5 cm) with a capacity of approximately 125 ml (obtained from United Glass, London,

England) have been found to be suitable. Before use the bottles are soaked for 24 hours in 2–5% Decon 90, followed by 10 rinses with tap water, 2 rinses with double-distilled water, and heat sterilization. After culture, the bottles can be cleaned and reused without loss of their ability to support hemopoietic cell proliferation.

Pyrex culture bottles and plastic (T60-Falcon) tissue culture flasks were found to be poor in their ability to sustain stem cell proliferation.

B.   Establishing Liquid Cultures

Two types of culture are regularly used. In the first case, thymus and bone marrow cells are cocultured (Dexter et al., 1973; Dexter and Lajtha, 1974); in the second case, bone marrow cells are added to an established bone marrow monolayer culture (Dexter and Lajtha, 1975).

1.   THYMUS AND BONE MARROW COCULTURES

Thymuses are removed aseptically from BDF_1 mice (4–6 weeks old), placed in Fischer's medium, and minced coarsely with scissors. The coarse cell suspension is then forced through needles of various diameters until a more or less single-cell suspension is produced. This is then added to growth medium (Fischer's medium supplemented with 20% horse serum) to give a final concentration of 10^6 nucleated cells per milliliter, and 10 ml of the suspension are inoculated into each culture bottle. The cultures are gassed with a mixture of air + 5% CO_2 and incubated at 37°C for 24 hours. They are then inoculated with 10^7 nucleated bone marrow cells (in a volume of 0.25 ml) obtained from the pooled femora of 6–8-week-old BDF_1 mice (the femoral marrow cells are collected in ice cold Fischer's medium using the technique reported in Section II,A,1). The thymus plus bone marrow cocultures are gassed with air + 5% CO_2 and incubated at 37°C. After several days, such cultures consist of an attaching and a nonattaching population of cells. The attaching cells consist of "fibroblastic," "epithelioid," and "phagocytic" mononuclear cells. The nonadhering cells contain mainly granulocytes in all stages of maturation. After culture for 1 week, the bottles are gently shaken to remove any loosely attaching cells; half (5 ml) of the growth medium (containing cells in suspension) is removed, and an equal volume of fresh growth medium is added. This is repeated at weekly intervals. The cells present in the growth medium removed are counted directly on a hemacytometer. (In this way the weekly production of suspension cells can be monitored.) The growth medium is then centrifuged (800 g for 10 minutes). The cell-free supernatant is poured off, and the cells are resuspended in a small volume of Fischer's medium and subsequently assayed for the presence of

CFU$_s$ and CFU$_c$, using techniques previously described (Sections II and III). Additionally smears are made for morphological examination, and the cell-free growth medium is assayed for CSA activity (as for "conditioned medium," Section III).

2. Bone Marrow Plus Bone Marrow Cultures

Pooled femoral bone marrow cells collected in Fischer's medium are added to growth medium (Fischer's supplemented with 20% horse serum) to give a final concentration of 10^6 nucleated cells per milliliter, and 10 ml are then inoculated into each culture bottle, gassed with air + 5% CO_2, and incubated at 37°C. After 1 week (and subsequently at weekly intervals), the bottles are gently shaken to disperse any loosely attaching cells; half (5 ml) of the growth medium is again removed, and an equal volume of fresh growth medium is added. It is found that the number of nonattaching cells in suspension, CFU$_s$ and CFU$_c$, rapidly declines, but a monolayer of cells develops in the cultures. After 3 weeks the cultures are reinoculated with growth medium containing a further 10^7 fresh bone marrow cells. Subsequently the cultures are fed and the cells are harvested as described for the thymus plus bone marrow cocultures.

C. Characteristics of Liquid Cultures

1. Thymus plus Bone Marrow Cocultures

Initially all cultures consist of granulocytes in various stages of maturation. After 2 week's growth however, two main types of culture are observed —those producing mainly granulocytes and those producing mainly phagocytic mononuclear cells. The different cultures can be easily recognized by observing the cells *in situ*. In the granulocytic cultures (G-cultures) the cells in suspension are round, of variable size, and have distinct cell membranes. In the cultures containing mononuclear cells (M-cultures) the cells are larger and have irregular indistinct cell membranes. In both types of culture the attaching population is heterogeneous, containing phagocytic, fibroblastic, and epithelial cells.

The ratio between M- and G-type cultures varies considerably in different experiments. On occasions all the cultures established at any one time (from the same pool of thymus and bone marrow cells) may be G-type, on other occasions all M-type. Generally, however, both M- and G-type cultures are found in the same experiment.

The results of a "typical" culture of thymus plus bone marrow cells is shown in Table VIII. Cultures were established from the same "pool" of

TABLE VIII

Stem Cell Maintenance in a "Typical" Coculture of Thymus plus Bone Marrow Cells (T+BM) in Long-Term Liquid Cultures

Weeks cultured[a]	Type of culture	Cell count (suspension) × 10^6	Total CFU$_s$ (suspension)	Total CFU$_c$ (suspension)	Morphology		
					Blasts	Granulocytes	Mono-nuclear cells
1	M + G	1.9	278 ± 20	1750 ± 180	1	85	14
2	M	0.8	160 ± 12	3930 ± 200	2	15	83
	G	1.8	700 ± 45	ND^b	2	96	2
4	M	0.9	20 ± 5	1170 ± 40	0	12	88
	G	1.2	330 ± 50	2080 ± 58	3	62	35
7	M	0.6	0	186 ± 14	0	0	100
	G	1.4	220 ± 10	1280 ± 64	0	42	58
10	M	0.5	0	0	0	0	100
	G	1.9	52 ± 4	890 ± 16	1	28	71
12	M	1.6	0	0	0	0	100
	G	1.8	21	ND	0	8	92
15	M	0.8	0	0	0	0	100
	G	1.1	0	0	0	0	100

^{a} Cell counts, CFU$_s$, CFU$_c$ assays were performed *every* week.
^{b} ND, no data.

thymus and bone marrow cells, cultured for 1 week, and separated subsequently into M- or G-type cultures and assayed independently.

In both types of culture, cell proliferation is obviously occurring since the numbers of suspension cells increase after each feeding (approximately doubling in number). During this time interval, the numbers of attaching cells remains fairly constant and the increase in cells occurring in the suspension between "feedings" does not represent simply detachment of the adhering cells.

The production of stem cells and the morphology of cells produced is distinctly different in the two cultures. In M-type cultures, phagocytic mononuclear cells comprise more than 80% of the cells present at 2 weeks, and "classical" granulocytes disappear. Similarly CFU_s progressively decrease and are absent at 5 weeks. CFU_c are maintained for somewhat longer (7 weeks), then disappear. In G-type cultures, granulocytes in all stages of maturation are maintained for 10 weeks—but even here the percentage of mononuclear cells steadily increases. Stem cells (CFU_s) are present for 12 weeks. This is not simply maintenance of stem cells since CFU_s are not detectable as an adhering population (Dexter et al., 1973) and between weeks 3 and 9 the numbers of such cells present in suspension increases after each successive depopulation (feeding). The CFU_s produced are apparently normal, forming both erythroid and granulocytic spleen colonies, and if sufficient numbers are injected they can reconstitute the hemopoietic system of potentially lethally irradiated mice and protect them from radiation-induced hemopoietic death. Several such reconstituted mice are alive and apparently well 18 months after reconstitution.

CFU_c are also being produced in G-type cultures for at least 10 weeks. To what extent these CFU_c are being produced from the CFU_s is not known (there is a possibility that CFU_c may have a limited self-maintenance capacity that could account, in part, for the production of these cells). The formation of these colonies is still dependent upon the presence of exogenous CSA (in this case mouse embryo or heart conditioned medium). In the absence of added CSA, colonies do not form. The colonies produced consist mainly of differentiated granulocytes, and only a few colonies consisting of mononuclear cells are found.

2. Bone Marrow Plus Bone Marrow Cultures

Unlike the thymus and bone marrow cultures, the culture of bone marrow cells on a developed bone marrow monolayer gives more uniform and consistent results. The results of a typical culture are shown in Table IX. Twelve culture bottles were established from the same "pool" of bone marrow cells. At weekly intervals after the addition of the second bone marrow

TABLE IX

STEM CELL MAINTENANCE IN A "TYPICAL" CULTURE OF
BONE MARROW PLUS BONE MARROW CELLS (BM + BM) IN LONG-TERM LIQUID CELLS

Weeks cultured	Cell count (suspension) $\times 10^6$	Total CFU_s (suspension)	Total CFU_c (suspension)	Morphology		
				Blasts	Granulocytes	Mononuclear cells
1	0.5	420 ± 20	ND[a]	5	93	2
2	0.4	ND	2030 ± 160	ND	ND	ND
3	1.8	480 ± 56	3470 ± 250	2	87	9
4	1.9	660 ± 70	3690 ± 220	10	85	5
5	1.9	160 ± 40	ND	20	62	18
6	1.4	ND	1800 ± 120	20	54	22
7	1.3	77 ± 10	ND	41	54	5
8	0.8	ND	92 ± 10	56	30	14
9	0.8	90 ± 8	ND	ND	ND	ND
10	1.0	ND	180 ± 40	47	14	39
11	1.0	45 ± 5	ND	66	2	32
12	0.8	ND	0	37	0	62
13	0.9	0	0	17	0	83
15	0.9	0	0	0	0	100
17	0.8	0	ND	1	0	99
20	1.0	ND	265 ± 30	2	0	98

[a] ND, no data.

population, the cultures were fed and the suspension cells were pooled and assayed for stem cell content.

Cell proliferation is again occurring, since the numbers of suspension cells increases after each feeding (during the course of the experiment the numbers of attaching cells remains fairly constant). Similarly, stem cells (CFU$_s$) are also produced for up to 11 weeks. Again, this is not simply maintenance of stem cells since the numbers increase after each successive feeding [and previous work has established that CFU$_s$ and CFU$_c$ are not detectable as an attaching population (Dexter *et al.*, 1973)]. These CFU$_s$ are "functionally" normal—forming both erythroid and granulocytic spleen colonies and are also able to protect mice from potentially lethal X-irradiation. CFU$_c$ are also maintained at high levels for several months. As in the thymus and bone marrow cocultures, the development of these colonies is dependent upon the addition of CSA.

Interestingly, between 11 and 18 weeks, CFU$_s$ and CFU$_c$ were not detectable, but at 20 weeks CFU$_c$ were again present in the suspension. These CFU$_c$ formed granulocytic agar colonies. Similar results have been observed in several other experiments, where both CFU$_s$ and CFU$_c$ have not been detectable for several weeks but have subsequently "reappeared." We have as yet no explanation for this phenomenon.

Morphologically, obvious differences exist between these cultures and thymus/bone marrow cultures. An initial period of granulopoiesis is followed by an accumulation of primitive blast cells. With further time in culture, the percentage of blast cells increases and granulocytes disappear. Later, mononuclear cells become apparent, and blast cells decrease. During this time there is no increase in cell density. It is worthy of comment that concomitantly with the blast cell accumulation there is production of apparently normal CFU$_s$ and CFU$_c$ in these cultures.

D. General Comments on the Liquid Culture System

The thymus plus bone marrow system produces two types of culture: those where mainly mononuclear cells are produced and stem cells disappear, and those where granulocytes are produced and stem cells are maintained. The efficiency with which the granulocyte (G) cultures can be established is variable and the particular factors necessary for the maintenance of proliferation and differentiation of the stem cells *in vitro* are not known. Using a developed marrow monolayer population as a "feeder" layer, however, gives more consistent results for stem cell maintenance, and it may well be that in the thymus/marrow coculture system the thymus acts by stimulating the development of a specific population of marrow adherent cells, which in turn facilitate stem cell proliferation. Further work is needed to explore this possibility.

Nonetheless, certain fundamental differences exist between thymus/ marrow and marrow/marrow cultures, namely, the blast cell accumulation that occurs in marrow/marrow cultures. A possible explanation for this blast cell accumulation may be in the difference in the ability of the different cultures to produce CSF. Indeed, when the cell-free growth medium from these two different culture types was tested as a "conditioned medium" for the development of normal bone marrow CFU_c, it was found (Table X) that thymus/marrow cultures contained appreciable amounts of CSA activity, whereas marrow/marrow cultures contained little, if any, stimulating activity. Consequently, it is possible that the long-term production of granulocytes and macrophages in the thymus/marrow system is facilitated by the fairly high CSA concentration present (since, in the agar culture system CSA stimulates the development of macrophage and granulocyte colonies) and that in the marrow/marrow system the deficiency in CSA leads eventually to a "maturation block" (hence the blast cell accumulation). This maturation block is readily overcome when the cultured cells are grown in agar medium supplemented with CSA (Table IX). Again, further investigations are necessary to clarify the role of CSA in these systems. Of particular importance will be the effects of adding CSA directly to the marrow/marrow cultures before and during the blast cell accumulation.

TABLE X

CSF ACTIVITY OF "CONDITIONED MEDIA"
OBTAINED FROM VARIOUS LIQUID CULTURES[a]

Weeks cultured	Source of conditioned medium	CSF activity in cell-free growth medium[b]	
		20%	50%
1	T + BM	49	280
	BM + BM	0	0
8	T + BM	15	340
	BM + BM	0	2

[a] The growth medium from either thymus/bone marrow cocultures (T + BM) or from bone marrow:bone marrow cultures (BM + BM) was centrifuged and the cell-free supernatant was assayed at 20% or 50% for its ability to stimulate normal bone marrow CFU_c in agar cultures.

[b] Colonies stimulated/10^6 normal bone marrow cells.

REFERENCES

Bradley, T. R. (1973). *In* "Proceedings of the Second International Workshop on Haemopoiesis in Culture" (W. Robinson, ed.), p. 77.

Bradley, T. R., and Metcalf, D. (1966). *Aust. J. Exp. Biol. Med. Sci.* **44**, 287.

Bradley, T. R., and Sumner, M. A. (1968). *Aust. J. Exp. Biol. Med. Sci.* **46**, 607.

Bradley, T. R., Telfec, P. A., and Fry, P. (1971). *Blood* **38**, 353.

Chan, S. H., and Metcalf, D. (1972). *Blood* **40**, 646.

Cline, E. A., Warner, N. L., and Metcalf, D. (1972). *Blood* **39**, 326.

Dexter, T. M., and Lajtha, L. G. (1974). *Br. J. Haematol.* **28**, 525.

Dexter, T. M., and Lajtha, L. G. (1975). *Proc. Int. Symp. Comp. Leuk. Res., 7th, 1975* (in press).

Dexter, T. M., Allen, T. D., Lajtha, L. G., Schofield, R., and Lord, B. I. (1973). *J. Cell. Physiol.* **82**, 461.

Duttera, M. J., Whang-Peng, J., Bull, J. M. C., and Carbone, P. P. (1972). *Lancet* **1**, 715.

Fisher, G., and Sartorelli, A. (1964). *Methods Med. Res.* **10**, 247.

Hahn, G. M., and Kersen-Bax, I. (1971). *C. R. Helid. Seances Acad. Sci.* 2338.

Ichikawa, Y., Pluznik, D. H., and Sachs, L. (1966). *Proc. Natl. Acad. Sci. U.S.A.* **56**, 488.

Iscove, N. N. (1971). *In* "*In vitro* Culture of Haemopoietic Cells" (D. W. van Bekkum and K. A. Dicke, eds.), p. 459. Radiobiol. Inst. TNO, Rijswijk, The Netherlands.

Iscove, N. N., Senn, J. S., Till, J. E., and McCulloch, E. A. (1971). *Blood* **37**, 1.

Metcalf, D. (1969). *J. Cell. Physiol.* **74**, 323.

Metcalf, D. (1971a). *In* "*In vitro* Culture of Haemopoietic Cells" (D. W. van Bekkum and K. A. Dicke, eds.), p. 449. Radiobiol. Inst. TNO, Rijswijk, The Netherlands.

Metcalf, D. (1971b). *J. Cell. Physiol.* **77**, 277.

Metcalf, D., and Moore, M. A. S. (1971). *Front. Biol.* **24**, 70–108.

Metcalf, D., MacDonald, H. P., Odartchenko, N., and Sardat, L. B. (1975a). *Proc. Natl. Acad. Sci. U.S.A.* **72**, 1744.

Metcalf, D., Warner, N. L., Nossal, G. J. V., Miller, J. F. A. P., Shortman, K., and Rabellino, E. (1975b). *Nature (London)*, **255**, 630.

Metcalf, D., Parker, J., Chester, H. M., and Kincade, P. (1975c) *J. Cell. Physiol.* **84**, 275.

Moore, M. A. S., Williams, N., and Metcalf, D. (1972). *J. Cell. Physiol.* **79**, 283.

Moore, M. A. S., Spitzer, G., Williams, N., Metcalf, D., and Buckley, J. (1974). *Blood* **44**, 1.

Pike, B. L., and Robinson, W. A. (1970). *J. Cell. Physiol.* **76**, 77.

Pluznik, D. V., and Sachs, L. (1966). *Exp. Cell. Res.* **43**, 553.

Robinson, W. A., Kurnick, J. E., and Pike, B. L. (1971). *Blood* **38**, 500.

Schredni, B., Rozenszajn, L. A., Kalechman, Y., Michlin, H. (1975). *Abstr., Int. Soc. Exp. Haematol. 1975.*

Seridan, J. W., and Metcalf, D. (1973). *J. Cell. Physiol.* **81**, 11.

Stephenson, J. R., Axelrad, A. A., McLeod, D. L., and Shreeve, M. M. (1971). *Proc. Natl. Acad. Sci. U.S.A.* **68**, 1542.

Stow, J. (1969). M.Sc. Thesis, University of Manchester.

Testa, N. G., and Dexter, T. M. (1975). *Abstr., Assoc. Clin. Pathol. 1975.*

Testa, N. G., and Lajtha, L. G. (1973). *Br. J. Haematol.* **24**, 367.

Testa, N., Lord, B. I. (1973). *Cell Tissue Kinet.* **6**, 425.

Testa, N. G., Hendry, J. H., and Lajtha, L. G. (1973). *Biomedicine* **19**, 183.

Till, J. E., and McCulloch, E. A. (1961). *Radiat. Res.* **14**, 213.

Worton, R. G., McCulloch, E. A., and Till, J. E. (1969). *J. Cell. Physiol.* **74**, 171.

Subject Index

CONTENTS OF PREVIOUS VOLUMES

Volume I

Volume IV

Volume VII

Volume X

Volume XIII

A 6
B 7
C 8
D 9
E 0
F 1
G 2
H 3
I 4
J 5